Active Tectonics

Prentice Hall Earth Science Series

Active Tectonics
Earthquakes, Uplift, and Landscape
Second Edition

Edward A. Keller
University of California, Santa Barbara

Nicholas Pinter
Southern Illinois University, Carbondale

Prentice Hall
Upper Saddle River, New Jersey 07458

Library of Congress Cataloging-in-Publication Data

Keller, Edward A.
    Active tectonics : earthquakes, uplift, and landscape / Edward A. Keller, Nicholas
Pinter.—2nd ed.
        p. cm.—(Books in the Prentice Hall earth science series)
    Includes bibliographical references and index.
    ISBN 0-13-088230-5 (pbk.)
    1. Plate tectonics. 2. Earthquakes. 3. Geomorphology.I. Pinter, Nicholas. II. Title.III.
Series
QE511.4.K45 2002
551.8—dc21   2001036837

Senior Editor: Patrick Lynch
Production Editor/Page Layout: Kim Dellas
Senior Marketing Manager: Christine Henry
Manufacturing Manager: Trudy Pisciotti
Assistant Manufacturing Manager: Michael Bell
Cover Designer: Jayne Conte
Cover Photo: Dr. Jeff Freymueller, Geophysical Institute, University of Alaska, Fairbanks
Art Editor: Adam Velthaus
Illustrator: David Crouch
Editorial Assistant: Sean Hale

© 2002, 1996 by Prentice-Hall, Inc.
Upper Saddle River, New Jersey 07458

Printed in the United States of America

10 9 8 7 6 5 4 3 2 1

ISBN 0-13-088230-5

Pearson Education LTD., *London*
Pearson Education Australia PTY, Limited, *Sydney*
Pearson Education Singapore, Pte. Ltd
Pearson Education North Asia Ltd, *Hong Kong*
Pearson Education Canada, Ltd., *Toronto*
Pearson Educación de Mexico, S.A. de C.V.
Pearson Education–Japan, *Tokyo*
Pearson Education Malaysia, Pte. Ltd

To Professor Marie Morisawa, SUNY Binghamton,
for her many contributions in geomorphology,
her assistance and guidance of students, friends and colleagues,
and her love of science and life.

Contents

6 Active Tectonics and Coastlines   191

7 Active Folding and Earthquakes   223

8 Paleoseismology and Earthquake Prediction 267

9 Mountain Building 313

Preface

Active tectonics is the study of dynamic tectonic processes that shape the landscape and have an impact on human society. Tectonic geomorphology is the part of active tectonics that is concerned with landforms produced by tectonic processes and the application of geomorphic principles to tectonic problems. Tectonic geomorphology increasingly has become one of the principal tools in a variety of applications, including identification of active faults, formation of geologic structures, seismic-hazard assessment, and the study of landscape evolution. Tectonic geomorphology has proven to be useful in these applications because tectonically produced landforms are created and preserved over time intervals ideal for recording landscape change.

This book requires a basic knowledge of geologic principles. It is appropriate for upper-division undergraduate students, graduate students, and others who work in the fields of geology, geomorphology, and earthquake studies. In universities, this book is appropriate for classes in active tectonics, tectonic geomorphology, earthquake geology, and geomorphology.

The field of active tectonics has expanded rapidly during the past decade or so, but it remains at the cutting edge of geologic research. Space-based positioning, analysis of digital topography, and new dating techniques are bringing a whole new class of information to studies of the dynamic Earth. Advances in topics such as buried reverse faulting, active fold growth, earthquake recurrence, climate change, isostasy, and long-term landscape evolution continue to refine and redefine our understanding of tectonic and geomorphic processes.

We hope the readers of this book will find it to be an up-to-date source of information, as well as a solid foundation for understanding future advances in the fields of active tectonics and tectonic geomorphology.

ACKNOWLEDGMENTS

The authors would like to thank colleagues who reviewed all or parts of the first or second editions of this book: Ronald L. Bruhn, University of Utah; Randel T. Cox. University of Memphis; Thomas W. Gardner, Pennsylvania State University; David R. Hickey, Graptolithics; John M. Holbrook, Southeast Missouri State University; William R. Lettis, Lettis & Associates, Inc.; Nancy Lindsley-Griffin, University of Nebraska-Lincoln; George W. Moore, Oregon State University; Karl J. Mueller, University of Colorado; Gomaa I. Omar, University of Pennsylvania; Frank Pazzaglia, Lehigh University; John B. Ritter, Wittenberg University; William A. Smith, Western Michigan University; Steven N. Ward, University of California, Santa Cruz; and John C. Weber, Grand Valley State University.

The authors are also pleased to acknowledge the assistance of the editors. Assistance from Ellie Dzuro (word processing), Dave Crouch (computer illustration), and Amy Selting (production assistance) is also greatly appreciated.

Active Tectonics

1

Introduction to Active Tectonics: Emphasizing Earthquakes

ACTIVE TECTONICS

In the geological sciences, the term **tectonics** refers to the processes, structures, and landforms associated with deformation of the Earth's crust. In a broader sense, it refers to the evolution of these structures and landforms over time. On a global scale, we are concerned with the origin of continents and ocean basins, which are the largest landforms produced by tectonics on Earth. At the regional scale, we are interested in the generation and evolution of mountain chains. At the local scale, we often study features such as small folds, which may form elongated low hills, or faulting, which may result in ground rupture and the production of relatively small (a few tens to thousands of meters long by less than 1 m to about 8 m high), steep slopes known as fault scarps.

The time scales of tectonics depend on the spatial scale at which the processes act. For example, it takes billions of years for continents to develop; hundreds of millions of years for large ocean basins; several million years for small mountain ranges; several hundred thousand years for small folds to produce hills; and fault scarps may be produced almost instantaneously during earthquakes. Rates of tectonic process are also extremely variable:

- Fault ruptures during earthquakes may propagate as fast as several kilometers per second. For example, during the 1983 Borah Peak earthquake in Idaho, eyewitnesses reported that a fault scarp approximately 1 m in height formed in less than 1 s [1]. During a major 1999 earthquake in Taiwan, a 7-m-high scarp formed at a rate of about 1 m per second (Ta-liang Teng, personal communication, 2000).
- When tectonic processes are averaged over years, rates generally range from fractions of millimeters to several millimeters per year for fault displacement to several centimeters per year for processes that move continents and ocean basins.

The term **active tectonics** refers to those tectonic processes that produce deformation of the Earth's crust on a time scale of significance to human society [1]. As such, we are most interested in processes likely to cause disruption of society within a period of several decades to several hundred years—the time period for which we plan the lifetimes of buildings and important facilities such as dams and power plants. However, in order to study and predict tectonic events over this time period, we must study these processes over a much longer time scale—at least several thousand years to several tens of thousands of years—because earthquakes on particular faults may have long return periods (time between events). Depending on when the most recent event occurred, faults may produce earthquakes in the next several decades or the next several thousand years. Another viewpoint is that the time frame necessary to study active tectonics is more like several millions of years [2]. Understanding tectonic processes over several millions of years is necessary to fully understand active tectonics and mitigate associated geologic hazards such as earthquakes [2].

Although active tectonics includes the slow disruption (warping or tilting) of the Earth's crust that may cause damage to human structures, we are most concerned with active tectonic processes capable of producing catastrophes. A **catastrophe** is defined as any situation in which the damage to people, property, or society in general is sufficiently severe that recovery or rehabilitation, or both, are a long, involved process [3]. One active tectonic process likely to produce a catastrophe is a great earthquake. However, moderate-sized earthquakes also can produce catastrophes, particularly if they occur in densely populated areas where buildings are constructed of materials that cannot withstand shaking (homes constructed of unreinforced cement blocks, bricks, or stones are particularly hazardous) or buildings constructed on thick layers of unconsolidated sediment (particularly those sediments with a high water content). Catastrophic earthquakes in history include:

- A sixteenth-century event in China that reportedly claimed approximately 850,000 lives
- A 1923 earthquake near Tokyo that claimed 143,000 lives
- A 1976 earthquake in China that claimed several hundred thousand lives
- A 1985 earthquake originating in rocks below the Pacific Ocean off Mexico that sent seismic waves to Mexico City (several hundred kilometers from the source) and caused approximately 10,000 deaths

- A 1989 earthquake on the San Andreas fault system south of San Francisco, California, that killed 62 people and caused $5 billion in property damage
- A 1994 earthquake in the Los Angeles, California, urban area (Northridge) that killed 61 people and caused more than $40 billion in damage
- A 1995 earthquake in Japan (Kobe) that killed more than 6000 people and caused about $100 billion in damages
- A 1999 earthquake in Taiwan that caused 2300 deaths and $7 billion in damages
- A 1999 earthquake in Turkey that killed over 30,000 people with damages of $10 to $20 billion

It has been estimated that a great earthquake in a densely populated part of Southern California could do $100 billion in damage and kill several thousand people [4]. Thus, the 1994 event, as terrible as it was, was not "the big one". Because earthquakes have the proven potential for catastrophic damage, much of our research in active tectonics is to better understand earthquake processes, the damage likely to occur, and the ways to minimize loss of life and property damage.

Research in active tectonics at a variety of scales, from regional to local, can be useful to society. Figure 1.1 is a generalized flowchart from data input (at the regional or local levels) to output and possible social impacts of active-tectonic information such as improving regional planning and site-specific land use, establishing building codes, and planning for earthquake-hazard reduction. At the data input stage, we make measure-

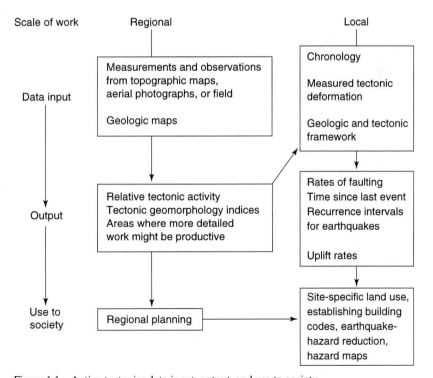

Figure 1.1 Active tectonics data input, output, and use to society.

ments and observations from topographic maps, aerial photographs, and field work that help define relative tectonic activity and areas where more detailed work is needed to better understand the earthquake hazard. This reconnaissance-level work delineates areas where detailed evaluation may provide rates of active-tectonic processes, recurrence intervals of earthquakes, and rates of crustal deformation. This information is necessary for society to develop regional planning strategies and site-specific strategies for earthquake-hazard reduction [1]. With this general introduction, the remainder of this chapter discusses selected aspects of earthquakes and related phenomena at local to global scales.

GLOBAL TECTONICS

Tectonic processes are driven by forces deep within the Earth that deform the crust, producing external forms as large as ocean basins, continents, and mountains. Collectively these processes are known as the **tectonic cycle**. Earth scientists, through detailed study of the ocean basins and continents, have established that the **lithosphere**—the outer layer of the Earth (crust and uppermost mantle) that contains the continents and ocean basins—is relatively strong and rigid compared with deeper material and ranges in thickness from several tens of kilometers beneath parts of ocean basins to greater than 100 km beneath parts of continents. The lithosphere is not a continuous uniform shell but is broken into several large pieces called lithospheric plates that move relative to one another [5]. Plates can include both continents and portions of ocean basins or ocean basins alone (Figure 1.2).

The lithospheric plates move over the **asthenosphere**, which is thought to be a more-or-less continuous, hot, and plastically flowing layer of relatively weak rock below the lithosphere (Figure 1.3). This motion causes the continents to change their relative positions on the surface of the Earth [6]. The idea that continents move is not new, but it is only in the last 30 years that this hypothesis has been tested intensively enough to gain the status of a unified theory of the Earth. Alfred Wegener, a German scientist, first suggested in the early twentieth century that the continents were moving or drifting. His evidence was based in part on the good fit of the continents, such as between South America and Africa, but most importantly on the similarities in rock types, geologic structures, and paleontological (fossil) evidence now found on opposite sides of the Atlantic Ocean. However, it was not until the late 1960s, when the process of sea-floor spreading (shown on Figure 1.3) was discovered, that a plausible mechanism for continental drift was provided. We now know that lithosphere plates move at rates that range from about 1 to 15 cm/yr—in general, about as fast as fingernails grow. The most recent global episode of continental drift and sea-floor spreading started about 200 million years ago, the time when Wegener hypothesized the superintendent called Pangea broke up. Figure 1.4 shows Pangea as well as the present positions of the continents and ocean basins. Sea-floor spreading in the past 180 million years separated Africa and Eurasia from North and South America, South America from Africa and Antarctica, and Australia and India from Antarctica. The Tethys Sea closed, leaving the small remnant of it today known as the Mediterranean Sea. About 50 million years ago, India crashed into Eurasia, producing the Himalayan Mountains and Tibetan Plateau. That collision is still happening today.

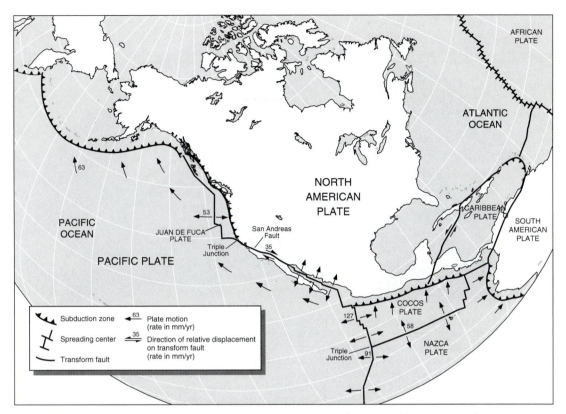

Figure 1.2   Boundary between the North American and Pacific Plates. (Courtesy of Tanya Atwater.)

The boundaries between lithosphere plates are areas of geological activity where most earthquakes and volcanic activity occur. The three types of boundaries between the plates are divergent, convergent, and transform [7]. **Divergent boundaries** occur at spreading ridges where new lithosphere is produced and plates move away from each other (Figures 1.3 and 1.4). **Convergent boundaries** occur where one plate dives ("subducts") beneath the leading edge of another plate, and thus are also known as subduction zones. However, if both leading edges are composed of relatively low-density continental material (average composition of granite), it is more difficult for subduction to start, and a special type of convergent plate boundary, called a **continental-collision boundary**, may develop. This type of boundary condition produces some of the highest linear mountain systems on Earth, such as the Alps and the Himalayas. **Transform boundaries** occur where one plate slides past another, displacing spreading ridges. This type of boundary is most common in oceanic crust, but it also occurs on land (for example, along the San Andreas fault in California; Figure 1.2). At some locations, three plates meet, and these areas are known as triple junctions. One example of a **triple junction** is where the Juan de Fuca, North American, and Pacific Plates meet. Another example of a triple junction is located west of South America north of the equator (Figure 1.2), where the three spreading ridges between the Pacific, Cocos, and Nazca Plates meet.

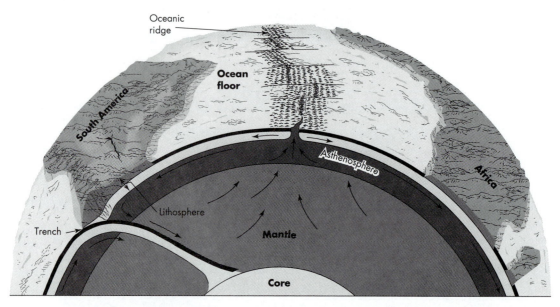

Figure 1.3 Model of plate movement, sea-floor spreading, and mantle convection (Modified after Grand, 1994, *Journal of Geophysical Research*, 99:11,591–11,621). The outer layer (lithosphere) is approximately 100 km thick and is stronger and more rigid than the deeper asthenosphere, which is a hot and slowly-flowing layer of relatively low-strength rock. The oceanic ridge is a spreading center where the plates pull apart, drawing hot, buoyant material into the gap. After these plates cool and become dense, they descend at oceanic trenches (subduction zones). This process of spreading produces ocean basins, and mountain ranges often form where plates converge at subduction zones. (Modified after Hamblin, 1992. *Earth's Dynamic Systems*, 6th ed. New York: Macmillan.)

The good correlation between plate boundaries and earthquakes is dramatically shown on Figure 1.5, which is a map of global seismicity from 1963 to 1988. The locations of earthquakes clearly are related to the major plate boundaries and, in fact, can be used to map the boundaries. It is important to recognize that several large and damaging earthquakes also have occurred far from plate boundaries. However, these events are the exception and not the rule.

EARTHQUAKES AND RELATED PHENOMENA

Our discussion of global tectonics established that the Earth is a dynamic, evolving system. Earthquakes are a natural consequence of the dynamic processes forming the ocean basins, continents, and mountain ranges of the world. As new lithosphere is produced at oceanic ridge systems, older lithosphere is consumed at subduction zones, or as plates slide past one another, **stress** (force per unit area on a specified plane, in a material such as rock) is produced, and **strain** (deformation, such as change in length or volume, or rupture resulting from stress) builds up in the rocks. When the stress exceeds the strength of the rocks, the rocks fail (rupture), and energy is released in the form of an earthquake. As a result, faults are considered **seismic sources**. Identification of seismic sources in an area is the first step in evaluating the earthquake risk.

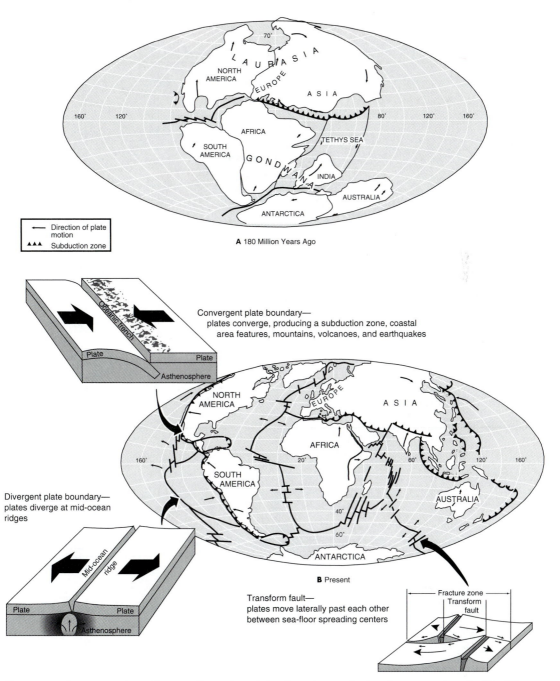

Figure 1.4   The supercontinent Pangaea (Laurasia and Gondwana) started to break up approximately 200 million years ago. Shown here are the inferred positions of the continents at 180 million years ago (a) and at present (b). Arrows show directions of plate motion. See text for further explanation of the closing of the Tethys Sea, the collision of India with Asia, and the formation of mountain ranges. (From Dietz and Holden, 1970. *Journal of Geophysical Research*, 75: 4939–4956. Copyright by the American Geophysical Union; Modifications and block diagrams from Christopherson, 1994. *Geosystems*, 2nd ed. New York: Macmillan.)

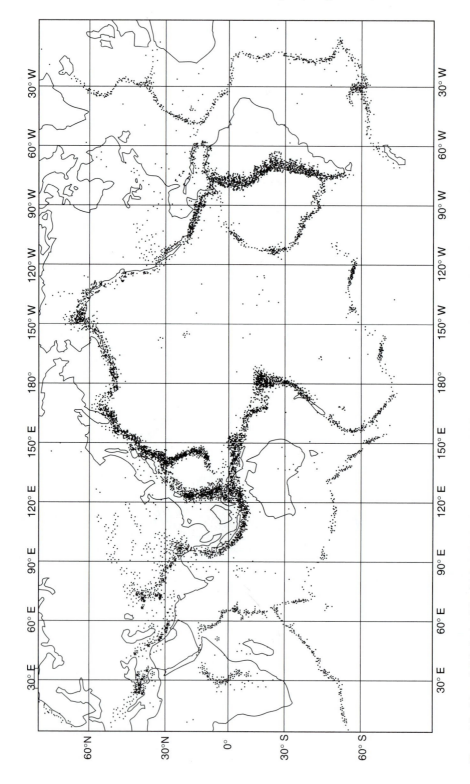

Figure 1.5   Map of global seismicity (1963–1988, Richter magnitude M ≥ 5), delineating belts of earthquake activity that define plate boundaries. Compare with Figure 1.2. (Courtesy of U. S. National Earthquake Information Center.)

The process of **faulting** can be compared with sliding two rough boards past one another. Friction along the boundary between the boards may temporarily slow their motion, but rough edges break off and slip occurs along the boundary. This process is analogous to what happens at plate boundaries where one plate slides past or overrides another. The rocks undergo strain and, if the stress continues, the rocks eventually break, forming a **fault**. A fault is defined as a fracture or fracture system along which rocks have been displaced; that is, rocks on one side of the fault have moved relative to rocks on the other side. Figure 1.6 shows the major types of faults based on sense of relative displacement. Figure 1.7 shows selected aspects of the three major types of faults—**normal**, **reverse** (a low-angle reverse fault, dip less than 45°, is called a thrust fault), and **strike-slip**—and how they may look at the surface. Chapter 2 provides a detailed discussion of how faulting shapes the landscape.

Most of the faults shown on Figure 1.6 displace the surface. However, some faults are buried; that is, fault rupture during earthquakes does not propagate to the surface even during large earthquakes (for example, the 1994 Northridge earthquake). **Buried faults** are commonly reverse faults associated with folding of rock. This important class of faults is discussed further in Chapter 7.

Faults almost never occur as a single trace or rupture. Rather, they form fault zones. A **fault zone** is a group of related fault traces that are subparallel in map view and often partially overlap in en echelon or braided patterns. Fault zones vary in width, ranging from a meter or so to several kilometers wide.

Most long faults such as the San Andreas fault are **segmented**, with each segment having an individual style and history, including earthquake history. Rupture during an earthquake is thought to stop at the boundaries between two segments; however, great

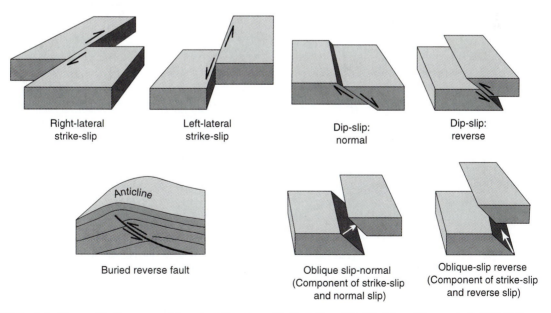

Figure 1.6   Types of fault movement based on the sense of fault motion. (Modified from Wesson et al., 1975. U. S. Geological Survey Professional Paper 941A.)

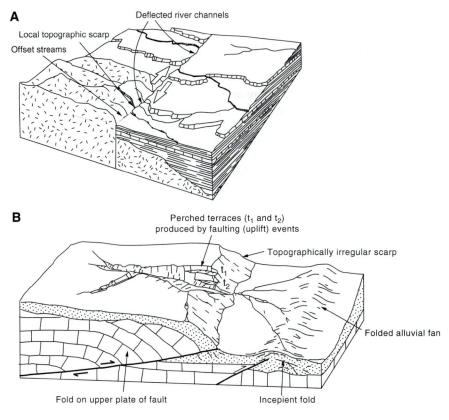

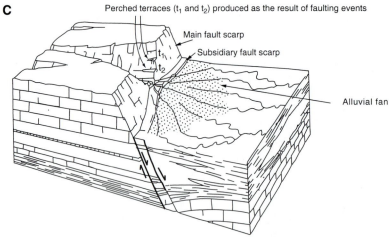

Figure 1.7   Idealized diagrams showing types of topographic expression possible with different types of faults. (a) A strike-slip fault showing a local topographic scarp, deflected river channels, and offset streams. (b) Thrust fault (low-angle reverse fault) with a fold above the fault tip. Topography shows an irregular scarp, perched terraces (t_1 and t_2) produced by faulting (uplift) events, and a folded alluvial fan. (c) Normal fault, with topography showing main and subsidiary fault scarps, perched terraces (t_1 and t_2) produced by faulting (uplift) events, and an alluvial fan. (After Ramsay and Huber, 1987. *Modern Structural Geology*, Vol. 2. New York: Academic Press.)

earthquakes may involve several segments of a fault zone. When the earthquake history of a fault zone is unknown, individual fault segments may be characterized based on changes in fault-zone morphology or geometry. It is preferable, from an earthquake-hazard evaluation point of view, to define fault segments based on seismic activity and paleoseismic evaluation (see Chapter 8). Considerable research is being conducted to better understand the geology and processes that govern fault segmentation and the earthquakes generated on individual segments. Fault segmentation is discussed in later chapters.

Because most earthquakes are concentrated near plate boundaries, most large U.S. earthquakes are in the West, particularly near the North American and Pacific Plate boundaries. However, large, damaging earthquakes also have occurred far from plate boundaries; these are termed **intraplate earthquakes**. For example, during the winter of 1811–1812, a series of particularly strong earthquakes struck the central Mississippi Valley, nearly destroying the town of New Madrid, Missouri, and killing an unknown number of people. These earthquakes rang church bells in Boston, over 1600 km away, and produced intensive surface deformation over a wide area from Memphis, Tennessee, north to the confluence of the Mississippi and Ohio Rivers. During the earthquakes, forests were flattened, fractures opened so wide that people had to cut down trees to cross them, and the land sank several meters in some areas, causing flooding. It was reported that the Mississippi River actually reversed its flow during the shaking [8]. The earthquakes occurred along a seismically active structure known as the New Madrid Seismic Zone (NMSZ), which underlies the geologic structure known as the Mississippi Embayment. The Mississippi Embayment is a downwarped rift in the crust where the lithosphere is relatively weak; it has broken repeatedly because compressional stress is transmitted from the distant boundaries of the plate. The recurrence interval for large earthquakes along the embayment is estimated to be at least several hundred years [8].

The possibility of future damage demands that the earthquake hazard be considered in design and construction of facilities such as power plants and dams, even in the "stable" interior of the North American Plate. A recent study utilizing deformed (folded) Holocene sediments with radiocarbon (^{14}C) age of 2300 ± 100 years [9] concluded that active folding above a buried reverse fault associated with a restraining bend of an active right lateral strike-slip fault in the NMSZ is very young (perhaps less than 10,000 years). Rates of reverse faulting (movement along the fault plane) of 5 to 6 mm/yr and strike slip of 1 to 2 mm/yr are sufficient to generate infrequent large earthquakes. These rates are also great enough to produce, over a period of several hundred thousands of years, significant topographic relief. The fact that relief in the Mississippi River floodplain area is very minor and structural relief (as a result of folding) of deformed sediments is only a few meters supports the hypothesis that active tectonics in the modern or reactivated NMSZ is very recent [9]. Another example of a large intraplate earthquake is the 1886 event that hit Charleston, South Carolina, which claimed 60 lives and caused $23 million in property damage.

MAGNITUDE AND INTENSITY OF EARTHQUAKES

The point within the Earth where earthquake rupture starts and seismic energy is released is called the **focus** [10]. The **epicenter** is the point on the surface of the Earth directly above the focus (Figure 1.8). News media usually report the location of the

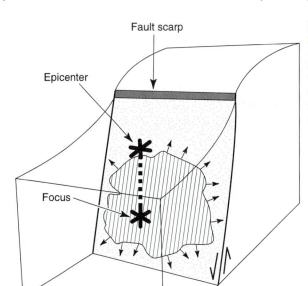

Fault scarp

Epicenter

Focus

Fault slip

Fault plane

Spreading rupture surface

Figure 1.8   Block diagram of a fault plane and rupture area associated with an earthquake. Also shown are the focus, epicenter, and the rupture surface spreading during an earthquake. When the rupture reaches the surface, a fault scarp is produced.

epicenter, but scientific reporting includes the location of the epicenter and the depth to the focus. The depth of an earthquake's focus may vary from a few kilometers to almost 700 km. The deepest earthquakes occur along subduction zones, where slabs of brittle oceanic lithosphere sink to great depths. However, most earthquakes are relatively shallow, with foci less than about 60 km.

However, the depth to the focus in that 60 km can make a tremendous difference. The 1994 Northridge earthquake, with magnitude 6.7, was the most costly in U.S. history, with damages exceeding $40 billion. In 2001 the magnitude 6.8 earthquake (Nisqually Earthquake) that struck the Seattle, Washington region only caused about $2 billion in damages. The difference in damages was due, in large part, to the depths of the foci of the two events. Northridge had a focus of 18 km, whereas in Seattle the depth was 52 km. Buildings above the Northridge earthquake were close to the source of the earthquake energy, and many were extensively damaged. The buildings above the Seattle (Nisqually) earthquake were three times the distance from the source of the energy release (at a depth of 52 km), and experienced less intense shaking and less damage. Thus the depth of the focus of an earthquake is an important parameter in predicting damages.

In Southern California, most earthquakes have foci of about 10 km to 15 km depth, although deeper earthquakes do occur. The magnitude 7.6 event in 1992 on strike-slip faults in the Mojave Desert near Landers, California, had a focus of less than 10 km.

Figure 1.9   Ground rupture from the Landers earthquake, 1992. (Photograph by E. A. Keller.)

The Landers event caused extensive ground rupture for about 85 km, with local vertical displacement exceeding 2 m and extensive lateral displacements as much as about 5 m [11] (Figure 1.9). If the Landers earthquake had occurred in the Los Angeles Basin, it would have caused extensive damage and loss of life.

The **Richter magnitude** (M) of an earthquake has been a useful tool in comparing earthquakes [10]. The Richter magnitude is determined from the largest amplitude (in thousandths of millimeters) of seismic waves recorded on a standard **seismograph** (an instrument for recording earthquake waves) at a distance of 100 km from the earthquake epicenter [10]. The amplitude of the shaking is converted to a Richter magnitude using a logarithmic scale; for example, a M 7 earthquake produces a displacement on the seismograph 10 times larger than does a M 6. The Richter magnitude of an earthquake can be estimated using graphical solutions, as illustrated on Figure 1.10a. Records from a minimum of three seismographs in a region are necessary to precisely locate the epicenter (Figure 1.10b).

In recent years there has been a change from the Richter magnitude (M) to the **moment magnitude** (M_W) scale. The moment magnitude is considered by seismologists to be a natural progression to a more quantitative and physically based scale. The moment magnitude scale is based on the **seismic moment** (M_O), with units of N · m (Newton meters), which is defined as

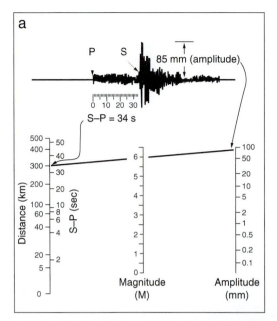

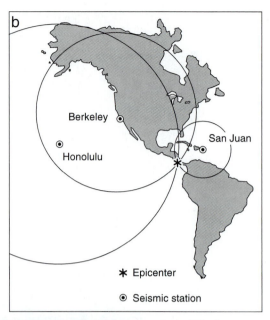

Figure 1.10 (a) Idealized diagram showing one procedure for determining the Richter magnitude (M) of an earthquake. In this example, the maximum amplitude (85 mm) is measured from the seismic record. The difference in arrival time between the S and P waves (34 s) is also taken from the seismic record, and this value also indicates the distance between the epicenter and the recording station (300 km). The approximate magnitude of the earthquake (in this example, M = 6) is obtained by placing a straight line between the amplitude in millimeters and difference in arrival time in seconds, as shown on the diagram. (b) Generalized concept of how the epicenter of an earthquake is located. Distance to event from at least three seismic stations is determined (Figure 1.10a) and plotted. The intersection of the arc distances (circle radii) defines the epicenter. For the diagram, the epicenter in Central America is located from data supplied by three seismic stations. Accurate location of the epicenter is not always as simple as the hypothetical example. (Part a) After Bolt, 1993 [10].)

$$M_O = S \cdot A \cdot \mu \qquad (1.1)$$

where S is the average amount of slip (m) on the fault that produced the earthquake, A is the area (m^2) on the fault plane that ruptured, and μ is the shear modulus of the rocks that failed (resistance to distortion by shear stress; about 3×10^{10} N $\cdot$ m^{-2} for the crust) [12]. In practice, seismic moment may be estimated for an earthquake by examining the records from seismographs, determining the amount and length of rupture, and applying the shear modulus (strength) of rocks at the fault. The moment magnitude (M_W) is then determined from the mathematical relationship

$$M_W = 2/3 \log M_O - 6.0 \qquad (1.2)$$

where M_O is the seismic moment (SI units are N $\cdot$ m), and 6.0 is a constant [12]. The moment magnitude scale has a more sound physical basis and is applicable over a wider range of ground motions than is Richter magnitude. Therefore, its use has been encouraged in reporting earthquake statistics. Earthquake magnitudes reported in this

10 Largest Earthquakes in the World Since 1900

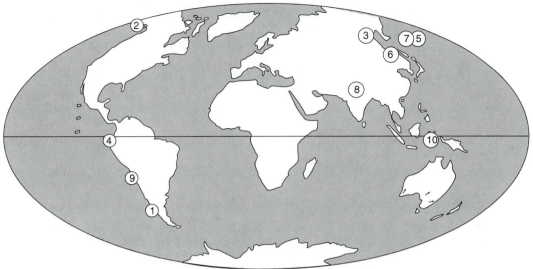

1.) Chile 05/22/1960 9.5 Mw 38.2 S 72.6 W   6.) Kuril Islands 11/06/1958 8.7 Mw 44.4 N 148.6 E
2.) Alaska 03/28/1964 9.2 Mw 61.1N 147.5 W   7.) Alaska 02/04/1965 8.7 Mw 51.3 N 178.6 E
3.) Russia 11/04/1952 9.0 Mw 52.75 N 159.5 W   8.) India 08/15/1950 8.6 Mw 28.5 N 96.5 E
4.) Ecuador 01/31/1906 8.8 Mw 1.0 N 81.5 W   9.) Argentina 11/11/1922 8.5 Nw 28.5 S 70.0 W
5.) Alaska 03/09/1957 8.8 Mw 51.3 N 175.8 W   10 .) Indonesia 02/01/1938 8.5 Mw 5.25 S 130.5 E

Figure 1.11   Locations and moment magnitudes of the 10 largest earthquakes in the world (1900–2000). (U.S. Geological Survey, 2000. Accessed 1/4/00 at http://wwwneic.cr.usgs.gov)

book are mostly moment magnitudes estimated by the U.S. Geological Survey and listed on several Web sites (see citation for Figure 1.11). Figure 1.11 shows the locations with moment magnitudes for the 10 largest earthquakes in the world from 1900 to 2000. Eight of the ten events rim the Pacific Ocean, and half are in the Alaska region.

The energy released in an earthquake (SI [Système Internationale] units are joules) is proportional to the magnitude (Table 1.1), but a one-unit change in magnitude increases the energy released by about 32 times. For example, an M_W 5 earthquake releases about 32 times more energy than an M_W 4 event. Therefore, about 33,000 ($32 \times 32 \times 32$) shocks of M_W 5 are required to release as much energy as a single earthquake of M_W 8. Considering the entire Earth, there are about 1 million earthquakes per year that can be felt by people. However, only a small percentage of these can be felt very far from their source. In a general sense, the magnitude is related to the damage expected from an earthquake. A M_W 8 or above is considered to be a great earthquake, capable of causing catastrophic damage; an earthquake with M_W 7–7.9 is a major earthquake also capable of causing catastrophic damage; and an earthquake of M_W 6–6.9 is a strong earthquake capable of causing extensive to catastrophic damage (Table 1.2).

Table 1.1

RELATIONSHIPS BETWEEN MAGNITUDE, GROUND MOTION, AND ENERGY OF EARTHQUAKES

Magnitude Change	Ground Motion Change (Displacement)	Energy Change
1.0	10.0 times	about 32 times
0.5	3.2 times	about 5.5 times
0.3	2.0 times	about 3 times
0.1	1.3 times	about 1.4 times

Source: U.S. Geological Survey 2000 earthquakes facts and statistics. Accessed on 1/3/00 at http://www.neic.cr.usgs.gov

EARTHQUAKE INTENSITY

A qualitative way of comparing earthquakes is to use the **Modified Mercalli Scale** (abridged), which has 12 divisions of intensity based on observations concerning the severity of shaking during an earthquake (Table 1.3). Intensity reflects how people perceived and structures responded to the shaking. An *instrumental intensity* can be obtained from data recorded from a dense network of high-quality seismographs. Although a particular earthquake has only one Richter or moment magnitude, different levels of intensity may be assigned to the same earthquake at different locations, depending on proximity to the epicenter and local geologic conditions. Figure 1.12 is a map showing the spatial (aerial) variability of instrumental intensity for the 1994, M_W 6.7 Northridge earthquake.

GROUND ACCELERATION DURING EARTHQUAKES

Ground acceleration is the rate of change of the horizontal or vertical velocity of the ground during an earthquake. It is measured relative to the acceleration of gravity, which is 9.8 ms^{-2}, or 1 g. An acceleration of 0.5 g would be 4.9 ms^{-2} and 0.1 g is 0.98 ms^{-2}. Esti-

Table 1.2

WORLDWIDE MAGNITUDE AND FREQUENCY OF EARTHQUAKES BY DESCRIPTOR CLASSIFICATION

Descriptor	Magnitude	Average Annually Number of events
Great	8 and higher	1
Major	7–7.9	18
Strong	6–6.9	120
Moderate	5–5.9	800
Light	4–4.9	6,200 (estimated)
Minor	3–3.9	49,000 (estimated)
Very Minor	>3.0	Magnitude 2–3; about 1000 per day
		Magnitude 1–2; about 8000 per day

Source: U.S. Geological Survey 2000 Earthquakes facts and statistics. Accessed on 1/3/00 at http://www.neic.cr.usgs.gov

Table 1.3

MODIFIED MERCALLI INTENSITY SCALE (ABRIDGED).

Intensity	Effects
I	Not felt except by a very few under especially favorable circumstances.
II	Felt only by a few persons at rest, especially on upper floors of buildings. Delicately suspended objects may swing.
III	Felt quite noticeably indoors, especially on upper floors of buildings, but many people do not recognize it as an earthquake. Standing motor cars may rock slightly. Vibration like passing of truck. Duration estimated.
IV	During the day felt indoors by many, outdoors by a few. At night some awakened. Dishes, windows, doors disturbed; walls make cracking sound. Sensation like heavy truck striking building; standing motor cars rocked noticeably.
V	Felt by nearly everyone; many awakened. Some dishes, windows, etc., broken; a few instances of cracked plaster; unstable objects overturned. Disturbance of trees, poles and other tall objects sometimes noticed. Pendulum clocks may stop.
VI	Felt by all; many frightened and run outdoors. Some heavy furniture moved; a few instances of fallen plaster or damaged chimneys. Damage slight.
VII	Everybody runs outdoors. Damage negligible in buildings of good design and construction; slight to moderate in well-built ordinary structures; considerable in poorly built or badly designed structures; some chimneys broken. Noticed by persons driving motor cars.
VIII	Damage slight in specially designed structures; considerable in ordinary substantial buildings with partial collapse; great in poorly built structures. Panel walls thrown out of frame structures. Fall of chimneys, factory stacks, columns, monuments, walls. Heavy furniture overturned. Sand and mud ejected in small amounts. Changes in well water. Disturbs persons driving motor cars.
IX	Damage considerable in specially designed structures; well-designed frame structures thrown out of plumb; great in substantial buildings, with partial collapse. Buildings shifted off foundations. Ground cracked conspicuously. Underground pipes broken.
X	Some well-built wooden structures destroyed; most masonry and frame structures with foundations destroyed; ground badly cracked. Rails bent. Landslides considerable from river banks and steep slopes. Shifted sand and mud. Water splashed (slopped) over banks.
XI	Few, if any, masonry structures remain standing. Bridges destroyed. Broad fissures in ground. Underground pipelines completely out of service. Earth slumps and land slips in soft ground. Rails bent greatly.
XII	Damage total. Waves seen on ground surfaces. Lines of sight and level distorted. Objects thrown upward into the air.

Source: Wood and Neuman, 1931. U.S. Geological Survey, 1974. *Earthquake Information Bulletin*; 6(5): 28

mated horizontal and vertical peak accelerations during an earthquake likely to occur in an area are useful information for designing buildings and other structures to withstand seismic shaking. It is the horizontal acceleration that is most likely to cause damage to buildings. Homes constructed of adobe, which are common in Mexico, South America, and the Middle East, can fail (collapse) under a horizontal acceleration as small as 0.1 g [10].

The vulnerability of unreinforced, prefabricated five-story apartments to seismic shaking was tragically illustrated on May 28, 1995, when an M_W 7.5 earthquake struck the town of Neftegorsk on the island of Sakhalin off the northeast coast of Russia. Seventeen five-story structures, constructed without consideration of potential earthquake hazard in the late 1960s as part of a development to support oil production, collapsed into rubble. Approximately 2000 of the town's 3000 people were killed.

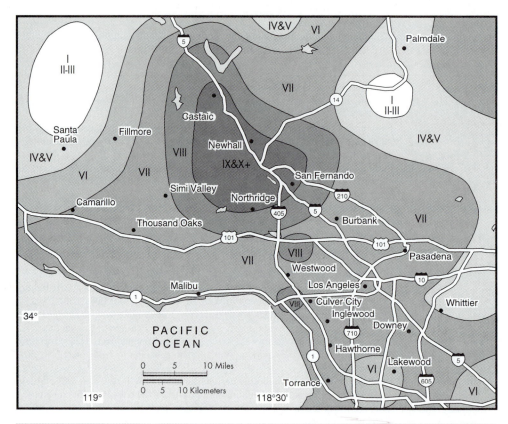

INTENSITY	I	II-III	IV	V	VI	VII	VIII	IX	X+
Shaking	not felt	weak	light	moderate	strong	very strong	severe	violent	extreme
Damage	none	none	none	very light	light	moderate	moderate/hvy	heavy	very hvy
Peak acc (%g)	<0.9	0.9–2.9	2.9–5.3	5.3–9.7	9.7–18	18–33	33–59	59–109	>109
Peak vel (cm/s)	<0.6	0.6–2.2	2.2–4.2	4.2–8	8-15	15–30	30–57	57–110	>110

Figure 1.12   Instrumental intensity map and peak ground acceleration (%g) for the 1994 M 6.7 Northridge, California earthquake. Map is based on observations and data from a dense network of high-quality seismographs. Epicenter is a star, and patterns match intensity of shaking shown on the table. (*Source*: U.S. Geological Survey, courtesy of David Wald.)

In 1999, two catastrophic earthquakes of comparable size occured, separated by less than one month. An earthquake in Turkey (August 17) killed over 30,000 people while an earthquake in Taiwan (September 21) killed far fewer people (2300). Both caused terrible loss of life, but a major reason why the Taiwan event claimed over 27,000 fewer deaths than the Turkey event was, in part, that during the past 30 years Taiwan has enforced building codes that are designed to produce buildings that resist earthquake shaking (Ta-liang Teng, personal communication, 2000). Turkey also has building codes, but the consensus is that they were not enforced, and as a result buildings suffered massive damage, killing thousands of more people than if the buildings had been properly constructed.

SEISMIC WAVES

Faulting breaks rock, and that movement along faults produces seismic waves that cause the ground to vibrate. Some of the seismic waves produced travel within the Earth and are known as **body waves**, whereas others travel along the surface (Figure 1.13). The two types of body waves generated by earthquakes are the primary, or **P waves**, and the secondary, or **S waves**. The P waves are the fastest of the two and, like sound waves, may travel through both solid and liquid materials. These waves push and pull in the direction of wave propagation with an alternating compression and dilation motion. The rate of propagation for P waves through rocks such as granite is approximately 5.5 km/s. Interestingly, P waves with frequency greater than about 15 Hz (cycles per second) are detectable to the human ear when propagated into the atmosphere, explaining why people sometimes hear earthquakes before feeling the shaking from the slower surface waves. S waves travel only through solid materials, and their speed through rocks such as granite is approximately 3 km/s. As S waves propagate, they produce a sideways shearing motion in rocks at right angles to the direction of propagation. This motion is similar to that produced in a clothesline by pulling down and letting go. Liquids (unlike rocks) are unable to spring back when subjected to sideways shear, explaining why S waves cannot move through liquids [10]. Surface waves cause much of the damage to buildings and other structures. Surface waves include **Love waves**, which consist of complex horizontal ground movement, and **Rayleigh waves**, with complex rolling motion (Figure 1.13). Both travel slower than body waves, but Love waves travel faster than Rayleigh waves, in general.

Because different types of waves and waves of different frequency travel at different speeds away from an earthquake source area, they become organized into groups of waves traveling at similar velocities. Near the source of the large earthquake, however, there is not time for this segregation of the waves to take place, and shaking may be severe and complex. Waves traveling through rocks are both reflected and refracted across boundaries between different earth materials and at the surface of the Earth, producing amplification that may enhance shaking and damage to buildings and other structures. Furthermore, as an earthquake occurs, the propagation of waves is also affected by the rupture along a fault, which may be in a particular direction and thus tend to focus earthquake energy in that direction. Finally, surface shaking can be further complicated and accentuated by local soil conditions and topography [10].

P and S body waves have a wide range of frequencies, but because of rapid attenuation (loss of higher frequencies with wave propagation), most body waves tend to have frequencies of 0.5 Hz to 20 Hz. The more complex surface waves have lower frequencies (less than 1 Hz). The **frequency** of an earthquake wave equals the number of waves passing a point of reference per second, expressed as Hertz units. The **period** equals the elapsed time (in seconds) between successive peaks of a wave (observed at a point). The frequency of an earthquake wave is equal to 1 divided by the period (i.e., the reciprocal of the period). For example, the frequency of a P wave with period of 2 s is 1 divided by 2, or 0.5 Hz. Buildings and other structures have natural frequencies of

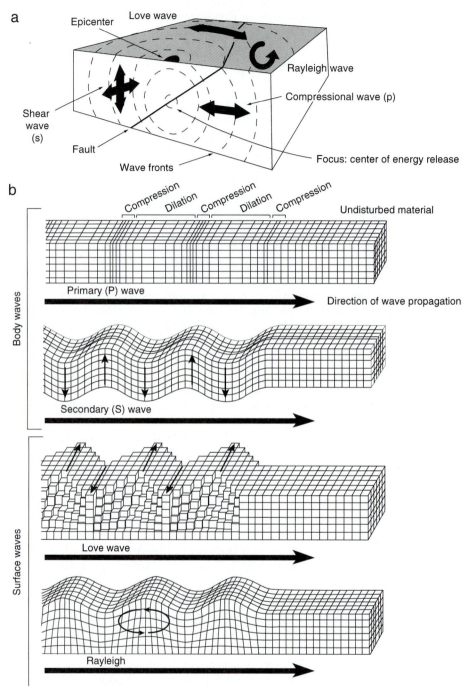

Figure 1.13   (a) Diagram of directions of vibration of body (P and S) and surface (Love and Rayleigh) waves generated by an earthquake associated with the illustrated fault. Also shown are the focus (center of energy release) and epicenter of the earthquake event. (b) Propagation of body and surface waves. (Part (a) From Hays, 1981 (13); Part (b) after Bolt, 1993 [10].)

vibration that may coincide with earthquake frequencies. Shaking of buildings is amplified when the frequency of earthquake waves is close to the natural frequency of the building. Low buildings have a higher natural frequency than taller buildings and, as a result, compressional and shear waves with relatively high frequencies tend to accentuate damage to low buildings. On the other hand, surface waves with lower frequencies tend to damage tall buildings.

High-frequency waves attenuate (die or diminish) much more quickly with distance from a generating earthquake than do low-frequency waves. Thus, tall buildings may be damaged at relatively long distances (up to several hundred kilometers) by large earthquakes [10, 13], whereas low buildings tend to sustain the greatest damage near earthquake epicenters. This principle was dramatically illustrated in 1985 when a M_W 8.1 earthquake several hundred kilometers away from Mexico City damaged or destroyed many of the taller buildings in the city.

MATERIAL AMPLIFICATION

Earth materials such as bedrock, sand and gravel, and silts and muds respond differently to seismic shaking. For example, the intensity of shaking of unconsolidated sediments may be much more severe than for bedrock (Figure 1.14). This effect is called **material amplification**. A major lesson from the 1985 earthquake affecting Mexico City was that buildings constructed on materials likely to amplify seismic shaking are extremely vulnerable to earthquakes, even if the event is centered several hundred kilometers away. Seismic waves from this earthquake, which occurred offshore of Mexico, initially contained many different frequencies, but the seismic waves that survived the several-hundred-kilometer journey to the city were those with relatively long periods of 1 to 2 s (frequencies of 1.0 to 0.5 Hz). It is speculated that when these waves struck the lake beds on which Mexico City is built, the amplitude of shaking may have increased at the surface by a factor of 4 to 5 times (Figure 1.15). The intense regular shaking caused buildings to sway back and forth, and eventually many of them collapsed or "pancaked" as upper stories collapsed onto lower ones. Most of the damage was to buildings with 6 to 16 stories, because these buildings had a natural frequency that nearly matched that of the arriving seismic waves [14].

The potential for amplification of surface waves to cause damage was again demonstrated with tragic results during the 1989 M_W 7.2 Loma Prieta (San Francisco) earthquake, when the upper tier of the Nimitz Freeway in Oakland, California, collapsed,

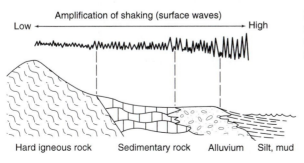

Figure 1.14 Generalized relationship between near-surface earth material and amplification of shaking during a seismic event.

a

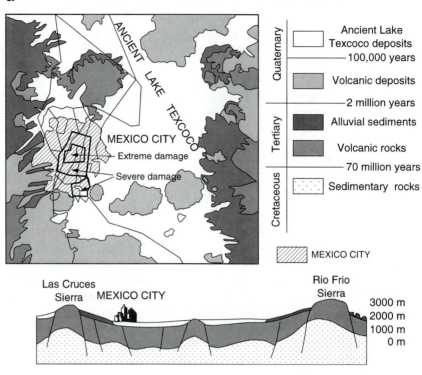

b

Figure 1.15 Earthquake damage, Mexico City, 1985. (a) Generalized geologic map of Mexico City showing ancient lake deposits where greatest damage occurred. (b) Multistory building, one of many that collapsed. (Map and photo courtesy of T. C. Hanks and D. Herd, U.S. Geological Survey.)

killing 41 people (Figure 1.16). Collapse of the tiered freeway occurred on a section of roadway constructed on bay fill and mud. Where the freeway was constructed on older, stronger alluvium, less shaking occurred and the structure survived. Extensive damage was also recorded in the Marina District of San Francisco (Figure 1.17), primarily in areas constructed on bay fill and mud, including debris dumped into the bay during the cleanup following the 1906 earthquake [15].

DIRECTIVITY

An earthquake may be considered as a process of rupture that starts from an initial point on the fault plane (the focus). Fault rupture does not occur instantaneously and it does not proceed in a uniform manner along the fault plane. For example, during the Northridge M_W 6.7 event, the earthquake ruptured the fault plane for approximately 8 seconds, during which time the earthquake the rupture propagated up and along the fault plane in a northwesterly direction at a speed of approximately 3 km/s. This process is known as **directivity**. The average slip across the fault was about 1 m, but the rupture propagation was not uniform and some parts of the fault plane experienced little or no slip while others experienced more than 3 m. Areas along the fault plane where slip changes are known as **asperities** and are the sources of pulses of earthquake energy that arrive at the surface at different times [16].

Directivity increases the amplitude of seismic waves in the direction of fault rupture. As a result the direction of rupture can greatly affect the intensity of seismic shaking (Figure 1.18). In the direction of propagation of fault rupture, the amplitude of the resultant wave may be as much as 10 times the amplitude of the waves in the reverse direction. This suggests that damages from seismic shaking may be much greater in the direction of fault rupture (propagation) than in the opposite direction (Figure 1.19).

ACTIVE FAULT ZONES

Most geologists would consider a fault to be **active** if it has moved during the past 10 ky[1] (Holocene Epoch). The Quaternary Period (the past 1.65 M.y.) is the most recent period of geologic time, and most of our landscape has been produced during that time. Any fault that has moved during the Quaternary Period may be classified as **potentially active** (Table 1.4). Faults that have not moved during the past 1.65 M.y. are generally classified as **inactive**. However, it is often difficult to prove the activity of a fault in the absence of historical earthquakes. To prove that a fault is active, it is necessary to determine its past earthquake history (paleoseismicity) based on the geologic record. This involves identifying faulted earth materials and determining when the most recent displacement occurred. The preceding definition of an active fault is used in the state of California for **seismic zoning**. However, other agencies have more conservative definitions for fault activity. For example, when considering seismic safety for nuclear power plants, the U.S. Nuclear Regulatory Commission defines a fault as **capable** if the fault has moved

---

[1] 1 ky = 1000 yrs; 1 ka = 1 ky before present (5 ka means an age of 5,000 yrs). 1 M.y. = 1,000,000 yrs; 1 Ma = 1 M.y. before present.

a

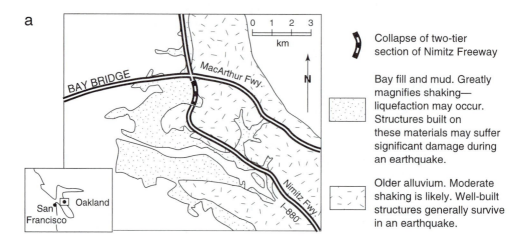

) Collapse of two-tier section of Nimitz Freeway

Bay fill and mud. Greatly magnifies shaking— liquefaction may occur. Structures built on these materials may suffer significant damage during an earthquake.

Older alluvium. Moderate shaking is likely. Well-built structures generally survive in an earthquake.

b

Figure 1.16 (a) Generalized geologic map of part of San Francisco Bay showing bay fill and mud and older alluvium. (b) Collapsed freeway. (Part (a) modified from Hough et al., 1990. *Nature*, 344:853–855. [copyright] Macmillan Magazines Ltd., 1990. Used by permission of the author. Part (b) courtesy of John K. Nakata, U.S. Geological Survey.)

Figure 1.17   Damage to buildings in the Marina District of San Francisco resulting from the 1989 earthquake. (Photograph courtesy of John K. Nakata, U.S. Geological Survey.)

at least once in the past 50 ky or more than once in the past 500 ky. These criteria provide a greater safety factor, reflecting increased concern for the risk of siting nuclear power plants.

SLIP RATES AND RECURRENCE INTERVALS

Our discussion of faults and earthquakes involves two important concepts: slip rates on faults, and recurrence intervals, or repeat times, of earthquakes. **Slip rate** on a fault is defined as the ratio of slip (displacement) to the time interval over which that slip occurred. For example, if a fault has moved 1 m during a time interval of 1 ky, the slip rate is 1 mm/yr (1 m/ky). The **average recurrence interval** on a particular fault is defined as the average time interval between earthquakes, and it may be determined by three methods:

1. *Paleoseismic data*: Averaging the time intervals between earthquakes recorded in the geologic record (see Chapter 8).
2. *Slip rate*: Assuming a given displacement per event and dividing that number by the slip rate. For example, if the average displacement per event is 1 m (1000 mm) and the slip rate is 2 mm/yr, then the average recurrence interval would be 500 yr.
3. *Seismicity*: Using historical earthquakes and averaging the time intervals between events.

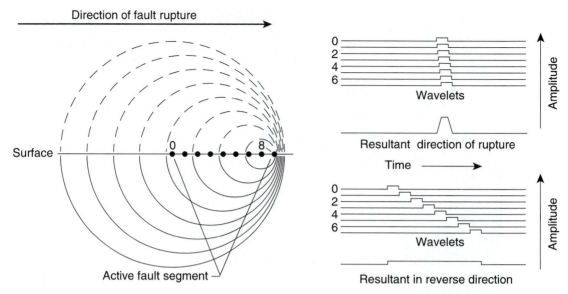

Figure 1.18   Concept of directivity increasing the amplitude of seismic waves in the direction that rupture propagates. Rupture begins at point 0, but expands asymmetrically to the right (1, 2 ... 8). Circles indicate the distance that seismic energy released at each time interval has traveled. Because the rate of rupture propagation is about equal to the speed at which the seismic waves travel, the waves from different times can coincide and amplify in areas to the right. This process is analogous to the Doppler effect of increasing pitch of sound waves from approaching train whistle. (After Benioff, 1955. *California Division of Mines Bulletin*, 171:199–202.)

Defining the terms *slip rate* and *recurrence interval* is easy, and the calculation is straightforward, but the underlying concepts are far from simple. Fault slip rates and recurrence intervals tend to be variable in time, casting suspicion on rates averaged over long periods of time. For example, it is not uncommon for earthquake events to be **clustered** in time and then be separated by relatively long periods of low activity. Both slip rate and recurrence interval will vary depending on the time interval for which data are available. The topics of slip rates and recurrence intervals will be discussed repeatedly in this book. They are introduced here to facilitate later discussions.

TECTONIC CREEP

Tectonic creep is the process of displacement along a fault zone that is not accompanied by perceptible earthquakes. The process can slowly damage roads, sidewalks, building foundations, and other structures. Tectonic creep has damaged culverts under the football stadium of the University of California at Berkeley, and periodic repairs have been necessary. Movement of approximately 3.2 cm in 11 years was measured (Figure 1.20) [17]. More rapid rates of tectonic creep have been recorded on the Calaveras fault zone, a segment of the San Andreas fault near Hollister, California. At one location, a winery located on the fault is slowly being pulled apart at about 1 cm/yr [18]. Damages resulting from tectonic creep generally occur along narrow fault zones subject to slow, continuous displacement. However, creep may also be discontinuous and variable in rate.

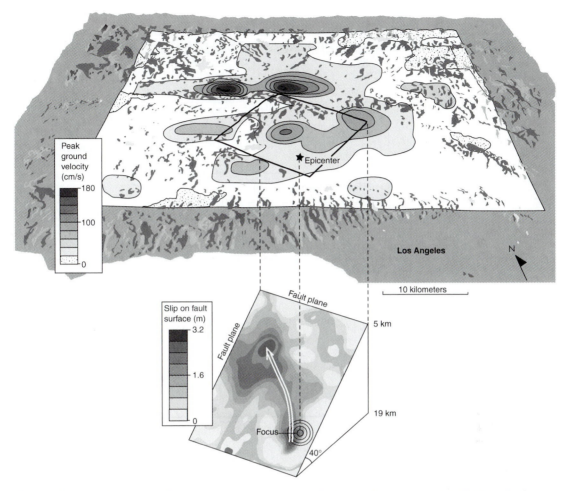

Figure 1.19   Aerial view of the Los Angeles region from the south showing the epicenter of the 1994 Northridge earthquake with peak ground motion in centimeters per second and the fault plane in its subsurface position. The fault rupture apparently began at the focus in the southeastern part of the fault plane and proceeded upward and to the northwest, as shown by the arrow. The area that ruptured is approximately 430 km^2, and the fault plane dips at approximately 40° to the south-southwest. Notice that maximum slip and peak ground velocities both occur to the northwest of the epicenter. (U.S. Geological Survey, 1996. U.S. Geological Survey Open-File Report 96–263.)

ESTIMATION OF SEISMIC RISK

Catastrophic earthquakes are devastating events. Historic earthquakes have destroyed large cities and taken thousands of lives in a matter of seconds. Table 1.5 lists some of the major historical earthquakes that have occurred in the United States.

Seismic risk maps have been prepared for the United States (Figure 1.21). One way of interpreting Figure 1.21 is that the darkest areas represent the regions of greatest seismic hazard, because those areas are most likely to experience the greatest seismic shaking (in this case, horizontal ground acceleration) in an average 50-yr interval.

Table 1.4

TERMINOLOGY RELATED TO DEGREE OF FAULT ACTIVITY.

Geologic Age			Years Before Present	Fault Activity
Era	Period	Epoch		
Cenozoic	Quaternary	Historic (Calif.) Holocene	— 200 —	Active
			— 10,000 —	
		Pleistocene		Potentially active
			— 1,650,000 —	
	Tertiary	Pre-Pleistocene		
			— 65,000,000 —	Inactive
Pre-Cenozoic time				
Age of the earth			4,500,000,000	

(After California State Mining and Geology Board Classification, 1973.)

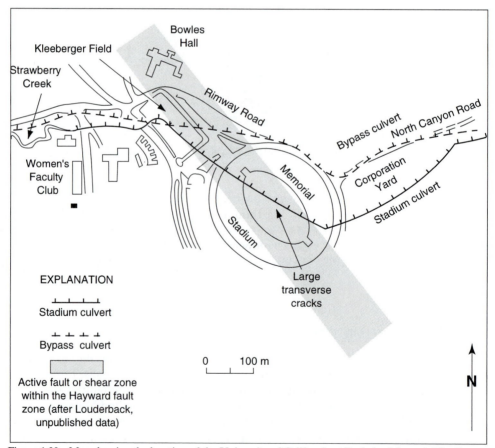

EXPLANATION

⊥⊥⊥⊥ Stadium culvert

+ + + + Bypass culvert

▨ Active fault or shear zone within the Hayward fault zone (after Louderback, unpublished data)

0   100 m

N

Figure 1.20   Map showing the location of the University of California at Berkeley Memorial Stadium, the active fault or shear zone within the Hayward fault zone, and the stadium culvert where major cracking has taken place. (After Radbruch et al., 1966 [17].)

Table 1.5

SELECTED MAJOR EARTHQUAKES IN THE UNITED STATES.

Year	Locality	Damage $ Million	Lives Lost
1811–12	New Madrid, Missouri	Unknown	
1886	Charleston, South Carolina	23	60
1906	San Francisco, California	524	700
1925	Santa Barbara, California	8	13
1933	Long Beach, California	40	115
1940	Imperial Valley, California	6	9
1952	Kern Country, California	60	14
1959	Hebgen Lake, Montana (damage to timber and roads)	11	28
1964	Alaska and U.S. West Coast (includes tsunami damage from earthquake near Anchorage)	500	131
1965	Puget Sound, Washington	13	7
1971	San Fernando, California	553	65
1983	Coalinga, California	31	—
1983	Central Idaho	15	2
1987	Whittier, California	358	8
1989	Loma Prieta (San Francisco), California	5,000	62
1992	Landers, California	27	1
1994	Northridge, California	40,000	61
2001	Seattle, Washington	2,000	1

(Modified after Hays, 1981 [13].)

The map is based on historical seismicity, frequency of earthquakes of various magnitudes, and slip rates on faults. Estimated ground accelerations assume firm rock conditions. Actual hazard at a particular site may vary as a result of material amplification or directivity of seismic shaking. Although regional earthquake hazard maps are valuable, considerably more data are necessary to evaluate hazardous areas more precisely in order to develop building codes and determine insurance rates.

In California, **conditional probabilities** (probability dependent on known or estimated conditions) of major earthquakes along segments of the San Andreas fault and related faults for a 30-yr period (1994–2024) have been calculated (Figure 1.22). The probabilities were calculated following synthesis of historical records and geologic evaluation of prehistoric earthquakes [19]. In 1988, this approach assigned a probability of about 30% for a major event on the San Andreas fault segment through the Santa Cruz Mountains, where the M_W 7.2 Loma Prieta earthquake occurred on October 17, 1989. Occurrence of this earthquake supports the validity of the conditional-probability approach. The probability of a large earthquake on the southern segment of the San Andreas fault is estimated to be close to 50% for the next 30 years. The M_W 7.6 Landers earthquake that occurred east of the San Andreas fault in 1992 was a surprising event. That event produced major right-lateral horizontal surface displacement of up to 5 m, and maximum Modified Mercalli intensity of VIII [20] on a fault system that was previously mapped but that had not received much attention. This large earthquake caused relatively little damage ($27 million) and one death, primarily because it occurred in a region with low buildings and few people.

Another large (M_W 7.1) right-lateral strike-slip earthquake, known as the Hector Mine earthquake, occurred in a sparsely populated part of the Mojave Desert about 40

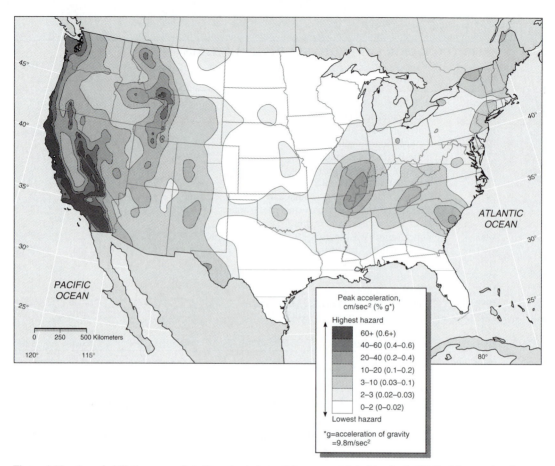

Figure 1.21   A probabilistic approach to the seismic hazard from ground shaking in the United States, showing ground accelerations having a 10% probability of being exceeded in a 50-year time period. (From U.S. Geological Survey 1999.)

km north of Joshua Tree, California, in October of 1999. Rupture length of the event, which was on the previously mapped north-northwest trending Lava Lake fault in the eastern Mojave Shear Zone, was 40 km with maximum right-lateral displacements of 4 to 5 m. The fault was mapped years ago by Thomas Dibblee, Jr. (a famous field geologist who mapped much of California) and reinforces the value of geologic mapping in recognizing faults. The Eastern Mojave shear zone was also the source for the 1992 (M_W 7.6) Landers earthquake. The shear zone evidently relieves some of the strain that builds up on the boundary between the North American and Pacific Plates. However, most strain is thought to be relieved along the boundary by the San Andreas fault system. It is speculated that Hector Mine (1999) and Landers (1992) events along with smaller events may represent a clustering of earthquakes. That is, the shear zone may produce several events within a relatively short time period of decades to a hundred or so years followed by several thousands years of seismic quiescence [21]. Of course, the events might be coincidental, but some connection is likely.

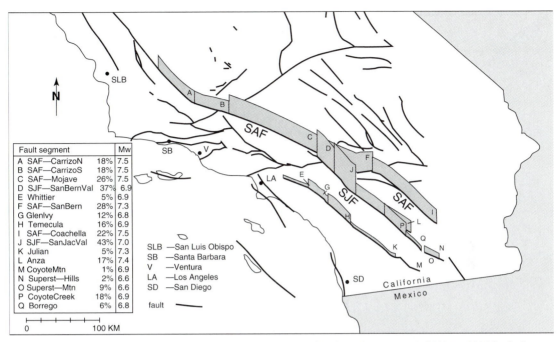

Figure 1.22   Rupture probabilities (%) and moment magnitudes (M_W) for the time period 1994 to 2024 for fault segments associated with the San Andreas fault (SAF), San Jacinto fault (SJF), and other related faults (listed) associated with SAF. (Modified after Working Group on California Earthquake Probabilities, 1995. Bulletin of the Seismological Society of America, 85:379–439.)

As more historical and geologic information is gathered, more detailed estimation of the probability of future earthquakes is possible. For example, based on research since the Loma Prieta (San Francisco) earthquake, it has been estimated for the San Francisco Bay region that at least one major earthquake with M_W 6.7 or larger has a 70 ± 10% probability of occurring between 2000 and 2030 [22]. Such an event is capable of causing widespread destruction and loss of life. If the event is centered in a highly urbanized area, damages and loss of life could, in a worst-case scenario, be similar to the M_W 6.9 event that occurred in Kobe, Japan, in 1995, which killed more than 6000 people and caused damages of about $100 billion. A M_W 6.7 event was assumed in the probability analysis because that is the magnitude of the 1994 Northridge (Los Angeles) earthquake, which caused 61 deaths and more than $40 million in damage.

Earthquake probabilities for specific faults for one or more M_W 6.7 events from 2000 to 2030 are shown on Figure 1.23. The 70 ± 10% probability for a large damaging earthquake in the San Francisco Bay region in the next 30 years was derived by analyzing several processes [22]:

- Motion of the Pacific and North American tectonic plates
- Slip on faults that mostly occur during earthquakes
- How strain from current plate motion of 3.8 cm/yr (measured from Global Positioning System, GPS; see Chapter 3) is distributed into the individual faults in the region

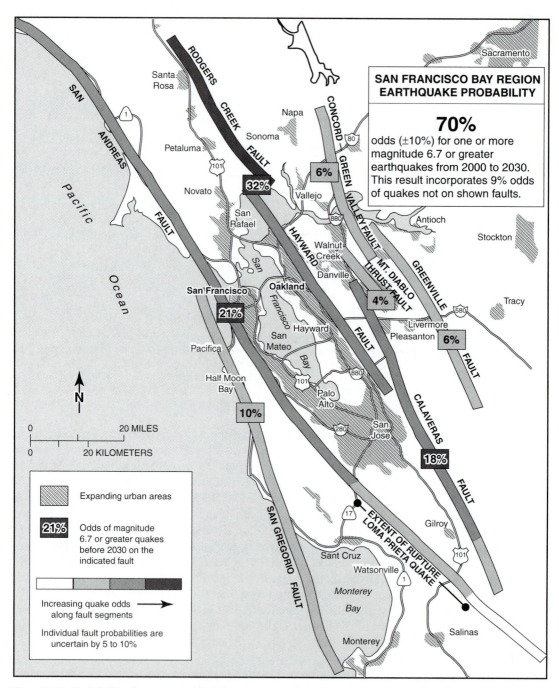

Figure 1.23 Probability of at least one M_W 6.7 or greater earthquake occurring from 2000 to 2030 on specific faults in the San Francisco Bay region. Probability of $70 \pm 10\%$ is the combined probability for the entire region and is not the simple sum of individual probabilities for faults shown. (Working Group on California Earthquake Probabilities, 1999. U.S. Geological Survey Fact Sheet 152–99.)

- Slip on fault that does not accompany earthquakes, known as tectonic creep (discussed previously)

Results of the San Francisco Bay region study emphasize the importance for all communities in the region to continue to prepare for earthquakes [22].

EFFECTS OF EARTHQUAKES

Primary effects of earthquakes are caused directly by the earthquake and can include violent ground-shaking motion accompanied by surface rupture and permanent displacement. For example, the M_W 7.7 1906 earthquake at San Francisco produced 6.5 m of horizontal displacement and a maximum Modified Mercalli intensity of XI [10]. Such violent motions can produce surface accelerations that snap and uproot large trees and knock people to the ground. This motion may shear or collapse large buildings, bridges, dams, tunnels, and pipelines, as well as other rigid structures [23]. The great 1964 Alaskan earthquake (M_W 9.2) caused extensive damage to railroads, airports, and buildings. The 1989 Loma Prieta (San Francisco) earthquake, with a M_W 7.2, was much smaller than the Alaska event and yet caused about $5 billion in damage. The 1994 Northridge earthquake, with M_W 6.7, was one of the most expensive disasters ever in the United States. The Northridge event caused so much damage because there was so much there to be damaged—the Los Angeles region is highly urbanized and has a high population density.

Short-term secondary effects of earthquakes include liquefaction, landslides, fires, seismic seawaves (tsunami), and floods (following collapse of dams). Long-term secondary effects include regional subsidence or emergence of landmasses and regional changes in groundwater levels.

LIQUEFACTION

Liquefaction is defined as the transformation of water-saturated granular material from a solid to a liquid state. During earthquakes, this may result from an increase in pore-water pressure caused by compaction during intense shaking. Liquefaction of near-surface water-saturated silts and sand causes the materials to lose shear strength and flow. As a result, buildings may tilt or sink into the liquefied sediments, and tanks or pipelines buried in the ground may float to the surface [24].

LANDSLIDING

Intense earthquake shaking commonly triggers landslides (a comprehensive term for several types of hillslope failure) in hilly and mountainous areas. Landslides can be extremely destructive and cause great loss of life, such as during a M_W 7.7 1970 earthquake in Peru. In that event, more than 70,000 people died; of this total, 20,000 were killed by a giant landslide that buried several towns. Both the 1964 Alaskan earthquake and the 1989 Loma Prieta earthquake caused extensive landslide damage to buildings, roads, and other structures.

FIRE

Fire is a major secondary hazard associated with earthquakes. Shaking of the ground and surface displacements can break electrical power and gas lines and ignite fires. In individual homes and other buildings, appliances such as gas heaters may be knocked over. The threat from fire is doubled because firefighting equipment may be damaged and water mains may be broken. Earthquakes in both Japan and the United States have been accompanied by devastating fires. The San Francisco earthquake of 1906 has been called the "San Francisco Fire"; in fact, 80% of the damage from that event was caused by firestorm that ravaged the city for several days. The 1989 Loma Prieta earthquake also caused large fires in the city's Marina District. Perhaps the most lethal earthquake-induced fire occurred in 1923 in Japan. The earthquake killed 143,000 people, and 40% of them died in a firestorm that engulfed an open space where people had gathered in an unsuccessful attempt to reach safety [23]. The 1995 M_W 6.9 Kobe, Japan, earthquake ruptured gas lines, and fires devastated parts of the city. Ruptured water lines and damaged roads prevented firefighters from reaching and extinguishing fires.

TSUNAMI

Tsunami, or seismic sea waves, can be extremely destructive and present a serious natural hazard. Most of the lives lost in the 1964 Alaskan earthquake were attributed to tsunami (Figure 1.24). Fortunately, damaging tsunami occur infrequently and are usually

Figure 1.24 Tsunami damage to fishing boats at Kodiak, Alaska, caused by the 1964 earthquake. (Photograph courtesy of National Oceanic and Atmospheric Administration [NOAA].)

confined to the Pacific Basin. The frequency of these events in the United States is about one every 8 years [25]. Tsunami originate when ocean water is vertically displaced during large earthquakes, submarine mass movements, or submarine volcanic eruption. In open water the waves may travel at speeds as great as 800 km/h, and the distance between successive crests may exceed 100 km. Wave heights in deep water may be less than 1 m, but when the waves enter shallow coastal waters they slow to less than 60 km/h, and their heights may increase to more than 20 m.

A small town on the island of Okushiri, Japan, was extensively damaged from the July 12, 1993, tsunami produced by an M_W 7.8 earthquake in the Sea of Japan. Vertical run-up (elevation above sea level to which water from the waves reached) varied from 15 m to 30 m [26]. There was virtually no warning because the epicenter of the earthquake was very close to the island and the big waves arrived only 2 to 5 minutes after the earthquake. The tsunami killed 120 people and caused $600 million in property damage.

A magnitude M_W 7.1 earthquake on July 17, 1998, with an epicenter located 50 km off the coast of Papua New Guinea, caused a large tsunami with wave heights 10 to 15 m. A series of three waves arrived 10 to 20 minutes after the earthquake, killing over 2100 people. The wave height and run-up on shore was surprisingly high for a subduction zone earthquake of M_W 7.1. The tsunami probably resulted from a combined effect of the earthquake and a submarine landslide. The Papua New Guinea event emphasizes the potential devastating damage that can result from unusually large waves produced by a locally generated earthquake [27, 28].

Tsunami also can cause catastrophic damage thousands of kilometers from where they are generated. In 1960 an earthquake originating in Chile triggered a tsunami that reached Hawaii 15 hours later, killing 61 people. However, long travel times now allow many tsunami to be detected in time to warn the coastal communities in their path. Following an earthquake that produces a tsunami, the arrival time of the waves can often be estimated to within 1.5 minutes per hour of travel time. This information has been used to produce tsunami warning systems such as that shown for Hawaii in Figure 1.25.

Consideration is now being given to produce other tsunami warning systems for example, to warn residents in northwestern California of tsunami generated by Alaskan or Cascadia subduction-zone earthquakes. Travel time for a tsunami from the Aleutian Islands in Alaska to northern California is about 4 hours, and movement of the tsunami southward can be monitored from changes in water levels at coastal tide gauges. The plan involves placing such gauges on the bottom of the Pacific Ocean, four off Alaska, and three off the northwest coast of California.

The hazard from a tsunami at a particular site on the coast depends in part on local coastal and sea-floor topography, that may increase or decrease wave height [10]. Damage caused by tsunami is most severe at the water's edge, where boats, harbors and buildings, transportation systems, and utilities may be destroyed. The waves may also cause damage to aquatic and supratidal life in both near- and on-shore environments [25].

Waves caused by landslides may also have considerable effect and cause extensive damage. In 1958 an earthquake triggered a landslide into Lituya Bay, Alaska, causing a truly giant wave that produced run-up on land to an elevation of over 500 m above sea level [29].

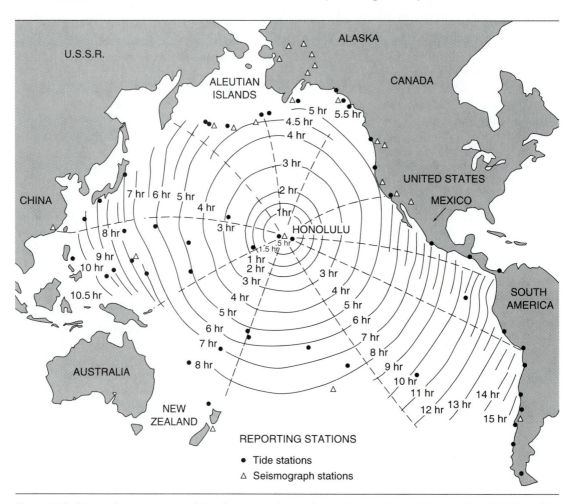

Figure 1.25 Tsunami warning system. Map shows reporting stations and tsunami travel times to Honolulu, Hawaii. (From NOAA.)

REGIONAL CHANGES IN LAND ELEVATION

Vertical deformation, including both uplift and subsidence, is another secondary effect of some large earthquakes. The great (M_W 9.2) 1964 Alaskan earthquake, with Modified Mercalli intensity of X–XI [10], caused vertical deformation over an area of more than 250,000 km^2 [30]. The deformation included two major zones of warping, each about 500 km long and more than 210 km wide (Figure 1.26), including uplift as much as 10 m and subsidence as much as 2.4 m. The effects of these regional changes in land level ranged from severely disturbing coastal marine life to changes in groundwater levels. As a result of subsidence, flooding occurred in some communities, whereas in areas of uplift, canneries and fishermen's homes were displaced above the high-tide line, rendering docks and other facilities inoperable. In 1992, a major earthquake (M_W 7.1) near Cape Mendocino in northwestern California produced approximately 1 m of uplift at the

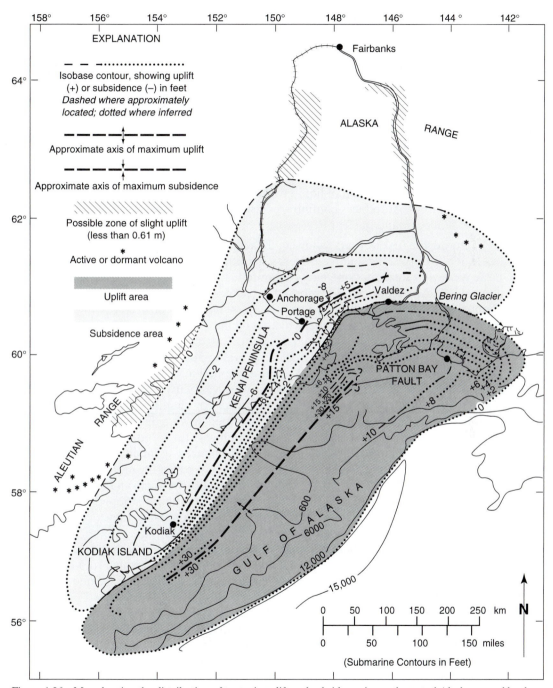

Figure 1.26   Map showing the distribution of tectonic uplift and subsidence in south-central Alaska caused by the Alaskan earthquake of 1964 (2 to 30 ft = 0.6 to 9.2 m). (From Eckel, 1970. U.S. Geological Survey Professional Paper 0546.)

shoreline, resulting in the deaths of communities of marine organisms exposed by the up-lift [31] (see Chapter 6).

EARTHQUAKES CAUSED BY HUMAN ACTIVITY

Several human activities are known to cause earthquakes or to increase earthquake activity. Damage from these earthquakes is regrettable, but the lessons learned may help control or stop large catastrophic earthquakes in the future. Four ways that the actions of people can cause earthquakes are [32]:

- Loading the Earth's crust, such as a result of building dams and impounding reservoirs (reservoir-induced seismicity)
- Deep-well injection of liquid waste
- Mining that reduces confining pressure of rocks above mined areas
- Underground nuclear explosions

During the first 10 years following the completion of the Hoover Dam on the Colorado River in Arizona and Nevada, several hundred local tremors occurred. Most of these were very small, but one had a magnitude of about 5, and two had magnitudes of about 4 [32]. An earthquake—attributed to reservoir-induced seismicity—of magnitude about 6 in India killed about 200 people following dam construction and filling of a reservoir. Evidently, faults may be activated by the increased load of water on the land and by increased water pressure in the rocks below the reservoir.

From April 1962 to November 1965, several hundred earthquakes occurred in the Denver, Colorado, area. The largest earthquake had a M 4.3 and knocked bottles off store shelves. The source of the earthquakes was eventually traced to the Rocky Mountain Arsenal, which was manufacturing materials for chemical warfare. Liquid waste from the manufacturing process was being pumped down a disposal well to a depth of about 3600 m. The rock receiving the waste was a highly fractured metamorphic unit, and injection of liquid increased the fluid pressure, apparently facilitating slippage along preexisting fractures and producing the earthquakes. Study of the earthquake activity revealed a strong correlation between the rate of injection of waste and the occurrence of earthquakes. When waste injection stopped, the earthquakes stopped [33]. These induced earthquakes in the Denver area were a milestone because they alerted scientists to the fact that earthquakes and fluid pressure are related.

Other human activities, including quarrying, mining, and withdrawal of petroleum, may induce shallow small earthquakes as a result of the removal of rock or fluid. For example, in April 1995 a magnitude 2.3 earthquake was felt at a diatomite processing plant and quarry in the westernmost Transverse Ranges near the city of Lompoc, California. The earthquake was recorded on seismograph stations as far as 185 km away. Several hours later, workers at the quarry found a 210-m-long reverse-fault scarp on the floor of the quarry. Other ruptures had been reported in 1981, 1985, and 1988, and investigation of the 1995 event suggests that the earthquake was the result of the removal of approximately 40 m of overburden in the quarry. Furthermore, it was concluded that the 40 m was a critical threshold value that reduced the normal stress (weight of the overburden) necessary to induce small earthquakes. Careful evaluation of the quarry floor and surrounding area suggests that the maximum observed rupture length was about 770 m and maximum scarp height was 18 cm. The Lompoc event illustrates co-seismic rock failure (faulting) at a scale intermediate between laboratory experiments and large damaging crustal scale earthquakes [34].

Numerous earthquakes with magnitudes as large as 5.0 to 6.3 have been triggered by underground explosions at the U.S. nuclear test site in Nevada [32]. Analysis of the aftershocks suggests that the explosions caused some release of natural tectonic strain. This led to discussions by scientists as to whether nuclear explosions might be used to prevent large earthquakes by releasing strain before it reached a critical point and caused a large earthquake. These discussions never resulted in serious consideration of actual application.

THE EARTHQUAKE CYCLE

Observations of the 1906 San Francisco earthquake led to a model known as the **earthquake cycle**. Important features of the hypothesis are related to drop in elastic strain following an earthquake and reaccumulation of strain prior to the next event. **Strain** was defined previously as deformation (displacement or change in shape or volume) resulting from stress, and **elastic strain** may be thought of as deformation that is not permanent, provided that the stress is released. If the strain is released, the deformed material returns to its original shape; for example, when a rubber band is stretched and released or when an archery bow is bent and released. During an earthquake, elastic strain drops because there is a stress drop when the rocks break and permanent displacement occurs (the rubber band or bow breaks). This process is referred to as **elastic rebound** (Figure 1.27). It takes time for sufficient elastic strain to accumulate again to produce another earthquake [12]. The earthquake cycle is discussed further in Chapter 3.

Although we have many empirical observations concerning physical changes in earth materials before, during, and after earthquakes, there is no general agreement on a physical model to explain the observations. One model, known as the **dilatancy diffusion model** [10, 35], assumes that the first stage in earthquake development is an increase of elastic strain in rocks that causes them to dilate, or undergo an inelastic increase in volume after the stress on the rock reaches one-half its breaking strength. During dilation, open fractures develop in the rocks, and at this stage, the first physical changes take place that might indicate a future earthquake. The model assumes that the dilatancy and fracturing of the rocks are first associated with a relatively low water pressure in the dilated rocks (stage 2, Figure 1.28), which helps to produce lower seismic velocity, more earth movement, higher radon gas emission (radon is a naturally radioactive gas that is dissolved in water and released as rocks fracture and dilate), lower electrical resistivity, and fewer minor seismic events. Water then enters the open fractures (stage 3, Figure 1.28), causing the pore pressure to increase (which increases the seismic velocity while further lowering electrical resistivity), thus weakening the rocks and triggering an earthquake (stage 4). After the movement and release of stress, the rocks resume many of their original characteristics (stage 5) [35].

There is considerable controversy concerning the validity of the dilatancy diffusion model. One aspect of the model gaining considerable favor is the role of **fluid pressure** (force per unit area exerted by a fluid) in earthquakes. As we learn more about rocks at seismogenic depths (the depth where earthquakes originate), it is apparent that a lot of

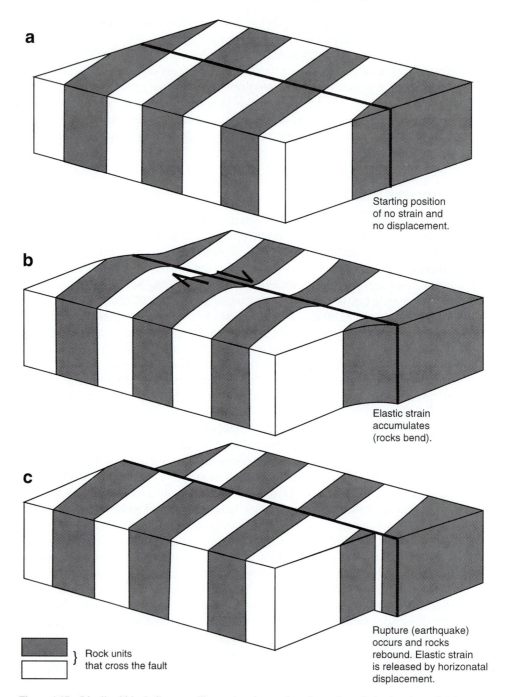

a

Starting position
of no strain and
no displacement.

b

Elastic strain
accumulates
(rocks bend).

c

Rupture (earthquake)
occurs and rocks
rebound. Elastic strain
is released by horizonatal
displacement.

} Rock units
that cross the fault

Figure 1.27   Idealized block diagrams illustrating the earthquake cycle and elastic rebound. (a) Beginning position with no strain or displacement. (b) After accumulation of elastic strain. (c) Following earthquake and rupture. (Courtesy of F. Duennebier.)

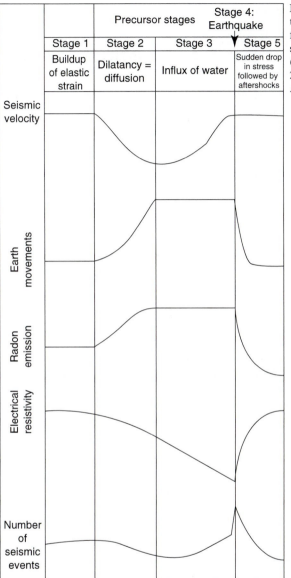

| | Stage 1 | Precursor stages | | Stage 4: Earthquake | |
		Stage 2	Stage 3	▼	Stage 5
	Buildup of elastic strain	Dilatancy = diffusion	Influx of water		Sudden drop in stress followed by aftershocks
Seismic velocity					
Earth movements					
Radon emission					
Electrical resistivity					
Number of seismic events					

Figure 1.28   Dilatancy diffusion model to explain the mechanism responsible for triggering earthquakes. The curves show the expected precursory signals. (After Press, 1975. Scientific American, 232(5): 14–23. © May 1975 by Scientific American, Inc. All rights reserved.)

water is present. Deformation of the rocks and a variety of other processes are thought to increase the fluid pressure at depth, and this lowers the shear strength. If the fluid pressure becomes sufficiently high, then this can facilitate earthquakes. A wide variety of data from several environments, including subduction zones and active fold belts, suggest that high fluid pressures are present in many areas where earthquakes occur. Thus, there is increasing speculation and interest in the role of fluid flow that affects fault displacement and is intimately related to the earthquake cycle. This process has been termed the **fault-valve mechanism** [36, 37]. The mechanism is a hypothesis in which fluid pressure rises until failure occurs, thus triggering an earthquake and discharging fluid upward.

Subsequent sealing of the rock matrix in the fault zone allows fluid pressure to reaccumulate, initiating another cycle.

PREDICTING GROUND MOTION

Engineering design of critical facilities such as power plants and dams requires careful evaluation of earthquake hazard. Of particular importance is prediction of **strong ground motion** due to earthquakes that may occur at or near facility sites. Seismographs provide information about the amplitude of seismic shaking, as illustrated in Figure 1.10a. Instruments known as **accelographs** measure and record vertical and horizontal accelerations produced by earthquakes. By measuring the vertical and horizontal components of acceleration in both the north-south and east-west directions, a three-dimensional picture of ground acceleration is created [10]. Another important parameter is the duration of shaking. For ground accelerations measured from an accelograph, the duration of strong shaking is defined as the **bracketed duration**, which is the time in which the acceleration is above a minimum value, often 0.05 g. For the example shown on Figure 1.29, the duration of strong shaking is approximately 8 s.

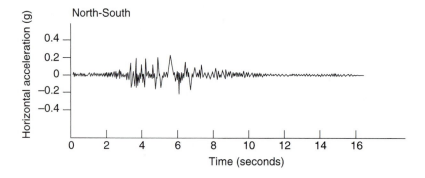

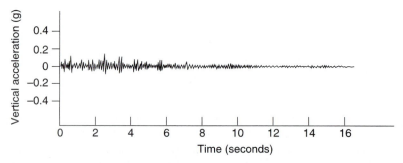

Figure 1.29   Hypothetical graph of vertical and horizontal accelerations from an earthquake with a magnitude M_W 6.5 at a distance of about 40 km from the center of energy release. Time "0" on the graph is the first arrival of the P waves. Vertical accelerations in this example are approximately 0.1 g. On the graph that shows the north-south horizontal acceleration, the S and L waves arrive approximately 4 s later than P waves with a maximum acceleration of approximately 0.25 g.

Mountain range and fault

Urban area

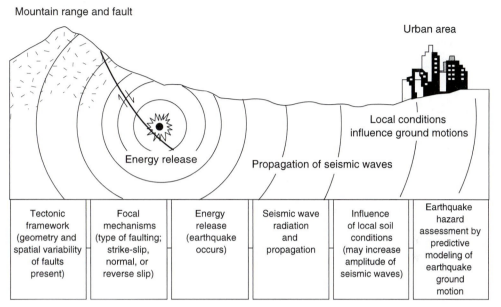

Local conditions
influence ground motions

Energy release

Propagation of seismic waves

Tectonic framework (geometry and spatial variability of faults present)	Focal mechanisms (type of faulting; strike-slip, normal, or reverse slip)	Energy release (earthquake occurs)	Seismic wave radiation and propagation	Influence of local soil conditions (may increase amplitude of seismic waves)	Earthquake hazard assessment by predictive modeling of earthquake ground motion

Figure 1.30   Assessment of earthquake hazard by modeling of ground motion. (After Vogel, 1988. In A. Vogel and K. Brandes (eds.), *Earthquake Prognostics*. Brannschweig/Weisbaden: Friedr. Vieweg & Sohn, 1–13.)

Assessment of earthquake hazard at a particular site starts with identification of the **tectonic framework** (geometry and spatial pattern of faults or seismic sources) in order to predict earthquake ground motion (Figure 1.30). Another major step in site assessment is to develop **time histories** (relationships between properties of seismic waves and time) of ground motion resulting from the largest earthquakes that could shake the site of interest. The process of predicting ground motion from a given earthquake may be illustrated by considering a hypothetical example. Figure 1.31 shows an example of a dam and reservoir site. The objective is to predict strong ground motion at the dam from several seismic sources (faults) in the area. The tectonic framework shown consists of a north-dipping reverse fault and an associated fold (an anticline) located to the north of the dam, as well as a right-lateral strike-slip fault located to the south of the dam. Figure 1.31b shows a cross section through the dam illustrating the geologic environment, including several different earth materials, folds, and faults. Assuming that earthquakes would occur at depths of approximately 10 km, the distances from the dam to the two seismic sources (the reverse fault and the strike-slip fault) are 42 km and 32 km, respectively. Thus, for this area, two focal mechanisms are possible: reverse faulting and strike-slip faulting.

The next step in the process is to estimate the largest earthquakes likely to occur on these faults. Assume that field work in the area revealed ground rupture and other evidence of faulting in the past, suggesting that on the strike-slip fault, approximately 50 km of fault length might rupture in a single event, with right-lateral strike-slip motion of 2 m. The field work also revealed that the largest rupture likely on the reverse fault would be 30 km of fault length, with vertical displacement of about 1 m. Given this in-

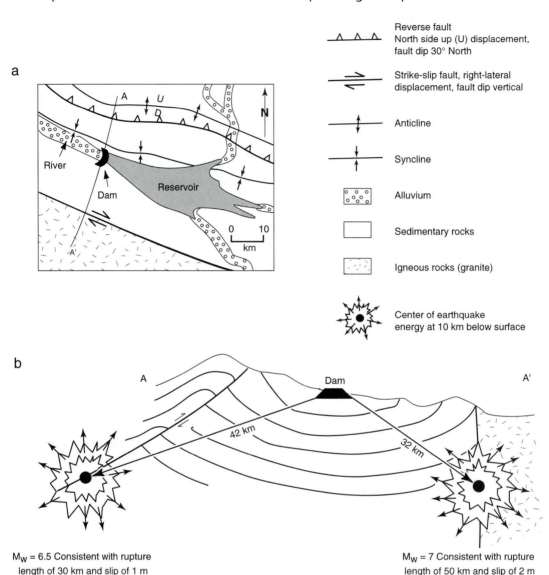

Figure 1.31   Tectonic framework for a hypothetical dam site. (a) Geologic map. (b) Cross section A-A', showing seismic sources and distances of possible ruptures to the dam.

formation, the magnitudes of possible earthquake events can be estimated from graphs such as those shown on Figure 1.32 [38, 39]. Fifty kilometers of surface rupture are associated with an earthquake of approximately M_W 7. Similarly, for the reverse fault with surface rupture length of 30 km, the magnitude of a possible earthquake is estimated to be M_W 6.5. Notice on Figure 1.32 that the regression line that predicts the moment magnitude is for strike-slip, normal, and reverse faults. Statistical analyses have suggested that the relation between moment magnitude and length of surface rupture is not sensitive to the style of faulting [39].

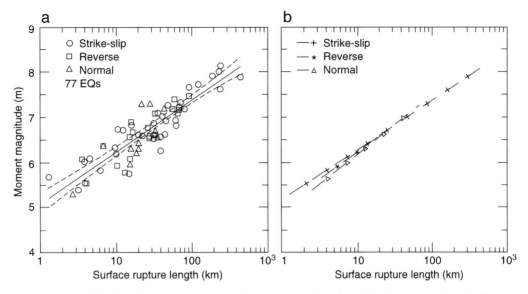

Figure 1.32   Relationship between moment magnitude of an earthquake and surface rupture length. Data for 77 events are shown in (a). Solid line is the "best-fit" or regression line, and dashed lines are error bars at the 95% confidence level. Graph in (b) shows individual lines for different types of faults. There is no significant difference between the lines. (After Wells and Coppersmith, 1994 [39].)

With the preceding information, the next task is to estimate the seismic shaking or ground motion expected from these events. These are referred to as **response spectra**, which are relationships between ground motion and period of earthquake waves [40]. There are two approaches available to estimate the response spectra: (1) empirical evaluation, and (2) simulation.

- Where ground motions have been recorded from previous events, **empirical evaluation** begins with identifying those records that most closely approach the conditions for the site of interest. The objective is to match as closely as possible the tectonic framework, rock types, type of faulting, and earthquake magnitude from a known event to known conditions at the site. The assumption is that the shaking and strong ground motion associated with a known earthquake will produce similar ground motion. Using our example of the dam site, an earthquake of M_W 6.4 on a reverse fault in Afghanistan might serve as a model for the reverse fault near the dam (Figure 1.31). Similarly, an earthquake of M_W 7.1 on the San Andreas fault in California may be used as a model to estimate the ground motion from such an event on the strike-slip fault. Of course, it is difficult to match exactly the conditions from a known event to those of the dam site, and so allowances are necessary to adjust for small differences in earthquake magnitude and distances to the predicted epicenters. If the match between a known event and possible event at the dam site is fairly good, then ground-motion parameters such as duration of shaking and average peak acceleration of ground motion may be estimated.

- The second approach is to develop theoretical or numerical models to estimate ground motions, including acceleration time histories for the various faulting scenarios [40, 41]. Such models are now commonly used in the seismic-risk evaluation of structures such as dams, bridges, and tall buildings.

Results from the modeling of ground motion are then compared with the empirical results and, if the agreement is good, the ground-motion parameters can be used to evaluate potential shaking, duration of shaking, and ground acceleration at the dam site. If this analysis is completed prior to the designing of a dam, then the parameters are useful to engineers designing the dam to minimize potential damage from earthquakes. If the evaluation is of an existing dam site, then the information is useful in evaluating whether additional engineering work is necessary to upgrade the structure to render it more resistant to earthquakes.

The preceding discussion of methods used to assess earthquake hazard outlines procedures for evaluating the seismic hazard at power plants, dams, and other critical facilities. Although this methodology suffers from limiting assumptions and shortcomings, it certainly is a valuable tool insofar as it allows estimation of strong ground motions likely to affect a particular site. As additional earthquake records are obtained and, in particular, strong ground motions close to the epicentral areas are recorded, then our understanding of how better to design structures to withstand earthquake shaking will improve.

SUMMARY

Tectonics refers to processes and landforms resulting from deformation of the Earth's crust, and active tectonics refers to those tectonic processes that produce deformation of the Earth's crust on a time scale that is significant to humans. Active tectonics includes slow disruption of the crust that may cause damage to human structures but it is most concerned with catastrophic events such as earthquakes that cause severe damage to people, property, and society.

A fault is a fracture or fracture system along which rocks have been displaced. A group of related faults or fault traces is known as a fault zone, and most major fault zones are segmented. Fault segments are recognized on the basis of changes in the fault-zone morphology or geometry as well as seismic and paleoseismic activity. Active fault zones are those for which it can be demonstrated that a fault has moved during the past 10 ky. Faults that have moved in the past 1.65 M.y. are considered potentially active, and those that have not moved during that period generally are classified as inactive.

Both magnitude and intensity of earthquakes are important in evaluating potential earthquake hazard. Although the Richter magnitude has been used for many years, it is now being replaced by the more physically based moment magnitude system. The Modified Mercalli Scale is based on observations concerning severity of shaking and response of structures to earthquakes. Effects of earthquakes include violent shaking, surface rupture, liquefaction, landslides, fires, tsunami, and regional changes in land elevation. Earthquakes also have been caused by human activities, such as reservoir construction, subsurface injection of liquid waste, and mining.

Estimation of seismic risk is an important endeavor and involves development of seismic risk maps and calculation of conditional probabilities of earthquakes occurring in the future. Slip rate on a fault is an important parameter in estimating the earthquake hazard for a particular fault. The average recurrence interval of earthquakes on a particular fault also is an important characteristic and can be determined by paleoseismic data, the ratio of assumed displacement per event to the slip rate, and historical seismicity.

An important aspect of earthquake-hazard reduction is prediction of strong ground motion. Assessment of the earthquake hazard starts with identification of the tectonic framework (geometry and spatial pattern of faults or seismic sources) followed by identification of possible focal mechanisms for earthquakes on faults present in the area being evaluated. Field work and other evidence are used to predict the magnitude of earthquake events that might be expected at a particular site.

REFERENCES CITED

1. Geophysics Study Committee, 1986. Overview and recommendations. In *Active Tectonics*. Washington, DC: National Academy Press, 3–19.
2. Davis, G. H., 1993. Basic science planning initative in "active tectonics." *EOS: Transactions, American Geophysical* Union, 74(43):59.
3. White, G. F., and J. E. Haas, 1975. *Assessment of Research on Natural Hazards*. Cambridge, MA: MIT Press.
4. Advisory Committee on the International Decade for Natural Disaster Reduction, 1989. *Reducing Disaster's Toll*. Washington, DC: National Academy Press.
5. Le Pichon, X., 1968. Sea-floor spreading and continental drift. *Journal of Geophysical Research*, 73:3661–3697.
6. Isacks, B., J. Oliver, and L. Sykes, 1968. Seismology and the new global tectonics. *Journal of Geophysical Research*, 73:5855–5899.
7. Dewey, J. F., 1972. Plate tectonics. *Scientific American*, 226(5):56–68.
8. Hamilton, R. M., 1980. Quakes along the Mississippi. *Natural History*, 89(8):70–75.
9. Mueller, K., J. Champion, M. Guccione, and K. Kelson, 1999. Fault slip rates in the modern New Madrid Seismic Zone. *Science*, 286:1135–1138.
10. Bolt, B. A., 1993. *Earthquakes*. San Francisco: W. H. Freeman.
11. Hart, E. W., W. A. Bryant, and J. A. Treiman, 1993. Surface faulting associated with the June 1992 Landers earthquake, California. *California Geology*, 46:10–16.
12. Hanks, T. C., 1985. The national earthquake hazards reduction program: scientific status. *U.S. Geological Survey Bulletin* 1659.
13. Hays, W. W., 1981. Facing geologic and hydrologic hazards: earth science consideration *U.S. Geological Survey Professional Paper* 1240-B.
14. Jones, R. A., 1986. New lessons from quake in Mexico. *Los Angeles Times*, September 26.
15. Hough, S. E., P. A. Friberg, R. Busby, E. F. Field, K. H. Jacob, and R. D. Borcherdt, 1989. Did mud cause freeway collapse? *EOS: Transactions, American Geophysical Union*, 70:1497–1504.
16. U.S. Geological Survey Staff, 1996. U.S.G.S. response to an urban earthquake, Northridge '94. *U.S. Geological Survey Open File Report* 96–263.
17. Radbruch, D. H., B. J. Lennet, M. G. Bonilla, B. J. Lennert, F. B. Blanchard, G. L. Laverty, L. S. Cluff, and K. V. Steinbrugge 1966. Tectonic creep in the Hayward fault zone, California. U.S. Geological Survey Circular 525.
18. Steinbrugge, K. V., and E. G. Zacher, 1973. Creep on the San Andreas fault. In R. W. Tank (ed.), *Focus on Environmental Geology*. New York: Oxford University Press, 132–137.
19. Working Group on California Earthquake Probabilities, 1995. Seismic Hazards in Southern California: Probable earthquakes, 1994–2024. *Bulletin of the Seismological Society of America*, 85:379–439.

20. Toppozada, T. R., 1993. The Landers–Big Bear earthquake sequence and its felt effects. *California Geology* 46 (Jan.–Feb.):3–9.

21. U.S. Geological Survey, 1999. Special Report: The Hector Mine earthquake, 10/16/99. Accessed @ http://www.socal.wr.usgs.gov/hector/report/html

22. Working Group on California Earthquake Probabilities, 1999. Major quake likely to strike between 2000 and 2030. USGS Fact Sheet 152–99.

23. Office of Emergency Preparedness, 1972. *Disaster Preparedness*: 1, 3: Washington, DC.

24. Youd, T. L., D. R. Nichols, E. J. Helley, and K. R. Lajoie, 1975. Liquefaction potential. In R. D. Borcherdt (ed.), Studies for Seismic Zonation of the San Francisco Bay Region. *U.S. Geological Survey Professional Paper* 941: A68–A74.

25. Office of Emergency Preparedness, 1972. *Disaster Preparedness*: 1, 2: Washington, DC.

26. Hokkaido Tsunami Survey Group, 1993. Tsunami devastates Japanese coastal region. *EOS: Transactions of the American Geophysical Union*, 74:417–432.

27. Geist, E. L., 1998. Source characteristics of the July 17, 1998 Papua New Guinea tsunami. American Geophysical Union, Fall Meeting, Abstracts, p. F571.

28. Koenig, R. 2001. Researchers target deadly tsunamis. *Science*, 293:1251-1253..

29. Slemmons, D. B., and C. M. DePolo, 1986. Evaluation of active faulting and associated hazards. In *Active Tectonics*. Washington, DC: National Academy Press, 45–62.

30. Hansen, W. R., 1965. The Alaskan earthquake, March 27, 1964: Effects on communities. *U. S. Geological Survey Professional Paper* 542-A.

31. Oppenheimer, D., G. Beroza, G. Carver, L. Dengler, J. Eaton, L. Gee, F. Gonzales, A. Jayko, W. H. Li, M. Lisowski, M. Magee, G. Marshall, M. Murray, R. McPherson, B. Romanowicz, K. Satake, R. Simpson, P. Somerville, R. Stein, and D. Valentine, 1993. The Cape Mendocino, California, earthquakes of April 1992: subduction at the triple junction. *Science*, 261:433–438.

32. Pakiser, L. C., J. P. Eaton, J. H. Healy, and C. B. Raleigh, 1969. Earthquake prediction and control. *Science*, 166:1467–1474.

33. Evans, D. M., 1966. Man-made earthquakes in Denver. *Geotimes*, 10(9):11–18.

34. Sylvester, A. G., and J. Heinemann, 1996. Preseismic tilt and triggered reverse faulting due to unloading in a diatomite quarry near Lompoc, California. *Seismological Research Letters*, 67(6):11–18.

35. Press, F., 1975. Earthquake prediction. *Scientific American*, 232(5):14–23.

36. Sibson, R. H., 1981. Fluid flow accompanying faulting: field evidence and models. In D. W. Simpson and P. G. Richards (eds.), *Earthquake Prediction, an International Review*. M. Ewing Series 4. Washington, DC: American Geophysical Union, 593–603.

37. Keller, E. A., and H. A. Loaiciga, 1993. Fluid-pressure induced seismicity at regional scales. *Geophysical Research Letters*, 20:1683–1686.

38. Slemmons, D. B., 1982. Determination of design earthquake magnitudes for microzonation. In M. Sherif (ed.), *Proceedings: 3rd International Earthquake Microzonation Conference*. Washington, DC: U.S. National Science Foundation, 119–130.

39. Wells, D. L., and K. J. Coppersmith, 1994. New empirical relationships among magnitude, rupture length, rupture width, rupture area and surface displacement. *Bulletin of the Seismological Society of America*, 84:974–1002.

40. Auersch, L., 1988. Seismic response of three-dimensional structures using a Green's function approach to the soil. *In* A. Vogel and K. Brandes (eds.), *Earthquake Prognostics*. Brannschweig/Weisbaden: Friedr. Vieweg & Sohn, 393–403.

41. Damrath, R., 1988. Modal response analysis of structures. *In* A. Vogel and K. Brandes (eds.), *Earthquake Prognostics*. Brannschweig/Weisbaden: Friedr. Vieweg & Sohn, 405–426.

2

Landforms, Tectonic Geomorphology, and Quaternary Chronology

TECTONIC GEOMORPHOLOGY

Landforms are surficial features, the collection of which constitutes the **landscape**. Examples of landforms include large features, such as mountains and plateaus, as well as smaller features, such as alluvial fans, hills, canyons, slopes, and sand dunes (see discussion of landscape scale in Chapter 9, especially Table 9.1). **Geomorphology** is the study of the nature, origin, and evolution of the landscape, focusing on physical, chemical, and biological processes that produce or modify landforms.

It is important to understand the relationships between the geologic environment and the landscape. Geologic factors are important because landform development is closely related to the underlying structure of the Earth. In geomorphology, structure is broadly defined to include rock and soil types, nature and abundance of fractures in the rocks, and faults and folds. Landform development also depends on the nature of surficial and geologic processes, including weathering (physical and chemical), fluvial (most of the landscape is produced by subaerial erosional and depositional processes related to streams and rivers), glacial, eolian (wind), mass wasting (including slope failures [landslide, mudflow, and earthflow]), tectonic (plate motion, faulting, folding, tilting, uplift, and subsidence), and volcanic.

Understanding geomorphology, and thus landscape evolution, is facilitated by the use of **process-response models**, which are qualitative and quantitative representations

49

of how processes influence landform development. For example, models have been developed to explain changes in deposition on alluvial fans that result from tectonic and fluvial processes and changes in climatic conditions [1]. Such models are tools for understanding tectonic processes because present-day alluvial-fan morphology can be used to infer tectonic activity in the past and predict activity in the future. The process-response model for alluvial fans is discussed in detail in Chapter 4.

Tectonic geomorphology may be defined in two ways: (1) the study of landforms produced by tectonic processes, or (2) the application of geomorphic principles to the solution of tectonic problems. The first definition implies that we are interested in the landforms themselves—their shape and origins—as functions of tectonic processes. The second definition has a utilitarian value; it allows us to use geomorphology as a tool to evaluate the history, magnitude, and rate of tectonic processes.

Geomorphology is a valuable tool in tectonic investigations because the **geomorphic record**, defined as the set of landforms and Quaternary deposits present at a site or in an area, generally encompasses the last few thousand to about 2 million years. Investigation of the geomorphic record provides the basic data necessary to understand the role of active tectonics in the development of a site or an area. For example, study of stream channels and their associated deposits offset by faulting may reveal the amount of displacement and timing of the last few earthquakes at a particular site—information that is critical in evaluating future earthquake hazards.

GEOMORPHIC CONCEPTS

The study of tectonic geomorphology requires investigation of geomorphic processes, tectonic processes, and earth materials, as well as an appreciation of how the landscape is formed and maintained and how it evolves through time.

Five important geomorphic concepts directly applicable to the study of tectonic geomorphology are as follows:

- Landscapes evolve through time, and changes in the assemblage of landforms are predictable.
- During landscape evolution, abrupt changes may occur as thresholds are exceeded [2].
- The interaction of landscape evolution with thresholds results in complex processes; this interaction is called **complex response** [2].
- A change in form implies a change in process.
- Normal, reverse, and strike-slip faulting each are associated with a characteristic set of landforms.

LANDSCAPE EVOLUTION

William Morris Davis put forward a simple model of landscape evolution in the late 1890s called the "Cycle of Erosion". His model was based on the assumption that relatively brief pulses of uplift are followed by much longer periods of tectonic inactivity and erosion [3]. Davis envisioned that during this cycle, the landscape would go from a brief period of **youth**, characterized by deep V-shaped valleys and an abundance of water-

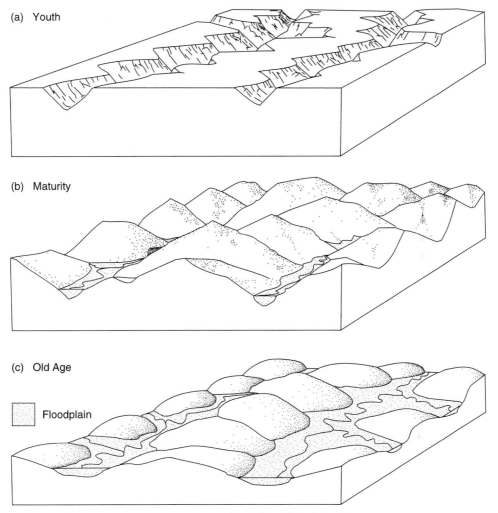

(a) Youth

(b) Maturity

(c) Old Age

Floodplain

Figure 2.1   The Cycle of Erosion, as envisioned by William Morris Davis.

falls and rapids; to a longer period of **maturity**, characterized by a great variety of land-forms, including meandering rivers and floodplains; followed by a long period of **old age**, in which the landscape might be worn down to an erosional surface of low relief that he called a **peneplain** (Figure 2.1). The time necessary for this hypothetical cycle is on the order of $\sim10^6$ years, and the cycle could be interrupted at any time by renewed uplift, called **rejuvenation**.

Over shorter periods of time (on the order of $\sim10^4$ to 10^5 years), landscapes may approach a **dynamic equilibrium**, in which the form of the landscape is nearly independent of time [4]. However, it has been argued that variables such as climate, tectonic processes, and geomorphic processes may change too rapidly for equilibrium to be reached for any appreciable length of time [5].

There is no doubt that landscape evolution is occurring over a range of time spans. Change is the norm, and sometimes that change may be very rapid. For example, less than 100 ky is necessary to produce a coastline characterized by several broad, uplifted, marine platforms (such as along the coasts of New Zealand, Japan, and California; see Chapter 6). Over a similar span of time, composite fault scarps up to 100 m high also may develop.

Fault scarps produced by individual earthquakes may be created and eroded away over periods of a few years to a few thousand years for small scarps on alluvial materials to a few hundred thousand years or more for large scarps on resistant rock. For example, an M_W 7.6 event known as the Chi-Chi earthquake of 1999 in Taiwan was associated with large thrust (vertical) motion as well as lesser left-lateral strike-slip displacements on the Chelungpu fault (Ta-liang Teng, personal communication, 2000). Total surface rupture length was about 100 km. Vertical displacements ranged from 1 to 8 m with horizontal displacements < 1 m. At one location the bed of the Tachia River was faulted, producing a new 8-m-high waterfall (Figure 2.2). This earthquake was the latest in a series of five large earthquakes during the twentieth century associated with the Taiwan Collision Zone, where the Philippine Sea Plate is moving northwest at 7 to 8 cm/yr relative to the Eurasian Plate [6].

Our discussion of geomorphic evolution of landforms, produced by tectonic processes (for example, folds and fault scarps) or isolated by tectonic processes (for example, stream or coastal terraces) and subsequently modified by erosion, concludes that the landscape is constantly changing over all scales of time and space. What is important is that the changes often are predictable, and this provides a mechanism for evaluating active tectonics.

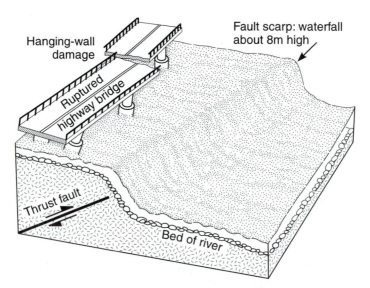

Figure 2.2   Eight meter high "new" waterfall produced by vertical displacement of the Tachia River bed where the thrust fault that produced the 1999 Chi-Chi earthquake in Taiwan crosses the river. Most damage including the highway bridge shown here was on the hanging wall of the fault.

THRESHOLDS

The life of a landscape must be like that of an airline pilot—long periods of routine work punctuated by brief periods of intense activity and change, like takeoff and landing or flying through an intense storm. Some landform changes reflect changes in **extrinsic variables**—external processes such as sudden earthquakes or the arrival of humans. Other changes result from changes in **intrinsic variables**, processes that are part of the everyday operation of the landscape system, for example, weathering, or movement of groundwater through deposits on a slope. Although a system may be held steady by a rough dynamic equilibrium, small changes may occur (for example, subsurface weathering that reduces the strength of rock and soil) until a **threshold** is surpassed and the balance collapses. When a threshold is exceeded, the effects may be both dramatic and quick. For example, in the earthquake cycle, the slow and steady deformation of rocks accumulates strain along a fault plane until the stress exceeds the strength of the rocks and an earthquake occurs.

Thresholds resulting from tectonic processes may change the operation of the system from deposition to erosion. For example, if a fault block containing a mountain and alluvial fans is slowly rotated or tilted, then the alluvial fans become steeper. Eventually a threshold slope is crossed, and streams on the alluvial fans downcut rapidly, moving the deposition of alluvium farther out from the mountain front. Thus, the processes near the mountain front change from deposition to erosion, and the deposition of alluvial-fan deposits is moved downstream [1].

An important consequence of understanding thresholds is that we are no longer limited to models in which long periods of slow change are interrupted by rapid external events (perturbations) such as climatic change and tectonic uplift. Rather, some changes in process and/or form, although they are abrupt, result from slower change that has occurred over a period of time within the system itself [2].

COMPLEX RESPONSE

We have established that landscape evolution occurs and that changes in landforms are expected and predictable. We also have suggested that some of these changes occur suddenly as thresholds are crossed. The interaction between these two can lead to a complex evolution of a particular landscape. This complexity is often called **complex response** [2]. For example, we once thought that if several river terraces were present in a river valley, then each terrace must reflect some external event, such as a climatic change or an uplift event. However, our understanding of geomorphic thresholds and complex response suggests that one perturbation or uplift event may cause a complex response in which several river terraces are produced. As a result, two drainages side by side may have different forms and histories. One basin may have several river terraces, while the adjacent basin has only one.

Changes in climate also have profound influences on geomorphic systems. For example, greater precipitation or aridity may result in channel deposition or downcutting, producing terraces. Climate changes, however, should be reflected in drainage basins on at least a regional scale. The fact that this is not always the case suggests that something else is occurring. That something else is complex response. Figure 2.3 shows a

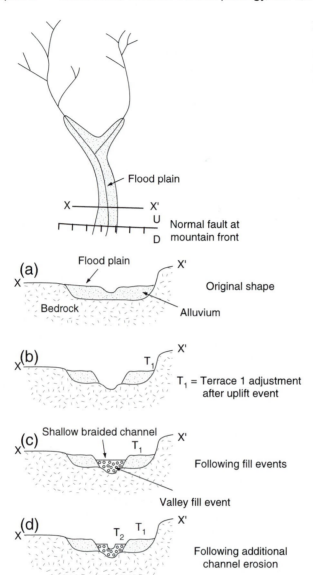

Figure 2.3 Idealized diagram showing the concept of the complex response. One tectonic uplift event produces two terraces. See text for explanation.

drainage basin, the lower part of which has a floodplain. Faulting at the mountain front initiates a complex sequence of incision followed by deposition. Figure 2.3a shows the original shape of the valley and floodplain prior to disturbance by faulting. Following faulting, which uplifts the basin by several meters, the stream responds by downcutting (erosion), producing terrace T_1 shown in Figure 2.3b. The downcutting and erosion produce sediment. As the erosion works its way headward, the channel in the lower part of the basin adjusts to the change prior to the upper part of the basin, where erosion is still occurring. By the time headward erosion in the channel reaches the upper part of the basin, additional sediment is transported downstream and deposited in the valley, perhaps causing a shallow, braided channel to form, as shown in Figure 2.3c. As deposition continues, the slope of the valley and the braided channel increases until a threshold in channel slope is crossed and downcutting occurs again, producing terrace T_2 (Figure 2.3d). Thus, in this simple example, two river terraces are produced by one tectonic disturbance.

Because no two river basins are exactly alike, the timing of erosion and deposition varies. In places, multiple terraces are produced by multiple tectonic events; elsewhere, multiple terraces might be related by complex response to a single tectonic or climatic perturbation. At the southern end of the San Joaquin Valley in California, San Emigdio Canyon crosses an axis of active uplift caused by a buried reverse fault (Figure 2.4). Upstream and downstream of the axis of uplift, there are two prominent Holocene stream terraces. Where the stream crosses the axis of uplift at the mountain front, however, there are several additional terraces, each separated by approximately a meter. The terraces upstream and downstream of the uplift probably reflect the more regional changes that have occurred. The additional terraces where uplift is centered may be related directly to earthquakes that have occurred at the mountain front [7]. A feature common to many geomorphic systems is that in areas with the highest rate of uplift, not only are the elevations of landforms greater, but the number of episodic events recorded also is greater.

RELATIONSHIP BETWEEN FORM AND PROCESS

A fundamental principle of geomorphology is that a change in landscape form often implies a change in landscape process. If you are walking up an alluvial fan with a uniform slope and you suddenly come to a sharp increase in gradient (perhaps to a near-vertical slope), followed by a return to the original slope, then this implies a change in process. In geomorphic terminology, the steep part of the slope is a **scarp**, possibly produced by differential erosion along a fault (forming a **fault-line scarp**) or directly by faulting (producing a **fault scarp**). The scarp also might be depositional in origin, representing the steep nose of a debris flow on the fan. Whatever the case, the change in form of the topography implies a change in the processes that formed it. Using this principle, geomorphologists observe landforms and look for anomalies that might reflect changes in process. When we are interested in tectonic geomorphology, we are often looking for landform surfaces that have been warped, tilted, uplifted, fractured, or otherwise deformed.

The principle that relates change in form to change in process is applicable at a variety of scales, from small fault scarps to mountain ranges. Just as the morphology of a fault scarp reflects a difference between erosion at its crest, producing a convex slope, and deposition at its base, producing a concave slope, so also is the morphology of an eroding mountain block fundamentally different from the morphology of the adjacent basin into which material eroded from the mountain range is deposited.

TECTONIC GEOMORPHOLOGY AND FAULTING

Active faulting causes a variety of landform features, including fault scarps, warped and tilted slopes, subsidence features such as sag ponds, and offset features such as stream channels. Each major category of faulting—strike-slip, normal, and reverse—may be

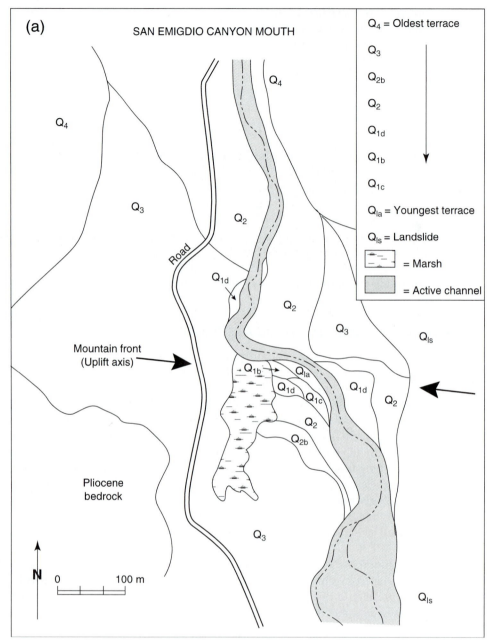

Figure 2.4 (a) Terraces near the mouth of San Emigdio Canyon, California. Multiple terraces have formed near the uplift axis associated with the buried Wheeler Ridge fault.

(b)

Figure 2.4 (*continued*) (b) Photograph of terraces at San Emigdio Canyon. (After Laduzinsky, 1989. M.A. thesis, University of California, Santa Barbara.)

discussed in terms of a characteristic assemblage of landforms. There is a fair amount of overlap between these different assemblages because many faults have oblique displacement, partially strike-slip and partially vertical. The balance between strike-slip and dip-slip displacement may vary significantly on a given fault system. Nevertheless, because specific processes tend to produce a particular set of responses, and therefore a particular assemblage of landforms, a generic classification of landforms is possible.

LANDFORMS OF STRIKE-SLIP FAULTING

A characteristic assemblage of landforms produced by active strike-slip faulting is shown in Figure 2.5. These features include the following:

- **Linear valleys** are troughs along main fault traces. These often develop because continued movement along recent fault traces crushes the rock, making it more vulnerable to erosion. Streams commonly follow these zones of weakness and flow for some distance along the troughs.
- **Deflected drainages** are stream valleys that enter a fault zone at an oblique angle and flow parallel to the fault for some distance before returning to the original orientation of flow. Streams may be deflected either in a right or a left sense.
- **Offset streams** are streams displaced by faulting; they indicate the direction of relative displacement. The offset may reflect cumulative offset of several earthquakes. Eventually the stream may erode a more direct route across the fault zone, producing a **beheaded stream** at the fault trace.

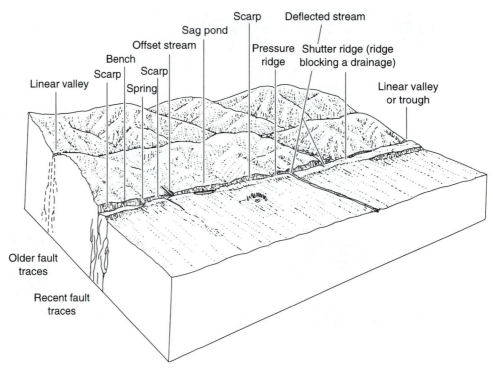

Figure 2.5 Assemblage of landforms associated with active strike-slip faulting. (After Wesson et al., 1975. U.S. Geological Survey Professional Paper 941A.)

- **Shutter ridges** (Figure 2.6) may form when fault displacements move ridges between drainages that are at a right angle to a fault zone or hills that are parallel to the fault zone to a position that blocks and diverts gullies or stream channels.
- **Scarps** can be produced by strike-slip motion by two possible mechanisms: (1) A small component of vertical displacement on individual fault strands results in local vertical separations, or (2) topographic relief on displaced landforms results in fault-parallel scarps.
- **Sag ponds** are often found in the fault zone and generally are related to downwarping between two strands of the fault zone (Figure 2.7).
- **Springs** are often found along the fault zone because pulverized, clay-rich crushed rock (known as **fault gouge**) associated with faulting can be an effective barrier to groundwater, forcing it to the surface. Along the right lateral southern San Andreas fault in the Coachella Valley, California, fault traces are often delineated by native palm trees, which utilize the shallow groundwater that is forced to the surface in this very arid desert environment. In some instances, the springs and palms form small oases along the fault (Figure 2.8).
- **Benches** are relatively small, flat elevated surfaces within strike-slip fault zones. These surfaces may be slightly warped or tilted and are usually are due to the displacements between several fault segments or strands in the zone.

Figure 2.6   Shutter ridges along the San Andreas fault. (Photograph by E. A. Keller.)

Figure 2.7   Sag pond (dry) along the San Andreas fault in central California. Note that the line down the center of the depression is a fence, not a fault-related feature. (Photograph by E. A. Keller.)

Figure 2.8   Palm trees mark the trace of the San Andreas fault in Indio Hills, California. (Photograph by E. A. Keller.)

- **Pressure ridges** are small warped areas produced by compression between multiple traces in a fault zone (Figure 2.9). Where a fault subsequently breaks through previous pressure ridges, shutter ridges may be formed.

Within strike-slip fault zones, **simple shear** (visualized and approximated by shearing a deck of playing cards) may produce a variety of structures and forms, including fractures, folds, normal faults, thrust faults, and reverse faults (Figure 2.10). In Figure 2.10, the dashed circle is an imaginary area that is deformed by simple shear into the ellipse shown. As a result of that deformation, both contraction and extension occur. Extension produces normal faults and **grabens** (fault-bounded basins), sometimes at a scale of only a few meters. Contraction produces reverse faults and folds of various sizes within the fault zone. Simple shear also produces **synthetic** and **antithetic shears** (faults with the same and opposite sense of displacement as the master fault, respectively) as well as what are called P shears. Orientations of those fractures are also shown in Figure 2.10. The landforms and structures associated with simple shear are most common within fault zones—which, for large faults, may be several kilometers wide—but in some instances may extend beyond the fault zone.

Localized uplift and subsidence are associated with bends and steps in a strike-slip fault (Figure 2.11). A left bend in a right-lateral fault system produces an area of uplift and thus is called a **restraining bend**. A right bend in a right-lateral fault produces an area of subsidence and thus is called a **releasing bend**. If two or more approximately parallel fault traces in a right-lateral strike-slip fault step to the left or right, then these are

Figure 2.9   Pressure ridges along the San Andreas fault in central California. (Photograph by E. A. Keller.)

called **restraining steps** and **releasing steps**, respectively, and also can produce areas of uplift or subsidence [8, 9]. Indio Hills, east of Palm Springs, California, is an area along the San Andreas fault where many fault-related landforms are present. Figure 2.12 shows an offset alluvial fan where a number of fault-related landforms can be seen. These include shutter ridges, deflected streams, beheaded streams, sags, and microtopography (consisting of a small **horst** [fault-bounded uplifted block] and **graben** [fault-bounded depression; in this case, small downdropped blocks produced by normal faulting]) with orientations consistent with the simple shear model of Figure 2.10. Figure 2.13 shows a low-sun-angle photograph of the offset alluvial fan and a 30-m-high scarp associated with a small left bend of the main fault trace (see Figure 2.12). Low-sun-angle photographs are very useful in recognizing tectonic landforms such as fault scarps. The left bend of the fault is thought to be responsible for the vertical component of motion at this point along the fault that produced the scarp.

Restraining and releasing bends also are found at more regional scales, producing much larger basins or areas of uplift. The general trend of the San Andreas fault is northwest-southeast, but near Los Angeles a section of the fault several hundred kilometers long is oriented closer to east-west (Figure 2.14). The western end of this anomalous east-west section of the fault is a sharp bend known as the "**Big Bend**". The geometry of the east-west section of the fault, sandwiched between the two northwest-southeast-trending sections, forms a large left bend or step about 100 km wide similar to that shown on Figure 2.11, and as such should be associated with contraction. In fact, the western Transverse Ranges lie along the Big Bend; they are east-west-trending mountains, in

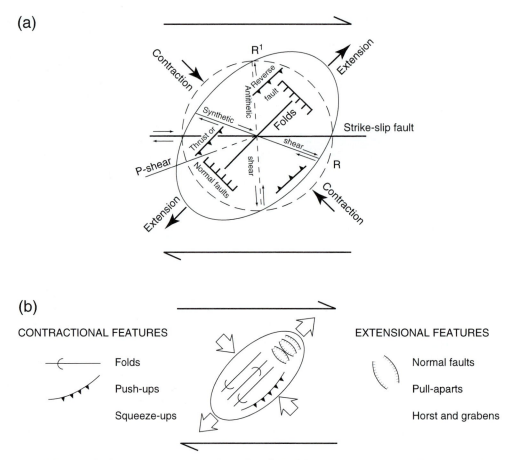

Figure 2.10   Simple shear associated with active strike-slip faulting produces a variety of fractures, faults, and folds with characteristic orientation (a), as well as extensional and contractional features (b). (After Wilcox et al., 1973. *American Association of Petroleum Geologists Bulletin*, 57:74–96; Sylvester and Smith, 1976. *American Association of Petroleum Geologists Bulletin*, 60:2081–2102.)

contrast to the northwest-southeast-trending Peninsular and Coast Ranges that characterize much of the California coast. Uplift of the Transverse Ranges can be explained in part as convergence (estimated to range from 5 to 10 mm/yr) caused by this large left bend of the San Andreas fault. It is also possible that the San Francisco Bay is a **pull-apart basin** (produced by releasing bends or steps) related to several strands of the San Andreas fault in that region. Discussion of some of the large-scale features associated with strike-slip faulting is a speculative but interesting area for future work.

Strike-slip faulting is not limited to the land. Figure 2.15 shows part of the Southern California borderland not far offshore from San Diego, where the San Clemente fault cuts the ocean floor [10]. This large, right-lateral strike-slip fault parallels the San Andreas fault system and has many of the landforms we see on land, including linear troughs, sags, fault scarps, offset or deflected channels, and large tectonic benches. Some of the higher scarps have relief of about 200 m.

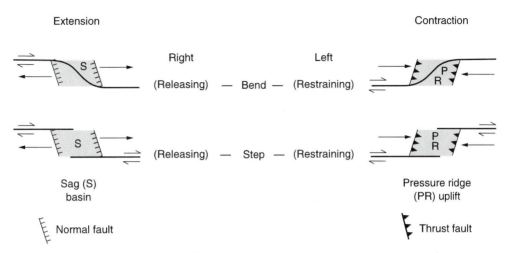

Figure 2.11   Bends and steps along strike-slip faults create zones of either extension or contraction, depending on the orientation of the bend or step and the sense of displacement on the fault. (After Crowell, 1974 [8]; Dibblee, 1977 [9].)

LANDFORMS OF NORMAL FAULTING

Some of the most remarkable topography on land and beneath the oceans is associated with crustal extension and normal faulting. The midocean ridges are the longest and most continuous mountain chains on Earth. The axes of oceanic ridges are marked by **rift valleys**, typically bounded by large normal faults. Rift valleys are also found on a smaller scale on continents (Figure 2.16). Perhaps the best example of continental rifting is the East African Rift system, which extends for thousands of kilometers (Figure 2.17). The major landform of the system is the rift valley, which is a graben. The morphology of the East African Rift system is similar to that found along some oceanic ridge systems, where overlapping segments of ridges have a basin between them. The fault scarps produced by normal faulting in rift valleys, both on land and beneath the ocean, are some of the largest fault scarps found on Earth. The East African Rift system, like the oceanic ridge systems, is a topographically high area with active volcanism, including the spectacular Mt. Kilimanjaro and other peaks. The rift valleys of the East African Rift are segmented along their lengths into a number of asymmetric basins (**half grabens**, in which the valley is downdropped along a primary set of normal faults on one side of the valley) with lengths on the order of 100 km. Some of these basins are the sites of large lakes, such as Lake Tanganyika and Lake Victoria. The tectonic geomorphology of the East African Rift valley is known in a gross sense, but much more basic research is necessary to understand the rates of tectonic processes and how they have produced this complex landscape. Understanding the geomorphic history of continental rift valleys may provide important information about how their larger cousins, the oceanic rifts, formed. The oceanic rifts represent the later stages of evolution of continental rift valleys, subsequent to the breakup of continents. Studies of the oceanic rifts are frustrated by the difficulty in obtaining topographic data and images comparable to those available on land.

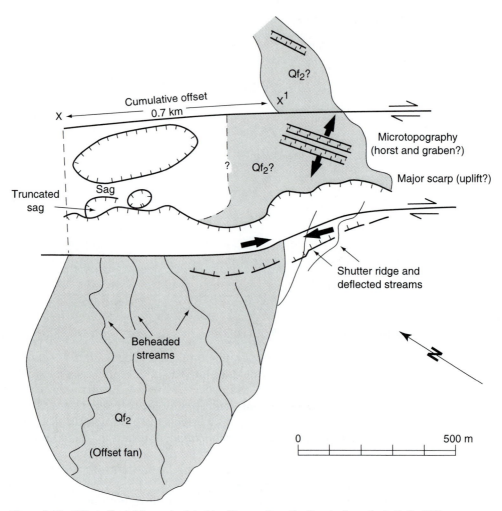

Figure 2.12 Offset alluvial fan and related landforms along the San Andreas fault, Indio Hills, California. (After Keller et al., 1982. *Geological Society of America Bulletin*, 93:46–56.)

In the United States, topographic expression of normal faulting is nowhere better expressed than in the Basin and Range province (Figure 2.18). The gross topography of the region has been compared to an army of caterpillars marching south from Idaho, or north from Mexico, depending on your bias. The caterpillars are long mountainous spines bounded by steep normal faults. The intervening basins (fault-bounded depressions) are grabens.

Two types of normal faults are shown in Figure 2.18b: high-angle normal faults that bound the basins and ranges (producing steep, linear mountain fronts) and a very-low-angle normal fault known as a **detachment fault**. Uplift of the ranges of the Basin and Range has been accompanied by **block tilting**, illustrated in Figure 2.18b. This tilting includes both down-to-the-east and down-to-the-west rotations, with regions of consistent tilt polarity separated by distinct "accommodation zones" [11]. Tectonic stress

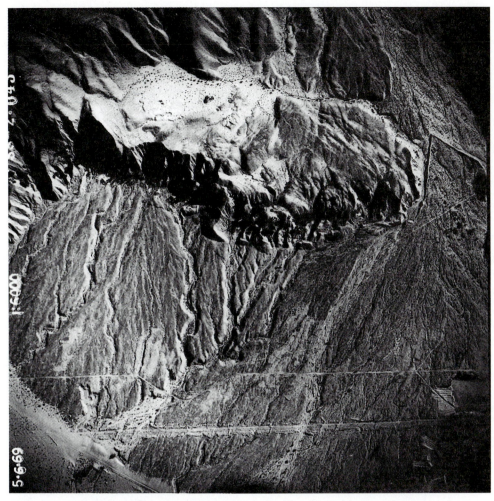

Figure 2.13   Offset alluvial fan, Indio Hills, California. See Figure 2.12. (Photograph courtesy of Woodward-Clyde consultants.)

and crustal thinning responsible for the normal faulting in the Basin and Range began several million years ago and are going on today.

Normal faults near the surface are generally steeply dipping (~60°). As a result, mountain fronts along active normal faults tend to be straight and steep. However, there are exceptions to this; for example, mountain fronts at Dixie Valley and Pleasant Valley, Nevada, are more sinuous because the normal fault zones are themselves sinuous. The same is true for the mountain front at Salt Lake City, Utah [12]. The combination of vertical motion on range-bounding normal faults and stream incision in the valleys results in the formation of **triangular facets**. Triangular facets are roughly planar surfaces with broad bases and upward-pointing apexes that occur between valleys that drain the mountains. Often there is a series of such features, sometimes called "flatirons", as shown in Figure 2.19. Triangular facets are characteristic of mountain fronts associated with active normal faults.

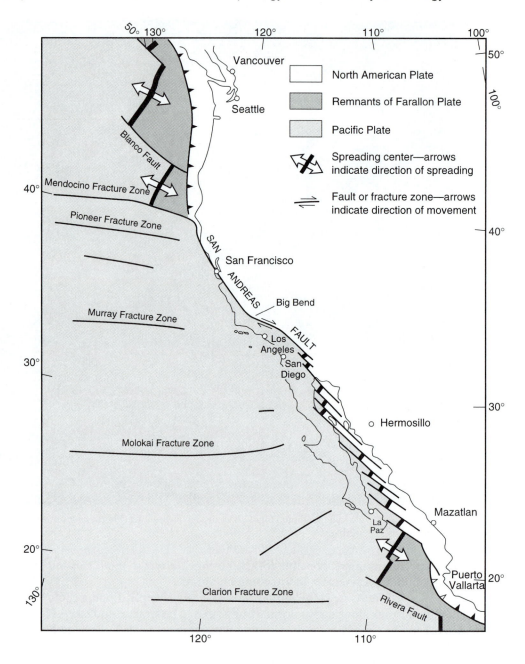

Figure 2.14   Tectonic framework of the West Coast of the United States. (After Irwin, 1990. *U.S. Geological Survey Professional Paper* 1515.)

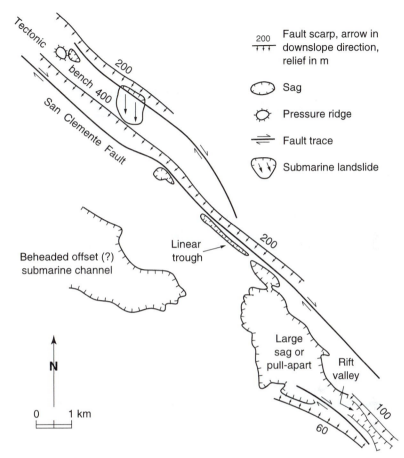

Figure 2.15   Idealized map of part of the San Clemente fault zone. Data are from Seabeam survey. (Courtesy of M. R. Legg, University of California, San Diego.)

The western boundary of the Basin and Range is the Sierra Nevada of California. A system of normal and strike-slip faults lies along the eastern flank of the range. Just north of the town of Bishop, California, hundreds of fault scarps produced by normal faulting cut what is known as the Volcanic Tableland. The Tableland was produced about 760 ka by a catastrophic volcanic eruption. Volcanic ash covered an area of at least 1000 km^2 to an average depth of 150 m. In the past 760 ky, numerous earthquakes have broken that surface (Figure 2.20), forming the fault scarps. The Volcanic Tableland is noteworthy because the durable surface records many faulting events so well, but also because faulting over such a broad area is not common in the Basin and Range. The pattern is attributed to flexure of the crust, accommodating westward tilting west of the Tableland and eastward tilting of the area to the east [13].

Normal faults result from crustal extension and thinning, and this environment also favors volcanic activity. Thin crust often is associated with high heat flow, and partial melting of rocks at depths of a few kilometers produces magma to feed volcanic activity. Active volcanism is common along active spreading ridges (midocean ridges), rift

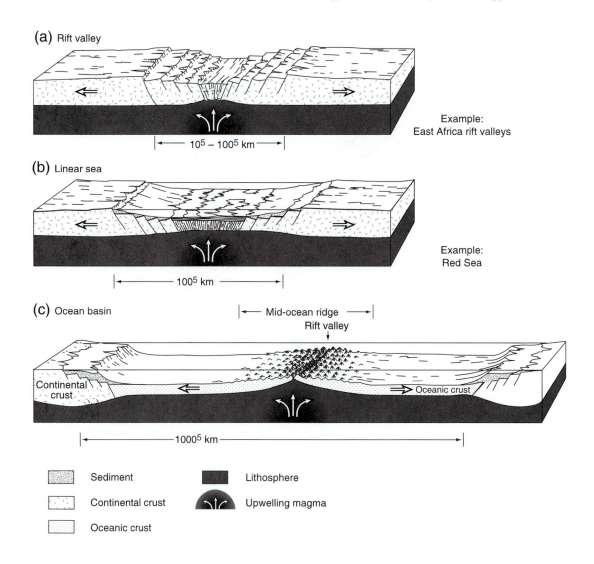

Figure 2.16 Landforms produced when the crust is pulled apart. Rifting and sea-floor spreading produce rift valleys, linear seas, and ocean basins. (After Lutgens and Tarbuck, 1992. *Essentials of Geology.* New York: Macmillan.)

valleys, and other locations associated with extensional tectonics (normal faulting), including the Basin and Range province. For example, in the Mono Basin, located north of the Volcanic Tableland, recent volcanism (within the last 40 ky) and intrusion of dikes reflect tensional strain previously taken up by normal faulting along the Sierra Nevada frontal fault system [14].

Normal faulting over several millions of years is one process that can form **escarpments** (long, nearly continuous clifflike slopes) that may dominate the topography of a region. For example, normal faulting has dominated landscape development in

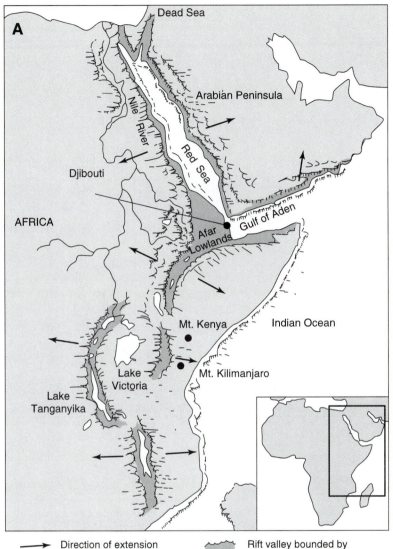

Figure 2.17   (a) East African Rift system. (From Lutgens and Tarbuck, 1992. *Essentials of Geology.* New York: Macmillan.)

⟶ Direction of extension   〰〰〰 Rift valley bounded by
                                              normal faults

much of eastern Greece and the Aegean Sea for at least the last several million years [15]. The Gulf of Corinth (Figure 2.21) is a graben bounded by normal fault escarpments with relief (difference in elevation from base to top of the escarpment) of several hundred meters. Two prominent escarpments south of Skinos climb 1100 m in two steps [15]. In February and March 1981, three moderate earthquakes (M 6.7, 6.4, and 6.4) produced considerable ground rupture on normal faults along both the north and south margins of the Gulf of Corinth (Figure 2.21). The major surface ruptures occurred along a 10-km-long normal fault segment trending east-west (east of Pision, Figure 2.21) and following the base of a very prominent escarpment produced by numerous earlier faulting events. Vertical displacements were as much as 1.5 m but mostly ranged from 0.5 to

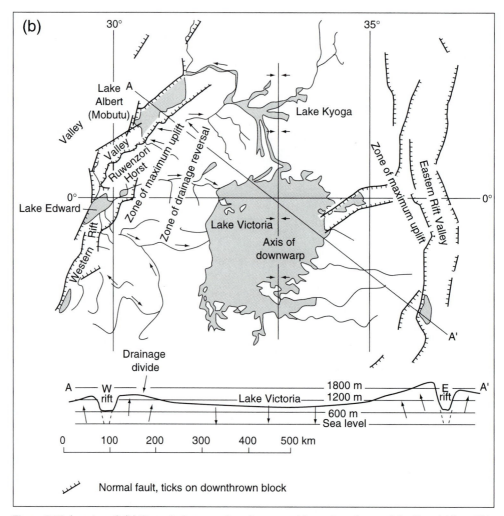

Figure 2.17 (*continued*) (b) Tectonic framework and topographic section of part of the East African Rift system. (After Bloom, 1991. *Geomorphology*, 2nd ed. Englewood Cliffs, NJ: Prentice Hall.)

0.7 m. Study of the ground rupture showed that this faulting was clearly the most recent displacement along the faults responsible for the major topographic features (escarpments) on the margins of the Gulf of Corinth [15].

In summary, the assemblage of landforms associated with normal faults includes steep, linear mountain fronts; fault scarps; horsts and grabens; escarpments; volcanic landforms (lava flows, cones, etc.); and at regional scales, rift valleys and axial rifts of oceanic ridge systems.

LANDFORMS OF REVERSE FAULTING

Reverse faulting generally is found in areas of crustal thickening, where mountains are being constructed. Some of the most spectacular scenery in the world is produced by uplift associated with reverse faulting. For example, at convergent plate boundaries, thrust

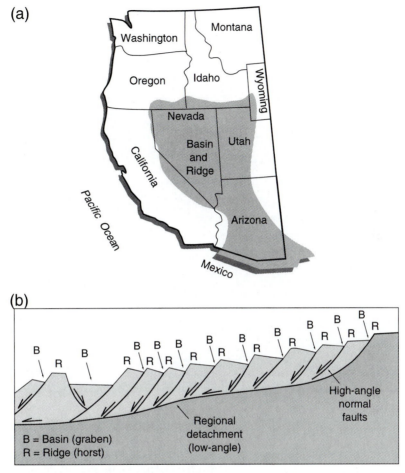

Figure 2.18   (a) Location of Basin and Range province, and (b) regional tectonic framework, including high-angle normal faults that produce the basins and ranges, and regional low-angle detachment faults.

faults (low-angle reverse faults) associated with subduction produce a variety of landforms, including uplifted coastal terraces, anticlinal hills (upwarps), and synclinal lowlands (downwarps). One of the best examples of this is the landscape near Eureka, in northwestern coastal California, where folds associated with the Cascadia Subduction Zone are exposed onshore. Landforms such as Humboldt Bay and other coastal lowlands are at sites of synclines, whereas intervening hills are anticlines (Figure 2.22) [16, 17].

The area of northern California just mentioned is an example of a **fold-and-thrust belt**. Thrust faults are often associated with folds, and the important subject of active folding and fold-related landforms is discussed in detail in Chapter 7. Active folding has produced spectacular folded topography in the Zagros fold belt of Iran (Figure 2.23) and the northern Apennine Mountains in Italy, as well as in other parts of the world. Landforms associated with reverse faulting include steep mountain fronts, fault scarps, fold

(c)

Figure 2.18 (*continued*) (c) Central Arrow Canyon Range in the Basin and Range province.
(Photograph © J. Shelton.)

scarps, extensional features, and landslides. The 1980, M_W 7.3, El Asnam earthquake in
Algeria produced a mean vertical displacement of 6 m on a 30-km-long reverse-fault system. One of the surprising discoveries following the earthquake was the amount of tensional (normal) faulting that accompanied the event. Compressional deformation caused
about 5 m of uplift of the anticlinal fold, but the most extensive surface-deformation
features caused by the earthquake were tensional, including normal fault scarps and
grabens [18]. These normal faults are all on the upper plate above the fault, which was
bent during faulting, producing the extension (Figure 2.24).

PLEISTOCENE AND HOLOCENE CHRONOLOGY

Geologists have subdivided geologic time in order to understand events in Earth history in their correct chronologic sequence and to correlate local sequences of rocks within their regional and global context (Table 2.1). For scientists and citizens concerned
with active tectonics, the period of most interest is the **Quaternary**, which is the most recent 1.65 M.y., consisting of the **Pleistocene** and **Holocene Epochs**. The Pleistocene is divided into three parts.

(a)

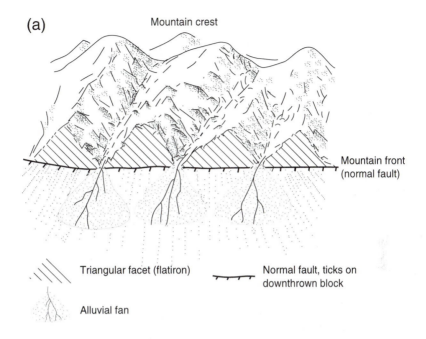

Mountain crest

Mountain front
(normal fault)

Triangular facet (flatiron)

Normal fault, ticks on
downthrown block

Alluvial fan

Figure 2.19   (a) Idealized diagram showing a mountain front, normal fault, and triangular facets (flatirons). (b) Features shown in (a) looking across Saline Valley at the Inyo Mountains and the snowy crest of the Sierra Nevada beyond. (Photograph © J. Shelton.)

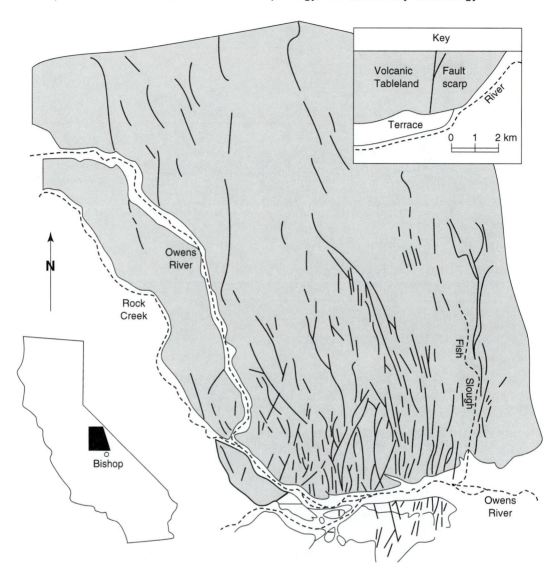

Figure 2.20 Fault scarps on the Volcanic Tableland, eastern California. The Tableland is a 760-ka volcanic sheet that has preserved extensive rupture across its surface.

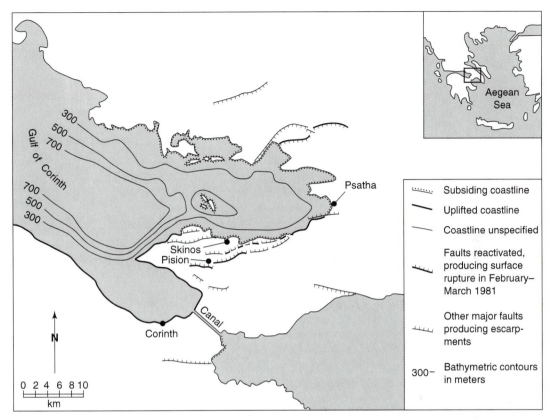

Figure 2.21   Normal faults in the Gulf of Corinth, Greece. Also shown are areas of surface rupture from the 1981 earthquakes, subsiding coastlines, and uplifted coastlines. (After Jackson et al., 1982. *Earth and Planetary Science Letters*, 57:377–397.)

- **Early Pleistocene**, from approximately 1.65 Ma to 780 ka. The boundary at 780 ka coincides with the time of the most recent reversal of the Earth's magnetic field.
- **Middle Pleistocene**, from 780 ka to 125 ka. The boundary at 125 ka marks the most recent major interglacial and associated highstand of sea level.
- **Late Pleistocene**, from 125 ka to 10 ka. The boundary at 10 ka marks the end of the Pleistocene and the beginning of the Holocene and reflects a major climatic change. Often in active tectonics we are interested in the part of the late Pleistocene from approximately 18 ka to 10 ka, known as the "latest Pleistocene". The boundary at 18 ka marks the maximum of the most recent glaciation.

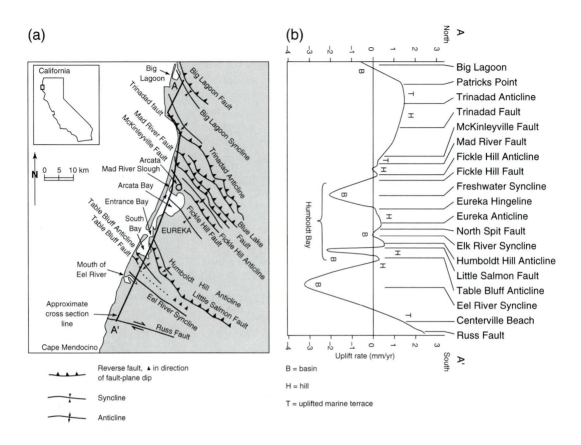

Figure 2.22   (a) Map of folds and thrust faults near Eureka, California. (b) Profile of uplift rates along A-A′. Uplift rates clearly match the topography. Anticlines and thrust faults produce hills, and uplifted coastal terraces and synclines produce bays (basins). (After Valentine, 1992 [17]; Clarke and Carver, 1992 [16].)

- **Holocene**, from 10 ka to present. The Holocene is the time period in which human civilizations burgeoned and grew. Across most of the planet, the climate stabilized (with some notable exceptions) to the conditions we know today. The **early** to **mid-Holocene** marks the Neolithic Revolution, which included such human advances as agriculture, settled communities, and development of the earliest complex societies. Active faults are defined as faults that have moved during the Holocene, and the study of the earthquake history during the Holocene allows us to estimate future earthquake hazards.

Dating is perhaps the most critical tool in assessing active tectonics, and it is often the most difficult part of a tectonic study. If there is no chronology, then we cannot calculate rates of tectonic processes ("no dates, no rates"). It is usually much easier to measure deformation resulting from tectonic activity than it is to accurately date when that deformation occurred. For example, we can easily measure the amount of vertical displacement associated with a past earthquake, but unless we are able to date the deposits

Figure 2.23   Oblique photograph of an anticline in the active Zagros fold belt, Iran. (Photograph courtesy of Aerofilms.)

associated with the displacement, we cannot calculate the rate of activity. If we are studying a sequence of alluvial deposits and they overlie and are not displaced by a fault below, then dating those deposits provides a minimum age since the last faulting (earthquake) event (Figure 2.25a). If the deposits are displaced, then dating them provides a maximum age for the most recent faulting event [19] (Figure 2.25b). Under ideal conditions, we would like to have several dates at a particular location—one or more for the overlying deposits that are not faulted and others for deposits that are faulted (Figure 2.25c). Given this information, we can bracket the time in which displacement occurred. New techniques are now in the geologist's toolbox for establishing the age of fault motion, including numerical dating of fault gouge and age of exposure of geomorphic surfaces.

Numerical age control for faulting events is vital in tectonic and earthquake studies in order to establish three parameters: the age of the most recent rupture; the typical duration of time between events (recurrence interval); and the long-term displacement rate on the fault. The age of most recent ruptures, in particular, has both

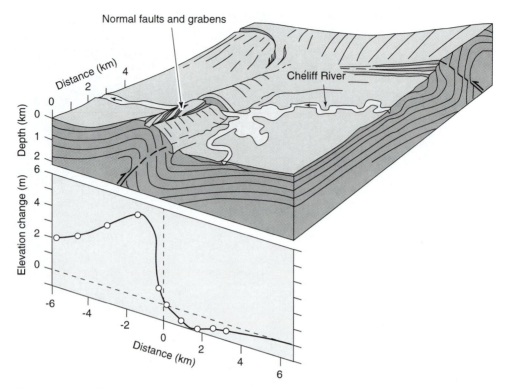

Figure 2.24   Idealized diagram of a fold deformed during the 1980 El Asnam earthquake (M 7.3) and graph of surface uplift produced by the event. The fold was produced by a sequence of such events. Note the extensional features at the crest of the fold. (After Stein and Yeats, 1989. *Scientific American*, 260(6):48–57.)

scientific and legal significance. In California, in order to call a fault **active** it must be demonstrated that the fault has moved in the past 10 ky [20]. For the U.S. Nuclear Regulatory Commission, a *capable* fault is one that has moved at least once in the past 50 ky (conveniently within the limits of conventional radiocarbon dating methods) or several times during the past 500 ky (see Chapter 1) [21].

At present, over 20 different methods of dating are useful for studying active tectonics. These methods may be categorized in terms of whether they are **numerical methods** (providing an absolute age in years), **relative dating** methods, or **correlation methods** [19]. Appendix A lists the more common methods used to date Pleistocene and Holocene materials and summarizes the age range over which each method can be applied, the basis for the method, and selected comments.

Historical records provide the most accurate chronology, but these are limited to only about 200 years in the United States. In other areas such as China, the historical record extends several thousand years. Given the brevity of human history compared with the great length of geologic history, we are usually forced to resort to analytical dating methods. All of the dating methods listed in Appendix A are applicable to active-tectonics studies under specific circumstances. Which method is used largely depends on the environment being studied. For example, if volcanic ash is present, then the logical

Table 2.1

THE GEOLOGIC TIME SCALE

Era	Period	Epoch	Millions of Years → Before Present
CENOZOIC	Quaternary	Holocene	0.01
CENOZOIC	Quaternary	Pleistocene	1.65
CENOZOIC	Tertiary	Pliocene	5.2
CENOZOIC	Tertiary	Miocene	23
CENOZOIC	Tertiary	Oligocene	35
CENOZOIC	Tertiary	Eocene	56
CENOZOIC	Tertiary	Paleocene	65
MESOZOIC	Cretaceous		146
MESOZOIC	Jurassic		208
MESOZOIC	Triassic		245
PALEOZOIC	Permian		290
PALEOZOIC	Pennsylvanian		324
PALEOZOIC	Mississippian		363
PALEOZOIC	Devonian		417
PALEOZOIC	Silurian		443
PALEOZOIC	Ordovician		495
PALEOZOIC	Cambrian		545
PRECAMBRIAN			
		Age of Earth	~4600

methods are **tephrachronology** and **potassium-argon** or **argon-argon analysis**. On the other hand, if carbonate rinds are found on clasts in soil horizons, they may be dated by using the **uranium-series method**. If the rinds are of Holocene or latest Pleistocene age, then the ^{14}C method may provide some age control. ^{14}C is the most often utilized of the dating techniques and also the most available, least expensive, and most thoroughly tested. Standard radiocarbon analyses can date material as old as about 50 ka (but many scientists trust the results only as far back as about 35 ka). Accelerator mass spectroscopy (AMS) is a variety of ^{14}C analysis that increases the temporal range that is datable, shrinks the error bars, and decreases the minimum sample size. Additional complications have been revealed by the extensive use of radiocarbon analyses. A fundamental assumption of the ^{14}C method is that the reservoir of ^{14}C in the atmosphere, to which all terrestrial organisms are equilibrated, has remained unchanged through time. Independent dating techniques reveal that this assumption is slightly in error, and radiocarbon ages are systematically too young by up to 3.5 ky at 20 ka, for example [22]. Independent corrections exist for this systematic error from tree rings [23], glacial varves (annual lake strata) [23], and uranium-thorium dates from corals [24].

In addition to systematic errors, random errors can be introduced into ^{14}C ages by sample contamination with carbon that is either too old or too young. Incorporation of 10% dead carbon (carbon with no ^{14}C left) into a sample will provide a date approximately 800 yr too old, and contamination of only about 0.5% recent carbon in any very

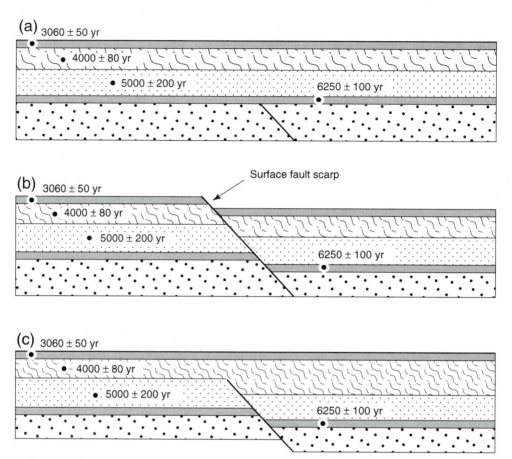

Figure 2.25   Principles of dating the most recent fault rupture. In all three cases, several horizons have been dated by [14]C. In case (a) the minimum age for the most recent faulting event is 6250 ± 100 yr. In case (b) the maximum age for the most recent faulting is 3060 ± 50 yr. In case (c) the most recent faulting occurred between 5000 ± 200 and 4000 ± 80 yr.

old (>50 ka) sample will produce a date of about 40 ka, even if the other 99.5% of the sample formed millions of years ago (see additional discusson in chapter 6)[19].

In addition to methods of dating geologic deposits, a battery of new techniques has been introduced to directly date the duration that a landform has been exposed at the Earth's surface (**exposure dating**). Research is underway to evaluate the use of beryllium-10, aluminum-26, chlorine-36, and other isotopes, all of which are produced when cosmic rays interact with the Earth's atmosphere or its surface [19, 25]. Because these isotopes accumulate in measurable quantities in surface materials such as soils, alluvial deposits, and rock surfaces, their concentration is a potential measurement of minimum time of exposure of those surfaces. Therefore, it is possible to measure the amount of beryllium-10 in the surface layer of a fault scarp in bedrock and determine the time the scarp has been exposed. This is the duration of time since faulting.

Lichenometry, the use of lichen growth rates to date exposure of rock surfaces, is emerging as a powerful tool to date earthquake-generated rockfall events during the past

500 years. Lichens are mosslike plants that grow on rock surfaces. They tend to grow in circular patches at variable rates on the order of a few millimeters per century. Surfaces may be dated ± 10 yr or better if sufficient care is taken in sampling and numerous (thousands) of lichen patches are measured. The method has been successfully used in California and New Zealand to date regional occurrences of rockfalls generated by large earthquakes, thus dating postseismic activity [26, 27].

Another new technique is being developed that may directly date the latest movement of a fault through analysis of the fault gouge. **Electron spin resonance** (ESR) takes advantage of the fact that ionizing radiation, from both cosmic rays and radioactive decay in surrounding material, occurs in the crystal structure of calcite or quartz [28]. Fault displacement can reset the ESR clock by either frictional heating or migration of fluids along the fault plane [29]. Thus any ESR signal measured in fault gouge should have accumulated only since the last fault rupture. That age can be estimated by measuring the signal and the in situ radiation dose.

The number of potential new dating techniques has increased in the past few years. These new methods are being studied and evaluated. Exposure dating is becoming common, and methods such as **optically stimulated luminescence** (OSL), based on exposure of sediment to ionizing radiation provided by radioactive elements including thorium, uranium, and potassium-40 following solar "bleaching", deposition and burial, are being advanced [30]. The preferable approach to establishing Quaternary chronology is to use as many potential methods as possible and summarize results to help estimate potential ages and chronology of Pleistocene and Holocene earth materials deformed by tectonic processes.

It is often difficult to collect useful dating material at a particular site where deformation from active tectonics occurs. As a result, it can be useful to establish a relative chronology based on methods such as soil-profile development or rock and mineral weathering. The development of a soil chronology is a complex process that requires the description and analysis of many soil profiles [31]. The tool is, however, a powerful one because the chronology may be established over an entire region and then carried to an area where independent age control is not available. Soils on alluvial surfaces of a variety of ages may be studied to produce what is known as a **soil chronosequence**, which is a series of soils arranged from youngest to oldest in terms of relative soil-profile development. Some of the soils may be dated by radiometric and other methods at some location in the region and then, through correlation of the soils (comparing dated profiles with profiles of unknown age), the chronology may be applied to other sites where absolute dates are unavailable. For example, the offset alluvial fan along the San Andreas fault in the Indio Hills near Palm Springs (Figure 2.12) has a relatively weak soil profile development with an estimated age of 20 to 30 ka. This age was determined by comparing soil-profile development at the site to soils of known age in the region. Soil dates based on correlation with dated soils are only rough estimates, commonly with uncertainties of up to ± 25%. Applying the estimated age range to the alluvial fan in the Indio Hills, which is offset about 700 m, provides a slip rate of approximately 23 to 35 mm/yr. As a second example of the use of soils in tectonic studies, consider the faulted terraces near Oak View, California, shown in Figures 2.26 and 2.27. Soils were described on the six river terraces. Development of a soil chronosequence allowed the faulted terraces to be correlated, and ^{14}C dates from charcoal in the terrace deposits for terraces Qt_{5a}, Qt_{5b}, and Qt_{6a} (along with measured displacement of the terraces) allowed calcu-

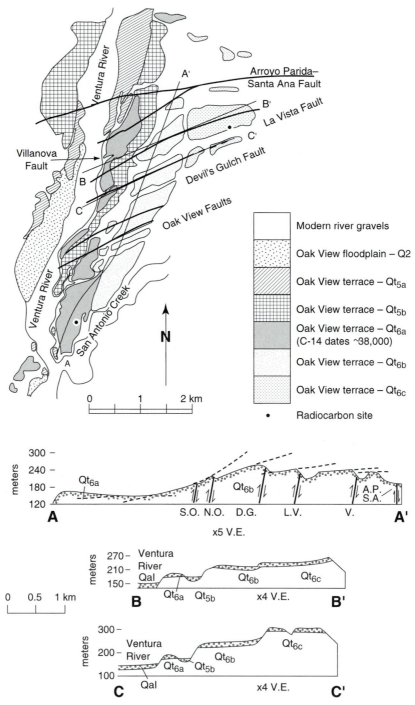

Figure 2.26   Faulted terraces of the Ventura River near Oak View, California. Cross section A-A' shows effects of faulting; cross sections B-B' and C-C' show the river terraces between faults. (After Rockwell et al., 1984 [32].)

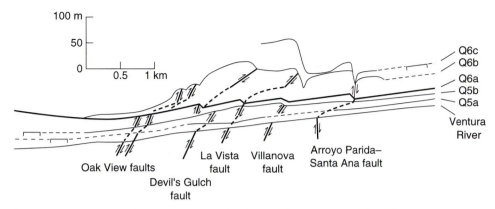

Figure 2.27 Longitudinal profiles of the modern Ventura River and faulted Late Pleistocene terraces. Locations of faults shown on Figure 2.26 (Modified after Rockwell et al., 1984 [32].)

lation of slip rates for the faults (ranging from 0.3 mm/yr to 1.1 mm/yr). Ages for the older terraces (Qt_{6b} and Qt_{6c}, Figure 2.26 were estimated by extrapolation from the younger dated terraces [32].

In summary, development of the Pleistocene and Holocene chronology is probably the most difficult part of an active tectonics study. Nevertheless, it is necessary if the history and rates of activity are to be determined. Vigorous research is developing new methods of dating that will be more reliable in estimating ages of materials, surfaces, and events.

SUMMARY

Tectonic geomorphology is defined as the study of landforms produced by tectonic processes or the application of geomorphic principles to the solution of tectonic problems. Five important geomorphic concepts applicable to tectonic geomorphology are (1) landscapes evolve through time, and changes in landforms are predictable; (2) abrupt changes in landscape evolution may occur as thresholds are exceeded; (3) interactions of landscape evolution with thresholds can produce complex response; (4) a change in form implies a change in process; and (5) active faulting produces a variety of landforms, and each major category of faulting may be discussed in terms of a characteristic assemblage of landforms.

Determination of the Pleistocene and Holocene chronology is critical to the solution of any tectonic-geomorphology problem. If there are no dates, there are no rates! Methods of establishing the chronology (dating) may be characterized in terms of whether they are numerical, relative, or based on correlation. Historical records provide the most accurate chronology, but these are limited to only about 200 years in the United States. Vigorous research is now underway to develop new methods of dating.

REFERENCES CITED

1. Bull, W. B., 1964. Geomorphology of segmented alluvial fans in western Fresno County, California. *U.S. Geological Survey Professional Paper* 352-E.

2. Schumm, S. A., 1977. *The Fluvial System*. New York: John Wiley & Sons. New York.

3. Davis, W. M., 1899. The geographical cycle. *The Geographical Journal*, 14:481–504.

4. Hack, J. T., 1960. Interpretation of erosional topography in humid temperate regions. *American Journal of Science*, 258A:80–97.

5. Bull, W. B., 1975. Allometric change of landforms. *Geological Society of America Bulletin*, 86:1489–1498.

6. Kuo-Fong, M., L. Chi-Tyi, T. Yi-Ben, S. Tzay-Chyn, and J. Mori, 1999. The Chi-Chi, Taiwan earthquake: large surface displacements on an inland thrust fault. *EOS: Transactions, American Geophysical Union*, 80:605–611.

7. Keller, E. A., D. B. Seaver, D. L. Laduzinsky, D. L. Johnson, and T. L. Ku, 2000. Tectonic geomorphology of active folding over buried reverse faults: San Emigdio Mountain front, southern San Joaquin Valley, California, 2000. *Geological Society of America Bulletin*, 112:86–97.

8. Crowell, J. C., 1974. Origin of late Cenozoic basins in southern California. *In* W. Dickinson (ed.), *Tectonics and Sedimentation*. Society of Economic Paleontologists and Mineralogists Special Publication, 22:190–204.

9. Dibblee, T. W., Jr., 1977. Strike-slip tectonics of the San Andreas fault and its role in Cenozoic basin evolvement. Late Mesozoic and Cenozoic Sedimentation and Tectonics in California, San Joaquin Geological Society Short Course.

10. Legg, M. R., and B. P. Luyendyk, 1982. Seabeam survey of an active strike-slip fault in the southern California continental borderland. *EOS: Transactions, American Geophysical Union*, 63:1107.

11. Thenhaus, P. C., and T. P. Barnhard, 1989. Regional termination and segmentation of Quaternary fault belts in the Great Basin, Nevada and Utah. *Bulletin of the Seismological Society of America*, 79:1426–1438.

12. Bruhn, R., 1994, personal communication.

13. Pinter, N., and E. A. Keller, 1995. Geomorphic analysis of neotectonic deformation, northern Owens Valley, California. *Geologische Rundschau*, 84:200–212.

14. Bursik, M., and K. Sieh, 1989. Rangefront faulting and volcanism in the Mono Basin, Eastern California. *Journal of Geophysical Research*, 94:15,587–15, 609.

15. Jackson, J. A., J. Gagnepain, G. Houseman, G. C. P. King, P. Papadimitriou, C. Soufleris, and J. Virieux, 1982. Seismicity, normal faulting and the geomorphological development of the Gulf of Corinth (Greece): the Corinth earthquakes of February and March, 1981. *Earth and Planetary Science Letters*, 57:377–397.

16. Clarke, S. H., and G. A. Carver, 1992. Late Holocene tectonics and paleoseismicity, southern Cascadia Subduction Zone. *Science*, 255:188–192.

17. Valentine, D. W., 1992. Late Holocene stratigraphy as evidence for late Holocene paleoseismicity of the southern Cascadia subduction zone, Humboldt Bay, California. Master's thesis, Humboldt State University, California.

18. King, C. G. P., and C. Vita-Finzi, 1981. Active folding in the Algerian earthquake of 10 October 1980. *Nature*, 292:22–26.

19. Pierce, K. L., 1986. Dating methods. In R.E. Wallace (ed.), *Active Tectonics*. Washington, DC: National Academy Press, 195–214.

20. State of California, 1973. California State Mining and Geology Board Classification (fault activity).

21. U.S. Nuclear Regulatory Commission, 1982. Appendix A. Seismic and geologic siting criteria for nuclear power plants. Code of Federal Regulations—Energy, Title 10, Chapter 1, Part 100. (App. A, 10, CFR 100).

22. Stuiver, M., and R. Kra, 1986. Calibration issues. *Radiocarbon*, 28:805–1030.

23. Stuiver, M., 1970. Tree ring, varve, and carbon-14 chronologies. *Nature*, 228:454–455.

24. Bard, E. B., R. G. Hamelin, and A. Zindler, 1990. Calibration of ^{14}C time scale over the past 30 ka using mass spectrometric U-Th ages from Barbados corals. *Nature*, 345:405–410.

25. Pavich, M. J., 1987. Application of mass spectrometric measurement of ^{10}Be, ^{26}Al, ^{3}He to surficial geology. In A. J. Crone and E. M. Omdahl (eds.), *Directions in Paleoseismology*. U.S. Geological Survey Open-File Report 87–673, 39-41.

26. Bull, W. B., and M. T. Brandon, 1998. Lichen dating of earthquake-generated regional rockfall events, Southern Alps, New Zealand. *GSA Bulletin*, 110:60–84.

27. Bull, W. B., 1996. Dating San Andreas fault earthquakes with lichenometry. *Geology*, 24:111–114.

28. Ikeya, M., 1975. Dating a stalactite by electron paramagnetic resonance. *Nature*, 255:48–50.

29. Fukuchi, T., 1988. Applicability of ESR dating using multiple centres of fault movement—the case of Itoigawa-Shizuoka tectonic line, a major fault in Japan. *Quaternary Science Reviews*, 7:509–514.

30. Aitken, M. J., 1998. *An Introduction to Optical Dating: The Dating of Quaternary Sediments by the Use of Photon-Stimulated Luminescence*. New York: Oxford University Press.

31. Birkeland, P., 1984. *Soils and Geomorphology*. New York: Oxford University Press.

32. Rockwell, T. K., E. A. Keller, M. N. Clark, and D. L. Johnson, 1984. Chronology and rates of faulting of Ventura River terraces, California. *Geological Society of America Bulletin*, 95:1466–1474.

3

Geodesy

INTRODUCTION

Geodesy is the science of the shape of the whole Earth and smaller parts of it and how these change over time. The first human attempts at accurate measurement of the land date to the first advanced civilizations. The Egyptians and Babylonians aspired to great works of architecture and engineering and required accurate surveys of the land to construct them. The sophistication and accuracy of surveying techniques have improved with the increasing ambitions of human civilization. In the modern world, networks of road and rail, canals, power lines, and telecommunication lines unite and bind the planet, spanning even the most forbidding terrain. These works are built, figuratively speaking, on networks of highly accurate geodetic lines.

Studies in active tectonics rely on geodesy for measurements of almost imperceptible changes in the surface of the Earth, changes that signal ongoing tectonic activity. The first applications of geodesy were not so subtle. In Classical and Postclassical Greece, there was considerable debate whether the Earth was a flat disk or, as Pythagoras and Aristotle postulated, a sphere. The father of geodesy was Eratosthenes, who lived between 276 and 195 B.C. and served as the librarian of the Museum of Alexandria, Egypt. Eratosthenes observed that, at noon on the day of the summer solstice, the sun

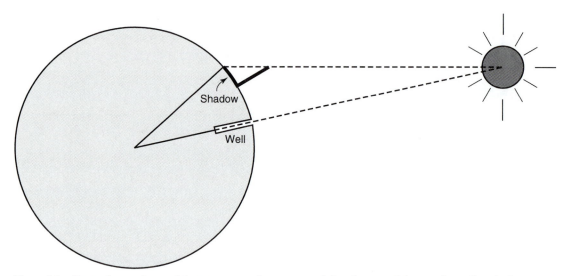

Figure 3.1.   Eratosthenes observed that at noon on the summer solstice, the rays of the sun shone directly down a well in Aswan, but a pole in Alexandria cast a shadow. By measuring the length of the shadow and the height of the pole, and knowing the distance between the two sites, he was able to calculate the radius of the Earth.

shone directly down into a well at Aswan, along the Nile valley. At the same moment in Alexandria, farther north at the Nile delta, the sun was not directly overhead, and a vertical pole cast a shadow (Figure 3.1). By measuring (1) the length of the shadow, (2) the length of the pole, and (3) the distance between Aswan and Alexandria (5000 Egyptian stadia, or 787.5 km), Eratosthenes calculated the radius of the Earth to be 6267 km [1, 2]. By this insightful observation and careful measurement, he not only proved that the Earth is roughly spherical in shape, but also was able to estimate its size—an estimate that was within 2% of the average radius of the Earth measured with today's sophisticated techniques (6371 km)!

The spherical model of the Earth remained the governing paradigm for nearly two millennia after Eratosthenes. It was the increasing accuracy of geodetic measurement that proved the model to be inadequate. In the early eighteenth century, it was found that a pendulum clock shipped from Paris to more equatorial latitudes ran slower near the equator. Sir Isaac Newton argued that this was evidence that the Earth was not truly spherical, but an **oblate ellipsoid**, bulging at the equator (Figure 3.2 a). The surface at the equator on an oblate body is further from the center than is the surface at higher latitudes. Gravity would be fractionally weaker, and a pendulum would swing more slowly. Never avoiding an opportunity to argue with the English, French scientists asserted that, on the contrary, the Earth was **prolate**—bulging at the poles (Figure 3.2b). The acid test of this dispute was to accurately measure the surface length of one degree of latitude, which would be greater near the poles on an oblate spheroid but greater near the equator on a prolate spheroid. The French Academy of Sciences sent out two teams of geodesists, one to Ecuador and the other to Lapland. The scientists assigned to the southerly expedition surely had no idea what was in store for them—they endured searing heat, icy cold, starvation, disease, and murder in a voyage that stretched more than

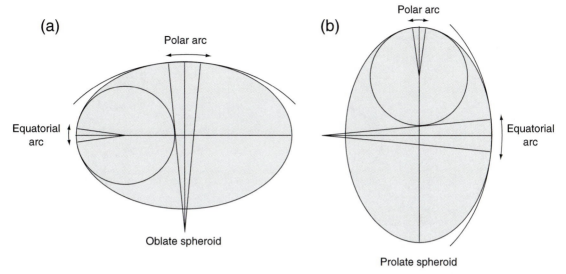

Figure 3.2.   Eighteenth-century theories on the Earth's shape. English scientists believed that the planet was an oblate spheroid (left)—the shape of a pumpkin sitting on the porch several weeks after Thanksgiving. French scientists believed it was a prolate spheroid (right)—egg shaped. On an oblate spheroid, the distance corresponding to one degree of latitude increases toward the poles; on a prolate spheroid, the distance diminishes from equator to poles. (After Fernie, 1991 [3].)

10 years. The northern expedition fared somewhat better—despite heavy seas, attempted piracy, and a subzero Scandinavian winter, they returned within 18 months. Through very careful geodetic and astronomical observation, both expeditions succeeded in their objective. The French found the length of 1° of arc in Ecuador to be 340,404 feet and in Lapland, 344,532 feet, proving that the the English were right—the Earth is indeed oblate [3]. Subsequent remeasurement found that the northern expedition had overestimated the distance in Lapland by about 600 feet, but the conclusion was correct nonetheless. We now estimate the radius of the Earth at the equator to be 1 part in 298.25, or about 21.4 km, greater than at the poles. From a purely spherical point of view, the Mississippi River runs uphill in that it is further from the center of the Earth at its delta than at its headwaters.

PRINCIPLES OF GEODESY

GEODETIC FRAMES OF REFERENCE

As the example of the Mississippi River illustrates, an important concept in geodesy is the frame of reference against which a position, including elevation, is measured. We now know that even the model of the Earth as an oblate spheroid is only an approximation (Figure 3.3). The global position of sea level, which approximates a surface of equal gravity potential known as the **geoid**, has an undulating shape that deviates from the oblate spheroid by as much as 100 m (Figure 3.3a). Elevation, as we commonly use the term, is a distance above or below the geoid, a distance more correctly called

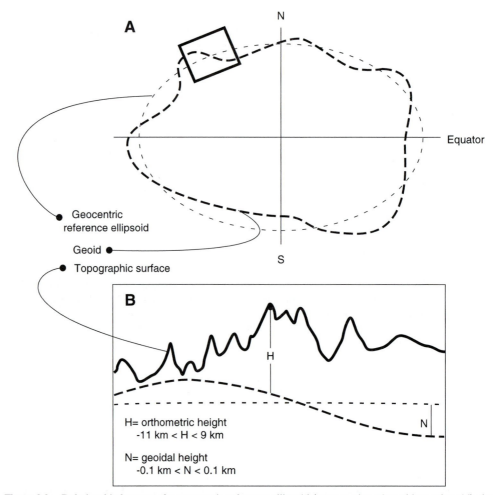

Figure 3.3. Relationship between the geocentric reference ellipsoid (cross section of an oblate spheroid), the geoid, and the surface topography of the Earth. Note that all deviations from a sphere are greatly exaggerated here. (After Wells, 1986. *Guide to GPS Positioning*. Fredericton, New Brunswick: Canadian GPS Associates.)

orthometric height (Figure 3.3b). Geodetic networks must be tied to the absolute vertical datum of the geoid, usually at several different locations. Mean sea level at a particular location is typically established by use of a **tide gauge** (see Chapter 6), which measures and averages the water level over long periods of time [4].

GEODETIC CHANGE

From a human perspective, the topography of the Earth is relatively stable through time. At most locations, once position and elevation have been accurately determined, they will not change much during a human lifetime. However, the precision of some geodetic measurements is so great that, in tectonically active regions, the motion can be measured instrumentally. All of the methodologies discussed in this chapter utilize the same fundamental approach—they precisely measure the relative positions of a network of points, wait for a period of time, and then remeasure the same network. Whether or not tectonic motion can be detected depends on three simple parameters: (1) the *precision* of the measurement technique, (2) the *rate* of tectonic motion, and (3) the *duration of time* between resurveys of the geodetic network. Along sections of the San Andreas fault, you could measure the northward motion of the Pacific Plate relative to North America with a ruler if you were patient enough to wait 30 years or more. With a geodetic positioning technique accurate to ±5 cm, for example, and a creeping fault that moves steadily at 10 mm/yr, at least 5 to 10 years between first and final measurements, and more likely 20 to 30 years, are required to say anything meaningful about the motion. With geodetic positioning techniques precise to less than 1 cm, motion across the same fault could be detected by resurveys in as little as 1 to 2 years.

GEODETIC TECHNIQUES

The technology used by modern geodesists has changed quite a bit since Eratosthenes and the French Academy of Sciences expeditions. The techniques that represent the current state of the art are informally subdivided as follows:

- Ground-based techniques

    Triangulation
    Trilateration
    Leveling

- Space-based techniques

    Very Long Baseline Interferometry (VLBI)
    Satellite Laser Ranging (SLR)
    Global Positioning Systems (GPS)
    Satellite Radar Interferometry (SRI)

The ground-based methods are a mature technology, the fundamentals of which have been around for centuries, even millennia. The space-based methods are relatively new, and their limitations and applications continue to evolve. Each of the different techniques must be evaluated in terms of the equipment required, its potential accuracy and precision, and its possible applications.

GROUND-BASED GEODESY

Ground-based geodesy takes on a number of forms, which range from localized surveying that you see construction and highway crews doing to three-dimensional positioning at points half a world apart. Until the 1970s and 1980s, ground-based geodetic measurements were the only tool available for detecting active tectonic deformation, and they may remain the most appropriate technique for certain applications today. Most high-precision surveying today is done with electronic distance measurement (EDM) systems, such as a geodolite (Figure 3.4). These systems bounce a laser pulse off a reflector positioned some distance away and detect the reflection of the pulse. The time interval

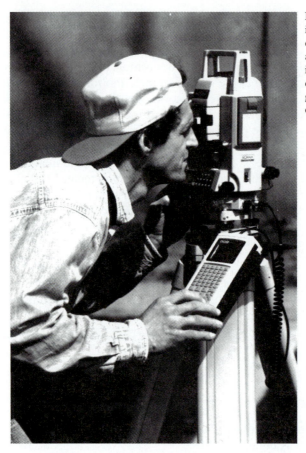

Figure 3.4.   A geodolite. This instrument and others like it are designed to measure distances and angles with great accuracy. These instruments are the most common devices for measuring ground-based geodetic networks. (Photo courtesy of Sokkia Corp.)

between sending the pulse and detecting its reflection is a measure of the distance between the laser source and the reflector. Depending on the equipment used, EDM systems can measure distances up to 10 to 20 km or more, although the greatest accuracy is achieved with much shorter lines. Under the best of conditions, measurement precision over baselines 1 to 35 km long using a geodolite has been found to be between 3 and 8 mm [5]. This precision was achieved by a series of short line sightings, temperature measurements at each station, and humidity measured by overhead aircraft.

Scales of geodetic measurements for tectonic applications are often subdivided into **near-field** and **far-field** approaches. Far-field geodesy refers to measurement over long distances, away from any one active structure in particular. In contrast, near-field geodesy involves measurement in the immediate vicinity, within meters to a few kilometers, of active features or features suspected to be active. Over the years, ground-based geodetic research has involved both far-field and near-field investigations of neotectonic deformation. Measurements have focused either on the precise horizontal positions of points (using "triangulation" or "trileration") or the precise vertical positions using **leveling**.

More than 700,000 km of leveling lines cross the United States, with accuracies measured in centimeters [3]. This control network is sensitive to local changes in topog-

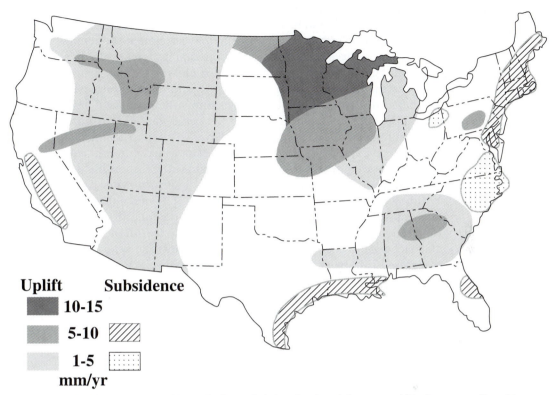

Figure 3.5.   The "neotectonic problem." Early geodetic leveling found that areas within the supposedly stable interior of the North American Plate appeared to be experiencing relatively rapid vertical motions, either uplift or subsidence. Either major secondary processes were at work within the plate interior, or there were systematic errors in the leveling measurements. (After U. S. Dept. of Commerce, 1972. News, 9/22/72. NOAA 72–122.)

raphy such as can be caused by ongoing tectonic activity. For example, one leg of the network crosses the Rio Grande rift in central New Mexico. Between surveys in 1911 and 1951, up to 20 cm of uplift occurred over what is now interpreted as an active magma body beneath the rift [6]. Also using leveling results, mid-twentieth-century geodesists identified broad regions of the United States that appeared to be subsiding and other regions that appeared to be gradually gaining in elevation (Figure 3.5). Because these areas are located in the supposedly stable interior of the North American Plate, many geologists of the day doubted the accuracy of these results, and the disagreement was sometimes called the "neotectonic problem". Subsequent geodetic work has basically confirmed the pattern illustrated in Figure 3.5, and geologic research has identified many of the mechanisms at work: for example, uplift of the Great Lakes area and New England is isostatic compensation following the removal of the latest Pleistocene ice sheets, and subsidence of the Gulf and Atlantic coastal plains is due to sediment loading on the continental shelves and the resulting lithospheric flexure. Some of the other regions of intraplate vertical motion are discussed in Chapter 9.

The geologists on the losing side of the "neotectonic problem" were not altogether amiss in their skepticism of the geodetic results of the day. Ground-based measurements, particularly precise elevation measurements over very broad areas, are subject to a number of sources of potential error. Small random errors such as due to thermal expansion of the leveling rods should be minor because they tend to average to zero over a large network. More problematic are errors that may accumulate systematically during the progression of a survey, particularly factors that systematically bias geodetic lines pushing into certain areas. For example, elevation and surface slope can directly and indirectly influence the refractive properties of light and thereby bias even the most careful ground-based survey into mountain ranges. Similarly, the mass of mountains and their associated lithospheric roots can perturb the geoid and add further systematic error to such measurements. Geodesists have, however, learned to compensate for most or all of these problems, and the United States has established different categories for classification of geodetic control of both horizontal and vertical coordinates: **first-order** surveys are the most precise, **second-order** the next, and **third-order** surveys are the least precise, although they still meet stringent criteria [7]. Some of the criteria for leveling surveys are listed in Table 3.1. Older ground-based geodetic measurements remain important data sources because they record much longer durations of time than possible with space-based techniques alone, and even newer ground-based measurements are being used side-by-side with space-based measurements to provide denser coverage and independent confirmation [e.g., 8–10].

In the immediate vicinity of faults or other geologic structures, near-field geodetic surveying can be used to detect fault creep, uplift or subsidence of the land, tilting, or the growth of active folds [11]. Several common near-field techniques (illustrated in Figure 3.6) are as follows:

- **Trilateration nets**. A number of stations are located on either side of a fault. Stations are arranged in a pattern of interlocking polygons so that the position of each point is known relative to every other. Repeated measurement of the orientations and lengths of the baselines between the stations results in a network sensitive to minute motion across the fault.

Table 3.1

SPECIFICATIONS FOR PRECISE LEVELING DONE BY THE U.S. COAST AND GEODETIC SURVEY.

	Order of Leveling			
	First	Second (Class I)	Second (Class II)	Third
Spacing of lines and cross-lines (km)	96.6	40.2-56.3	9.7	Not Specified
Average spacing of permanent benchmarks, not to exceed (km)	1.6	1.6	1.6	4.8
Length of sections (km)	0.8-1.6	0.8-1.6	0.8-1.6	Not Specified
Check between forward and backward runs between fixed elevations, not to exceed (L in km)	4 mm(L)0.5	8.4 mm(L)0.5	8.4 mm(L)0.5	12 mm(L)0.5

(After Brown and Oliver, 1976 Reviews of Geophysics and Space Physics, 14:13-35)

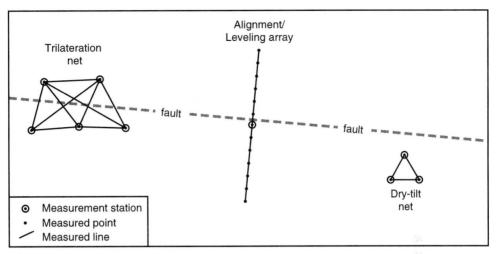

Figure 3.6.   Near-field geodetic techniques for measuring movement on an active fault. The trilateration net is sensitive to line length and orientation changes due to either strike-slip or dip-slip motion, the alignment array to strike-slip, the leveling array to dip-slip motion, and the dry-tilt net to tilting of the ground surface.

- **Alignment or leveling arrays**. Using a single base location, at which a measuring instrument is located, and a line of fixed points that cross the fault, any relative motion can be determined. Alignment arrays are designed to be sensitive to strike-slip motion, whereas leveling arrays are sensitive to vertical displacements.
- **Dry-tilt nets** (also called "spirit levels"). These points are generally arranged as an equilateral triangle located on one side of the fault or the other. Any changes in the relative elevations of the points indicate tilting of the ground [11].

Even in an age of space-based geodesy, near-field geodetic techniques are likely to remain in use because over short baselines, EDMs provide relative point positions at least as accurate as any space-based technology with far less measurement and processing time. In addition, techniques like triangulation are no longer in widespread use, but old ground-based surveys often provide a rich source of long time-scale data for comparison with recent EDM or GPS studies.

VERY LONG BASELINE INTERFEROMETRY (VLBI)

Very Long Baseline Interferometry crosses the line between geology and astronomy. This technique measures relative positions on the Earth's surface with millimeter-scale precision by observing quasars, which are among the most distant and most energetic features in the universe. Owing to their great distance from us, quasars are the most stationary beacons in the heavens. Closer celestial objects such as the stars shift positions at rates detectable by the most sensitive telescopes. Quasars also are desirable as beacons because their radio signals, far higher in frequency than visible light, are virtually undistorted by passage through the Earth's atmosphere [12].

Radio telescopes around the world track the positions of quasars and monitor the signals they emit. Each receiving station records this information, as well as the exact time at which it was received, calibrated by atomic clocks. For any one quasar, the radio signal detected at all of the telescopes is identical, except that the signal will be detected fractionally earlier at some stations than at others (Figure 3.7). The lag time between any two receivers is directly related to the distance between them parallel to the quasar's direction, as follows:

$$\text{Distance} = \text{rate} \cdot \text{time} = (\text{speed of light}) \cdot (\text{lag time}) = c \cdot \Delta t. \qquad \textbf{(3.1)}$$

By tracking several quasars simultaneously, a network of receiving stations can precisely determine their relative positions in all three dimensions. The precision of these VLBI distance measurements can be in the millimeter range [12]. In addition, because VLBI is measured against a celestial reference frame and not a terrestrial one, it provides other information, such as precise measurement of the Earth's rotation rate and the orientation of the rotation axis. These minute variations reflect processes such as tides, winds, seasonal redistribution of water and ice, and flow within the molten interior of the planet [12].

The principal disadvantage of the VLBI technique is that it requires a fully functional radio observatory, equipment that is neither easily portable nor available to the average geodesist. However, monitoring of existing VLBI stations provides a highly accurate, coarsely spaced framework for the shape and the rates of deformation of the surface of the Earth.

SATELLITE LASER RANGING (SLR)

In 1976, the United States launched the first fully dedicated laser geodynamics satellite (LAGEOS). LAGEOS is one of several satellites currently equipped for laser ranging, in which the satellite is used as a reference by which to locate ground positions precisely. Ground stations track the satellite's position in its orbit and bounce laser pulses off it. The distance between the station and the satellite is measured as half of the lag time between sending the laser pulse (t_{send}) and detecting its reflection (t_{detect}), multiplied by the speed of light through the atmosphere (c'):

$$\text{Distance} = \frac{1}{2} \cdot (t_{\text{send}} - t_{\text{detect}}) \cdot c'. \qquad \textbf{(3.2)}$$

At the ground station, three types of data are recorded: the distance to the satellite, the satellite's position in the sky above the observing station, and the times of the observations [2, 13] (Figure 3.8).

There are two ways to utilize the data recorded at an SLR ground station. The first is to compare the record with data from another station simultaneously tracking the laser-ranging satellite. In this fashion, relative position between the two points can be determined with great precision. The second method for locating SLR sites is to use measurements from all stations around the planet over a period of months to calculate a model of the satellite orbit (that model is known as an **ephemeris**). The sum of all ground measurements leads to an ephemeris that approximates the satellite's actual position within its orbit to within a few centimeters, and a single ground station can determine

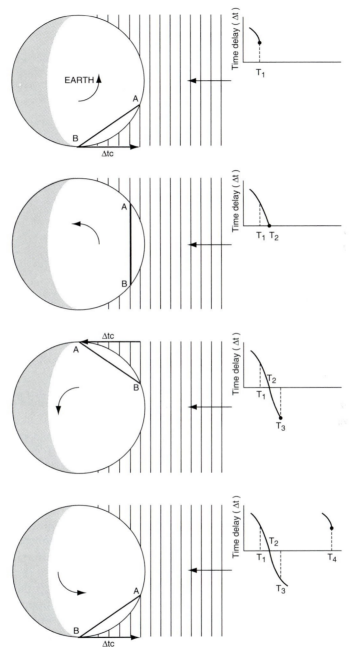

Figure 3.7.   VLBI measures the arrival times of signals from distant quasars, which can be viewed as a series of parallel wave-fronts. Two telescopes on the Earth at points A and B measure the arrival times (T_1, T_3) of a given wave-front at the two locations. The duration of the delay (T_3-T_1) is used to calculate the distance between the two sites (Δtc). (From Carter and Robertson, 1986 [12].)

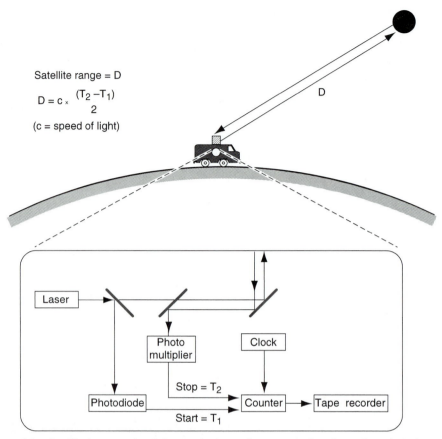

Satellite range = D

$$D = c \times \frac{(T_2 - T_1)}{2}$$

(c = speed of light)

D

Laser

Photo multiplier

Clock

Stop = T_2

Photodiode

Counter

Tape recorder

Start = T_1

Figure 3.8. Satellite laser ranging. A laser pulse is sent from a ground station, reflected off the satellite, and returned to the ground. An atomic clock measures the delay between the original signal and its reflection. The range to the satellite is calculated from this delay time. (After Wells, 1986. *Guide to GPS Positioning*. Fredericton, New Brunswick: Canadian GPS Associates.)

its own position relative to the model with similar precision. The accuracy of ground locations determined by satellite laser ranging is a few centimeters or better. SLR measurements are made with equipment that is increasingly compact and portable [14]. A particular weakness of the SLR technique, however, is that it requires favorable weather. Other space-based geodetic methods can operate regardless of weather (barring truly extreme conditions), but a cloudy day renders SLR inoperative.

In some applications of SLR, a very different type of satellite is used to reflect laser pulses back to the geodetic ground station—the moon. Mirrored reflectors were deployed by the Apollo 11, 14, and 15 missions and by two Soviet Luna landers [15] (Figure 3.9). Hitting the 46 cm², Apollo 11 reflector (Figure 3.9) has been compared with hitting a dime with a rifle fired from two miles away [15] and then catching the ricochet. Lunar Laser Ranging (LLR) reveals new information about the Earth–moon system (for example, that the drag of tides on Earth is causing the moon to slip away at a rate of 3.7 cm/yr). Furthermore,

Figure 3.9.   Lunar Laser Ranging reflector on the Sea of Tranquility, the moon. This is the first of three such devices emplaced by the Apollo astronauts (you see the footprints of Buzz Aldrin in the lunar dust). Laser ranging facilities on Earth bounce a pulse of light off one of the reflectors and measure the time it takes to detect the reflection. The detectors on Earth must be extremely sensitive because only a few photons will return to their source from a 2 billion watt laser emission. (From Faller and Dickey, 1990 [15].)

LLR has refined our understanding of the moon's orbit to the point that we can calculate the occurrence of solar eclipses back as far back as 1400 B.C. [15, 16]. With some discussion of a return of manned expeditions to the moon, there are plans for improved LLR capability that would improve precision by nearly two orders of magnitude [17].

GLOBAL POSITIONING SYSTEM (GPS)

The most extensively utilized space-based geodetic technology is the Global Positioning System [18]. The GPS system was developed in the 1970s, began being used as a tool for scientific and technical applications in the 1980s, and has exploded onto the commercial and consumer scene within the past decade or so. The heart of the GPS system is a constellation of 27 satellites that circle the Earth at an altitude of about 20,000 km

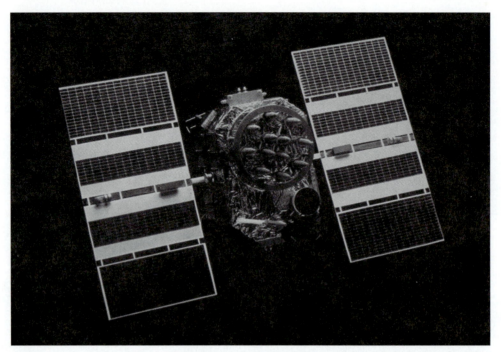

Figure 3.10. A GPS satellite. The current plan calls for 27 satellites in orbit around the Earth (24 operational satellites plus 3 spares), providing 24-hour, three-dimensional positioning capability around the globe. (Photo by Rockwell/Tsado/Tom Stack and Associates.)

and are operated by the U.S. Department of Defense (Figure 3.10). The GPS satellite array is continuously monitored by the U.S. Department of Defense using a string of permanent stations that gird the globe: on Hawaii, Kwajalein, Ascension Island, and Diego Garcia. Information on the satellite orbits, the health of each satellite, and clock accuracies is compiled at Colorado Springs, site of the central control station for the system. Civilian scientists now also operate a separate network of tracking stations.

Unlike the other space-based geodetic techniques discussed so far, operation of GPS receivers in the field is a passive procedure, made possible by the existing GPS infrastructure. Each satellite transmits a unique signal that gives its current position and the time of transmission according to on-board atomic clocks. Each satellite transmits this signal at two frequencies simultaneously, referred to as L_1 (1575.4 MHz) and L_2 (1227.6 MHz). The inexpensive hand-held GPS receivers that are now so common are **single-frequency receivers**, meaning that they utilize only one (the L_1) signal. Geodetic-quality GPS receivers generally are **dual-frequency** receivers, meaning that they receive and use both of signals. Because signal delays in the Earth's ionosphere are proportional to the frequency of the signal, this effect can be removed by comparing the arrival time of the two different frequencies [19, 20]. Finding the position of any GPS receiver on the ground is based on the principles of **ranging**—given the time of signal transmission, the current time, and the location of the satellite, it is possible to find the distance between the GPS receiver and each satellite it is tracking Figures 3.11 and 3.12. If three satellites are within range of the receiver, then its map position (*x*-

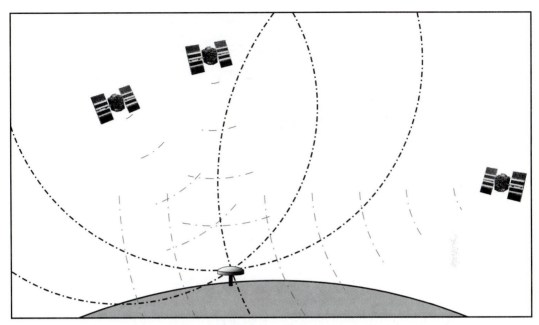

Figure 3.11.   GPS positioning is based on establishing the distances between the receiver on the ground and four or more of the orbiting satellites. Each satellite transmits time-coded signals that establish the lag time between signal transmission and signal reception. All together, the signals from the satellites allow calculation of the unique position of the receiver within the limits of the particular GPS technique being utilized. (After Herring, 1996 [21].)

and y-coordinates) can be determined; if four or more satellites are in range, then the position as well as the elevation (z-coordinate) can be calculated.

GPS positioning is possible in a large part because of the great precision of the satellite clocks; a clock error of as little as 1/1,000,000th of a second can result in a position error of 300 m [21]. Time at the receiver need not be so accurate because, in the process of tracking multiple satellites, there is a unique mathematical solution for the time at the receiver. The uncertainty in GPS positions measured on hand-held GPS receivers is generally cited as about ±20 m in the horizontal coordinates and ±40 m in the vertical [22]. Vertical precisions are generally estimated at one-half to one-third (i.e., imprecisions are twice to three times) that for the horizontal coordinates [21]. This results from the fact that the Earth blocks all the satellites located *beneath* beneath a receiver.

Until May 1, 2000 (8 P.M.), the estimated accuracies of hand-held GPS receivers was only ±100 m horizontal and ±200 m vertical. On that date, the U.S. Department of Defense ended encryption of the GPS signals, that had degraded GPS precision for civilian applications since 1990. The signals were encrypted so that U.S. military users would have an advantage over adversaries, but under pressure from scientific and commercial advocates, the military switched to a strategy of electronic jamming in battlefield situations [23]. The full implications of the end of signal encryption have not been fully explored at the present, but it seems likely that the benefits will be greatest for rough precision and will diminish for increasingly demanding applications. As outlined in the

Figure 3.12. GPS positioning has numerous applications in the geological sciences and beyond. Here, a steel rod mounted in the Antarctic ice is being measured in order to track ice thickening or thinning over time. (Photo courtesy of I. Whillans.)

following paragraphs, those applications have already developed methods for bypassing the military-imposed limitations. The relative precisions of several different GPS measurement strategies are outlined in Table 3.2 as well as some other details.

The most useful step in boosting the precision of GPS position measurements is to operate two or more receivers simultaneously (**relative positioning**). Uncertainty in receiver positions results from satellite clock errors and from ranging errors introduced by signal passage through the upper and lower atmosphere. Most of these errors are shared in common by multiple receivers operating in the same area. By comparing the signals recorded at one or more roving receivers with simultaneous measurements recorded at a receiver (a **base station**) fixed at a known location, such as a geodetic benchmark, the positions of the rovers can be determined much more precisely than from the same receivers operating in isolation. There currently are three distinct approaches to relative positioning:

- **post-processed relative positioning**
- **real-time differential GPS (DGPS)**
- **real-time kinematic (RTK) positioning**

Post-processing involves storing measurements obtained by roving receivers and base station and combining them at some later time using appropriate software. Post-processed relative positions can be the most accurate because they allow cumulative analysis of multiple satellite ranges at the same points and because the software algorithms generally are too intensive for the receivers themselves. The great disadvantage of post-processing is the lack of instantaneous gratification—results are not forthcoming until the number crunching is completed, at the end of the field day or much later. Alternatively, some of the additional precision can be gained using DGPS, in which roving GPS receivers are equipped with a supplementary radio receiver capable of picking up a correction signal from one of the DGPS beacons operated by the U.S. Coast Guard, the Corps of Engineers, or other sources. Each beacon consists of a permanently running GPS receiver operating at a point with precisely known coordinates and a radio trans-

Table 3.2

**PRECISION, APPLICATIONS, AND RELATIVE COST OF DIFFERENT GPS POSITION-
ING TECHNIQUES. [NUMBER OF DOLLAR SIGNS DENOTES RELATIVE COST OF
EQUIPMENT FROM INEXPENSIVE ($) TO VERY EXPENSIVE ($$$$$).]**

Method	Horizontal Precision	Applications	Relative Cost
Hand-held receiver, before 5/00 (with signal encrypted)	100 m	Recreational uses, approximate location of sample points	$
Hand-held receiver, after 5/00 (signal not encrypted)	20 m	Recreational uses, moderate precision sample locations	$
Real-time DGPS	1–5 m	Mapping within accuracy standards of most maps	$$
Postprocessed relative position, w/single-frequency receiver and base station	10 cm–1 m	Feature mapping, rough topographic measurements, GIS applications	$$$
Postprocessed relative position, w/dual-frequency receiver in "rapid-static" mode	2–5 cm	Point acquisition in ~10 min, precise topographical measurements, applications such as tectonic geomorph, volcanology	$$$$
Real-time kinematic (RTK)	1–10 cm	Real-time differential correction provides very rapid point acquisition and feedback	$$$$$
Postprocessed relative position, w/dual-frequency receiver w/24+ hr per site	2–5 mm	Plate motions; co-, post-, and interseismic deformation; volcanic hazard; glacial dynamics	$$$$

(After universities NAVSTAR Consortium, 1998 [24].)

mitter that sends out data about the difference between the position indicated by the
GPS and the actual position. Again, because errors are shared by nearby receivers, a
DGPS-equipped receiver operating close to a correction beacon can yield positions accurate to within a few meters in real time. Finally, RTK positioning utilizes a portable
base station-beacon combination that the user deploys at an appropriate location.

The next leap in GPS measurement precision is obtained using **carrier phase tracking**. Receivers capable of carrier tracking calculate the signal travel time from each satellite signal not only using the number of wavelengths that separate the time of
transmission from the time of reception, but also the *phase* of each incoming signal. This
procedure is computationally very intensive and generally requires extended data collection because a unique solution must first be obtained for the integer wavelength
delay of each signal so that remaining phase delay can be used to yield the added precision. The exact precision of a given measurement using carrier phase tracking depends
on the length of time over which receiver data are collected. Research in the California
Channel Islands demonstrated that dual-frequency phase-tracking GPS receivers could
be used in **rapid-static** mode, in which just 10 minutes of measurement time yielded pre-

cisions of ±2.5 cm horizontal and ±5 cm vertical [20]. Longer measurement times, typically 24 hours per site or longer, yield millimeter-level measurement precision [24]. This is the methodology and the precision that is used for the great bulk of geodynamic GPS research going on around the world today [e.g., 25–29].

There are two approaches to precisely measuring points used in geodynamical research: **campaign** and **permanent GPS**. In campaign-style data collection, positions are measured periodically, generally every 1 to 2 years or more depending on the application. These measurement sites must be clearly marked and must be engineered to be extremely stable and resistant to even the most minute shifting of the ground surface. It has been suggested that the precision of GPS measurements has, in some cases, outstripped the stability of the monuments being measured [30]. Monuments should ideally be in solid rock; among the best are stainless steel pins drilled into bedrock. In permanent GPS, a receiver is mounted at a fixed location and records data on an ongoing basis. The disadvantage of permanent GPS is that it requires the full commitment of a geodetic-quality receiver as well as nonstop data collection and analysis. The advantages of permanent GPS sites are that they achieve the maximum possible precision, they record any deformation continuously through time, and they provide a unified reference frame for the more numerous campaign sites in surrounding areas [31].

SATELLITE RADAR INTERFEROMETRY

In the new and rapidly evolving field of tectonic geodesy, satellite radar interferometry (also known as Interferometric Synthetic Aperture Radar, or InSAR) is the newest approach and one of the most exciting. Whereas the other space-based and ground-based techniques measure deformation fields at a finite number of discrete points, SRI is being used to image complete patterns of change associated with earthquakes or any other event or process that modifies the surface topography. The input data used in SRI are pairs of satellite radar images that bracket the deformation event or process in time, one before and one after. The two images are combined to produce an **interferogram** (Figure 3.13), which is a map of the study area showing changes in radar travel time between the time of the first image and the second. Some of this interferometric change may be the result of secondary factors, such as differences in satellite orbits, ionospheric activity, or even lower-atmosphere weather conditions, but all of these secondary effects can theoretically be identified and removed.

After such corrections, the remaining signal in an interferogram is a map of topographic change during the time interval bracketed by the two radar images. Figure 3.13 is an interferogram of the Okmok caldera in the Aleutian Islands constructed from images that bracketed an eruption in 1997. The figure illustrates both areas of seemingly random noise and areas of coherent signal marked by the concentric ringlike features. In this example, the areas of local noise correspond to new lava flows and ground covered by ash during this time span [32]; similar "loss of coherence" can also be caused by differences in vegetation or snow cover. In contrast, the concentric rings in Figure 3.13—called **interferometric fringes**—represent contours of systematic surface displacement along the satellite's line of sight. In this case, each ringlike cycle records 2.83 cm of co-eruptive subsidence, reaching a maximum of 1.40 m in the center of the caldera [32].

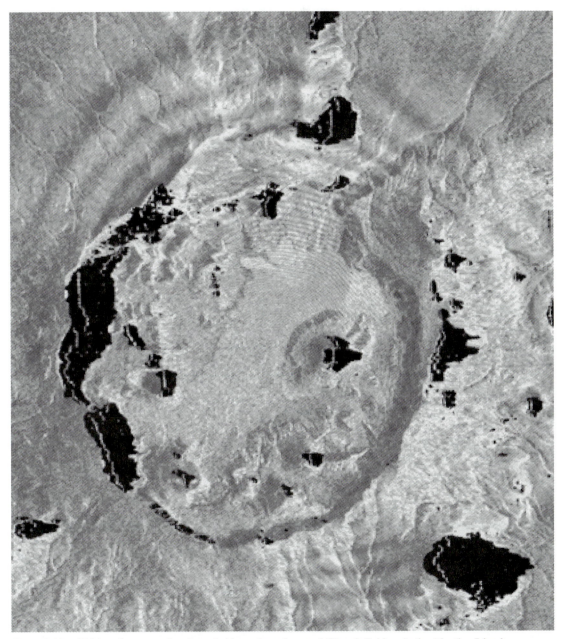

Figure 3.13.   SRI interferogram showing subsidence in and around Okmok Caldera in the Aleutian Islands associated with an eruption between February and March of 1997. As discussed in the text, an interferogram is a composite constructed from pairs of radar images taken at different times that shows change in the intervening time. This interferogram is interesting because it shows both the coherent concentric rings that are 2.83 cm contours of increasing subsidence approaching the center of the crater, as well as incoherent areas where new lava and ash were too irregular to allow matching of the "before" and "after" images. (From Lu et al., 1998 [32].)

In addition to volcanic inflation and deflation, SRI has been used successfully to image complex regional patterns of coseismic deformation [33, 34]. Such SRI images are increasingly being used for detailed modeling of fault rupture geometry and mechanics. Peltzer et al. [34] used an interferogram bracketing an M 7.6 earthquake that struck Tibet in November of 1997 to show that deformation was strongly asymmetrical across the fault, probably the result of significant but unrecognized heterogeneity in the crust. The full potential and range of applications of SRI to the field of active tectonics has not yet been realized at the time this book goes to press. New applications of SRI in other fields are sprouting seemingly every month. For example, the technique is being used to map the web of ice streams and tributaries that govern the motion and mass balance of ice sheet such as Antarctica [35]. In the field of active tectonics, SRI is likely to be a major source of new approaches in the near- and medium-term future. For example, simplification and acceleration of SRI processing could allow near-real-time monitoring of near-field strain accumulation. This, in turn, may prove to be a turning point in the search for warning signs of impending earthquakes.

SUMMARY OF GEODETIC TECHNIQUES

The reader can see that although the widespread use of geodesy in tectonics research is relatively new, a broad range of technologies have already proliferated. The utility and specific applicability of each of these techniques depend on the equipment required, the cost and availability of that equipment, the measurement precision achieved, and the distances over which those precisions can be obtained (Figure 3.14).

Even in the brief life span of tectonic geodesy so far, new technologies and evolving infrastructure have shifted the balance of which technology is most widely used for which application. Figure 3.15 presents a quick overview of recent trends in research publications utilizing the different geodetic approaches. Of the different space-based techniques, VLBI and SLR were the first to provide the measurement precisions necessary for tectonics applications, but the nature and cost of this equipment has limited much more widespread use. The technique that has proliferated most rapidly in recent years has been GPS. GPS positioning is even beginning to supplant traditional ground-based techniques used in topographic surveying. With the recent elimination of military encoding of the GPS signals, and with the cost of even the highest-end receivers still declining rapidly, there is no hint that the spread of GPS will soon be reversed. Of the geodetic techniques now being employed, the one that may represent the next revolution in tectonics research is SRI, with its capability of remotely imaging strain fields over entire regions. Some examples of recent applications of the various geodetic techniques are discussed in the following section, as well as some hints at future research directions.

APPLICATIONS

Geodesy and, in particular, the new space-based techniques represent the first tool in the toolbox of active tectonics capable of imaging regional deformation at the time scales of direct human observation. This tool presents a new opportunity to study the incremental motions, processes, and events that sum to comprise tectonic deformation over

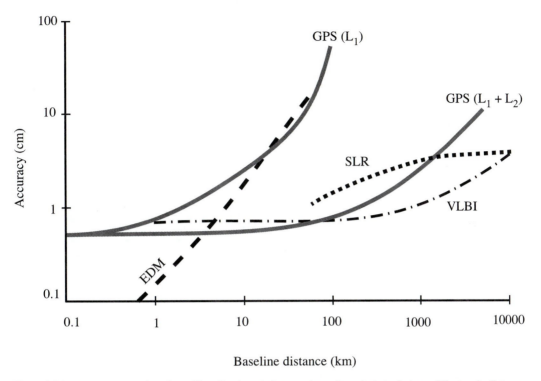

Figure 3.14.   Accuracy as a function of baseline length for a variety of geodetic techniques. Electronic distance measurement (EDM), for example, is most accurate over baselines up to a few kilometers in length but becomes decreasingly reliable (relative to the other techniques) over longer distances. The two curves for GPS positioning represent the accuracy for single-frequency GPS (L_1) versus dual-frequency GPS ($L_1 + L_2$). (Courtesy of D. Wells, Canadian GPS Associates.)

geological time. Looking at the range of recent work and speculating about near-future research directions, tectonic applications of geodesy can be loosely grouped into (1) assessment of and comparison with geological measures of plate motion, (2) measurement of earthquake mechanics, and (3) other applications.

GEODESY AND PLATE MOTION

Geologic evidence of the rates and directions of plate motions over geologic time scales ($\sim 10^5$ to 10^8 years) includes the following [36]:

- oceanic magnetic anomalies
- orientations of ocean transform fault
- offsets of geologic formations and features.

Additional information on relative motions comes from earthquake focal mechanisms at and near the plate boundaries. This evidence has been compiled into global models of plate motion, the most successful of which currently is the NUVEL model [37, 38]. Much can be learned about regional tectonics by first assessing the *agreement* between

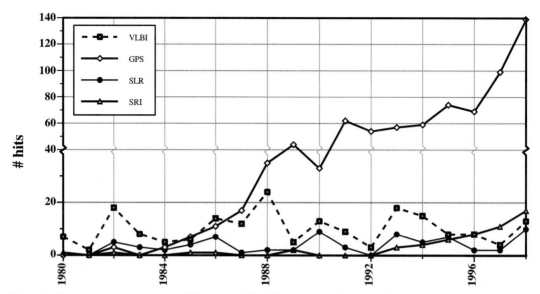

Figure 3.15. Overview of research publications, 1980–1998, using the different space-based geodetic techniques. Each data point represents the number of hits in the GEOREF database for a given year with that technique named in article titles, excluding abstracts (searches done in 7/2000). The plot informally shows the explosion of GPS as a research tool, and the more recent development of SRI.

models of long-term motions such as NUVEL and short-term geodetic measurements, and second by evaluating the *discrepancies* between the two sources of information.

Comparing the geologic and the geodetic data, the most striking conclusion is how close the estimates of plate motion are [39–41]. This is a testament first to the apparent precision of the new geodetic technology. More remarkable, the agreement between the two data sets demonstrates for the first time how steady plate motion is—with rates and directions that do not change whether measured over years or millions of years. We must conclude that, although they may stick and slip at their edges, the great bulk of the lithospheric plates rolls smoothly along with little or no variation in velocity [39].

Although the majority of geodetic measurements are consistent with global plate models, local disagreements are greatly improving our understanding of tectonic processes. Any such discrepancy could be explained in one or more of the following ways: (1) error in the GPS result, (2) an error in the geologic data or interpretation, (3) a change in plate motion over time, or (4) some local tectonic complexity not incorporated by the model. For example, GPS measurements have determined that recent relative motion between the Caribbean and North American Plates is about double the rate predicted by the NUVEL model [42]. Reasonable hypotheses to explain this difference include either that plate motion has changed in the past ~3 million years or that recent motion is a transitory phenomenon, perhaps catching postseismic relaxation following several historical earthquakes, but rigorous testing of these explanations suggests that NUVEL needs to be modestly revised in this area [43]. Similar tweaks are suggested in other areas as well [e.g., 44].

Perhaps the classical illustration of the stability of the plates and the potential for broadly distributed plate motion is what has been the "missing motion" on the San An-

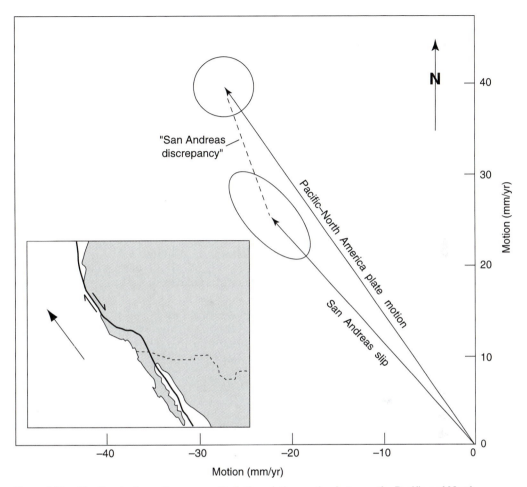

Figure 3.16.   The San Andreas discrepancy. Both the relative motion between the Pacific and North American Plates and slip on the San Andreas fault are displayed as vectors, illustrating directions and magnitudes of motion (including error ellipses). The "missing motion" on the San Andreas is represented by the gap between the two vectors, representing motion that must be accounted for on faults other than the San Andreas. (From Jordan and Minster, 1988 [45].)

dreas fault [45]. The far-field rate of motion between the Pacific and North American Plates is about 48 mm/yr [46], but the rate of motion on the San Andreas fault in central California is only 34 ± 2 mm/yr [47] (Figure 3.16). East of the San Andreas fault, the Sierra Nevada is separated from the core of stable North America by the Basin and Range province (Figure 3.17). The Basin and Range is characterized by deep basins and steep, fault-bounded mountain blocks formed by extension and right-lateral translation during the past 20 to 30 million years. VLBI measurements first determined that sites west of the Basin and Range are moving at a rate of 11 ± 1 mm/yr oriented at 36° ± 3° west of north relative to sites to the east [48]. When this vector is fit into the gap between local and regional measurements of Pacific–North American motion illustrated in Figure 3.16, it accounts for most of the "missing motion". Many of the details of this distributed motion in the Basin and Range are now being measured by GPS networks

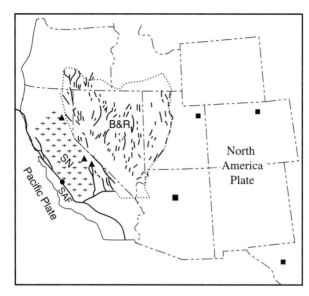

Figure 3.17.   Simplified tectonic map of the western United States, showing the Pacific Plate, the Sierra Nevada–Great Valley block (SN), the Basin and Range province (B&R), and the stable North American Plate. Solid squares are radio telescope sites on North America; triangles are radio telescopes on the Sierra Nevada block. (From Argus and Gordon, 1991 [48].)

across the region [49]. With Basin and Range faulting accounted for, there is a final residual of 6 ± 2 mm/yr of "missing motion" that is being taken up by additional distributed shear across the California margin [27, 50].

SEISMIC DEFORMATION CYCLE

Certainly one of the most important applications of active tectonics is to reveal the distribution and timing of earthquakes. The precision, distribution, and frequency of geodetic measurements contribute vital new information towards this application. Geodetic data are helping to answer questions in three main areas:

1. How are deformation and earthquakes distributed in the vicinity of plate boundaries and between them?
2. How does strain vary over time at the scale of individual earthquake cycles?
3. What are the present-day deformation rates across aseismic, locked faults, and what earthquake magnitudes and recurrences can be inferred from this deformation?

As discussed in the previous section, discrepancies between geodetic deformation measurements and global plate models can reveal important deviations from simple plate motion. Current plate models are simplified in that they assume that plate boundaries are sharp, concentrated features where all relative plate motion takes place, and that the plates are rigid bodies with no interior deformation (no **intraplate motion**) [39]. Detailed geodetic studies are increasingly showing that plate boundaries are not simple lines on the map, but rather broad zones across which relative motion is distributed. For example, in the South American Andes, GPS, SLR, and DORIS (a French GPS-like system) measurements (Figure 3.18) illustrate how convergence between the Nazca and South American Plates is distributed over a zone several hundred kilometers wide. These

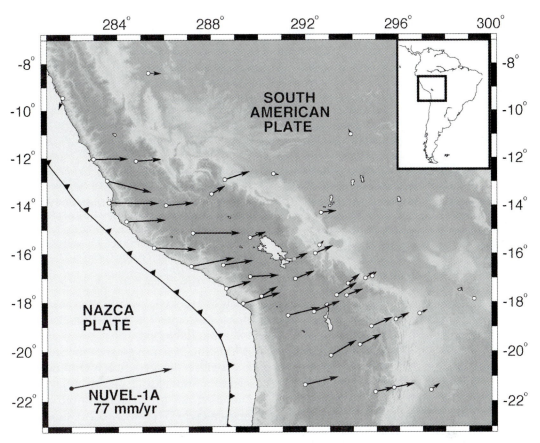

Figure 3.18.   GPS and other geodetic measurements along the Pacific margin of South America illustrate not only the convergence between the Nazca and South American Plates, but also the details of how that shortening is distributed over a very broad zone of deformation. Open circles are GPS measurement sites, and the arrows indicate the direction and relative velocity at those sites. Total convergence between the Nazca (NZ) and South American (SA) plates is about 77 mm/yr, but that relative motion varies gradually from west to east. (From Norabuena et al., 1998 [51].)

results have been modeled as total plate motion driving four distinct types of deformation: (1) strain accumulation on the locked plate interface, (2) a smaller but significant component of aseismic creep on that boundary, (3) regional shortening across the Andes, and (4) trench-parallel strike-slip motion [51] (Figure 3.19). Deformation in continent–continent collisional settings may be even more complex than at subduction zones, and geodetic measurements are now helping to detail the partitioning of strain in such areas [e.g., 52]. It has been estimated that such plate-tectonic complexities are widespread, and that the plate boundaries cover as much as 15% of the Earth's surface [36]. In addition, identifying the degree of motion within the supposedly rigid, stable plate interiors is an important goal for determining the likelihood and nature of earthquakes over the remaining 85% of the planet's surface. For example, careful examination of geodetic sites across the stable interior of North America suggests that the "North American" Plate is indeed moving as a single rigid unit, at least within the error margins (~1

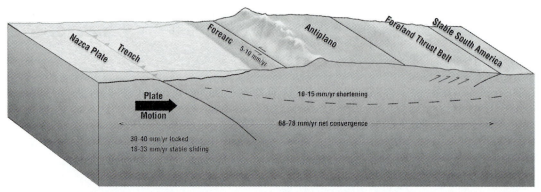

Figure 3.19.   Cartoon illustrating where the distributed deformation in Figure 3.18 goes. The reality is that this plate boundary is far from a simple line between two rigid plates. Instead, the total convergence of between 68 and 78 mm/yr is partitioned between interseismic strain accumulation at the plate boundary (30–40 mm/yr), aseismic creep (18–33 mm/yr), broadly distributed reverse faulting and uplift across the Andes (10–15 mm/yr), and trench-parallel shear (5–10 mm/yr). (After a figure from Universities NAVSTAR Consortium, 1998 [24].)

to 2 mm/yr)—an observation that is difficult to reconcile with earlier near-field geodetic measurements and paleoseismic evidence of frequent large earthquakes and rapid strain in the midcontinental New Madrid Seismic Zone [53].

In addition to the distribution of deformation in space, we are especially interested in how deformation near faults changes before, during, after, and between major earthquakes. The repeated pattern of coseismic motion during an earthquake, postseismic deformation, interseismic strain accumulation, any preseismic motion, and another earthquake is known as the **seismic deformation cycle** [54] (Figure 3.20). Geodetic studies are starting to image the various elements of this cycle to a degree not previously possible. In the years that high-precision geodetic networks have proliferated, coseismic motion associated with a growing number of large earthquakes has been measured [e.g., 55–60; see also Table 1 in 31]. As discussed previously, these geodetic measurements of coseismic deformation are often detailed enough to model the geometry of the fault on which the earthquake occurred and the rupture dynamics [e.g., 34].

Measuring the other elements of the seismic deformation cycle requires either multiple generations of geodetic observations or, as is increasingly becoming the case, the availability of permanent GPS receivers in areas of seismic activity and active deformation. Figure 3.21 nicely illustrates both the coseismic motion associated with a M 7.5 earthquake in Japan in 1994 and its M 6.9 aftershock as well as the significant postseismic slip following those events. Even more important, the nature of intraseismic deformation—the motion between major earthquakes—may prove to be an important tool in predicting the timing and magnitude of future earthquakes on locked faults. For example, GPS measurements of deformation in the Cascadia region of the U.S. Northwest confirm paleoseismic evidence for strain accumulation on a large plate-bounding megathrust; this same research also found residual motion going into northward translation that itself could be released in a M 7.6 earthquake every 100 years or in a larger number of smaller earthquakes [61]. Finally, the last element of the seismic deformation cycle that geodesy may be able to detect is perhaps the most intriguing—preseismic slip.

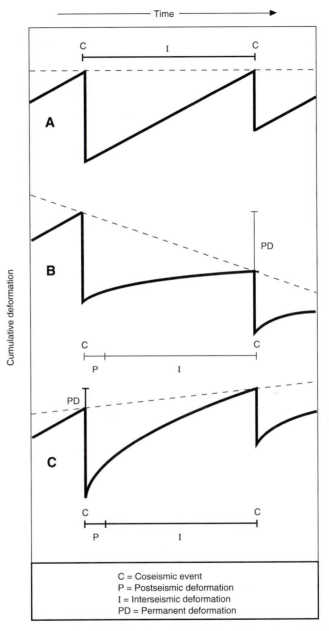

Figure 3.20.   Schematic illustrations of the seismic deformation cycle. (a)
Cycle with no permanent deformation—interseismic motion equals
coseismic slip. (b) Cycle with permanent deformation in the same direction
as the coseismic movement—the sum of the postseismic and interseismic
motion is less than the coseismic. (c) Cycle with permanent deformation of
the opposite sense as the coseismic movement—postseismic and
interseismic exceed the coseismic. (After Thatcher, 1986 [54].)

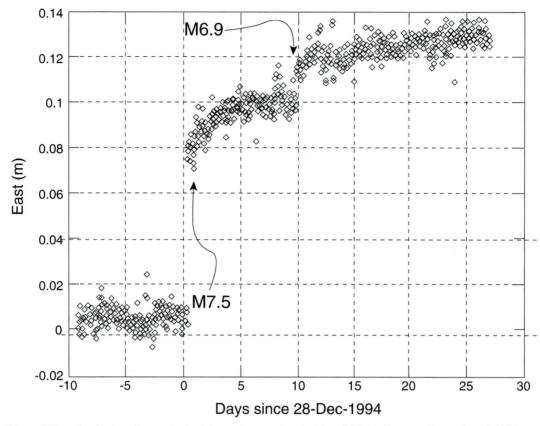

Figure 3.21.   Coseismic and postseismic deformation associated with an M 7.5 in Japan on December 28, 1994. The vertical axis here is the east-west location of a permanent GPS station on the island of Honshu computed relative to a stable frame of reference. Note that most displacement coincided with the main earthquake and its aftershock, but as much as an additional 40% of total motion occurred as gradual postseismic motion. (After F. Webb, unpublished data, shown in Segall and Davis, 1997 [31].)

With the coverage and frequency of geodetic observations increasing, there is the possibility that this approach may succeed where numerous others have failed—that is, in providing evidence of an imminent earthquake before it actually occurs. Before indulging in unbridled optimism, however, it is worth remembering that the failure of previous supposed earthquake precursors has shown us that rupture initiation is a complex process that may differ enormously from one fault to the next.

OTHER APPLICATIONS OF GEODESY

It is worth mentioning that there are many additional applications of the new geodetic technologies in other branches of the geosciences and beyond. For example, a related application of geodesy, which indeed some workers would consider an important part of active tectonics, is geodetic monitoring of volcanic hazard [62–65]. Volcanic eruptions are often preceded by ground-surface deformation caused by the injection of magma at depth [66]. A network of GPS receivers around Teishi Volcano in Japan detected significant deformation of the surface preceding an eruption in July of 1989 [67, 68]. A period of increased earthquake activity prior to the eruption was coincident with changes of 4 to 5 cm in relative elevation and about 13 cm in distance between two stations on either side of the volcanic center (Figure 3.22). Another seismic swarm occurred in 1997, and surface deformation was measured using a broad range of geodetic tools [9]. SRI interferograms created from repeated satellite passes over Yellowstone caldera reveal complex patterns of magma inflation and deflation and migration [10]. Volcano monitoring can be fully automated, with receivers permanently deployed at trouble spots and deformation measurements made almost in real-time. Although geodesy-based systems for earthquake warning remain just an optimistic hope at the present time, geodetic volcano monitoring is already a reality.

SUMMARY

Rapid advances in space-based geodesy are leading to a revolution in the study of active tectonics, global geology, glacial dynamics, astronomy, as well as numerous other applications. In active tectonics, it is now possible for the first time not only to reconstruct surface deformation, but actually to measure it directly. Repeated geodetic surveys can detect the directions and rates at which lithospheric plates move. An early result is that the relative motion of the centers of the plates is remarkably regular, but stress and strain are unevenly distributed near the plate boundaries. Geodetic measurements of coseismic deformation provide corroboration of focal mechanism solutions from seismologic records and allow modeling of fault geometry and rupture mechanisms. The density and frequency of geodetic measurements is beginning to allow discrimination of different phases of the seismic deformation cycle. As technology and detailed understanding of rupture mechanics improve, geodesy may provide important new tools for identifying preseismic motion that may warn of imminent earthquakes on some faults.

REFERENCES CITED

1. Torge, W., 1980 [translated by C. Jekeli]. *Geodesy: An Introduction*. New York: Walter de Gruyter.
2. Vaníček, P., and E. J. Krakiwsky, 1986. *Geodesy: The Concepts*. New York: North-Holland.
3. Fernie, J. D., 1991 and 1992. The shape of the Earth, Parts I, II, and III. *American Scientist*, 79:108–110; 79:393–395; 80:125–127.
4. Emery, K. O., and D. G. Aubrey, 1991. *Sea Levels, Land Levels, and Tide Gauges*. New York: Springer-Verlag.
5. Savage, J. C., and W. H. Prescott, 1973. Precision of Geodolite distance measurements for determining fault movements. *Journal of Geophysical Research*, 78:6001–6008.
6. Reilinger, R., and J. Oliver, 1976. Modern uplift associated with a proposed magma body in the vicinity of Socorro, New Mexico. *Geology*, 4:583–586.
7. U.S. Federal Geodetic Control Committee, 1984. *Standards and Specifications for Geodetic Control Networks*. Rockville, MD: National Oceanographic and Atmospheric Administration.
8. Liu, L., M. D. Zoback, and P. Segall, 1992. Rapid intraplate strain accumulation in the New Madrid seismic zone. *Science*, 257:1666–1669.
9. Aoki, Y., P. Segall, T. Kato, P. Cervelli, and S. Shimada, 1999. Imaging magma transport during the 1997 seismic swarm off the Izu Peninsula, Japan. *Science*, 286:927–930.

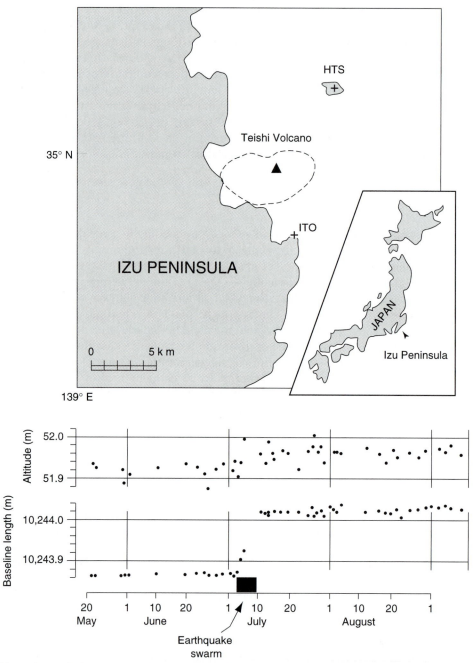

Figure 3.22.   GPS monitoring of Teishi Volcano, Japan. A swarm of earthquakes between July 3 and 10 preceded the eruption of July 12. GPS receivers at two locations across the volcanic feature, ITO and HTS, detected permanent changes in relative elevation and distance during the increased seismic activity. (After Shimada et al., 1990 [67].)

10. Wicks, C. Jr., W. Thatcher, and D. Dzurisin, 1998. Migration of fluids beneath Yellowstone Caldera inferred from Satellite Radar Interferometry. *Science*, 282:458–462.

11. Sylvester, A. G., 1986. Near-field tectonic geodesy. In Active Tectonics. Washington, DC: National Academy Press, 164–180.

12. Carter, W. E., and D. S. Robertson, 1986. Studying the Earth by very-long-baseline interferometry. *Scientific American*, 255(5):46–54.

13. Harrison, C. G. A., and N. B. Douglas, 1990. Satellite laser ranging and geological constraints on plate motion. *Tectonics*, 9:935–952.

14. Christodoulidis, D. C., D. E. Smith, R. Kolenkiewicz, S. M. Klosko, M. H. Torrence, and P. J. Dunn, 1985. Observing tectonic plate motions and deformations from satellite laser ranging. *Journal of Geophysical Research*, 90:9249–9263.

15. Faller, J. E., and J. O. Dickey, 1990. Lunar Laser Ranging. *EOS: Transactions, American Geophysical Union*, 71:725–726.

16. Morrison, D. C., 1989. An unsung legacy of the first lunar landing. *Science*, 246:447–448.

17. Bender, P. L., and J. E. Faller, 1990. Ranging goals for our return to the Moon. *EOS: Transactions, American Geophysical Union*, 71:475 (abstract #G21A-9).

18. Hofmann-Wellenhof, B., H. Lichtenegger, and J. Collins, 1994. *Global Positioning System: Theory and Practice*, 4th edn. New York: Springer-Verlag.

19. Prescott, W. H., personal communication.

20. Pinter, N., B. Johns, B. Little, and W. D. Vestal, 2001. Fault-related folding in the California Northern Channel Islands documented by rapid-static GPS positioning. GSA Today, 11(5):4-9.

21. Herring, T. A., 1996. The Global Positioning System. *Scientific American*, 274(2):44–50.

22. Holland, S., 5/1/2000. GPS gets better: Government unscrambles signal, boosting accuracy. ABCNews.com: http://abcnews.go.com/sections/tech/DailyNews/gps_000501.html.

23. Markoff, J., 3/5/96. Finding profit in aiding the lost: A civilian industry is built on the military's locator technology. *New York Times*: D1, D7.

24. Universities NAVSTAR Consortium, 1998. UNAVCO brochure, http://www.unavco.ucar.edu/community/brochure/.

25. Bilham, R., et al., 1997. GPS measurements of present-day convergence across the Nepal Himalaya. *Nature*, 386:61–64.

26. King, R. W., F. Shen, B. C. Burchfiel, L. H. Royden, E. Wang, Z. Chen, Y. Liu, X. Zhang, J. Zhang, and Y. Li, 1997. Geodetic measurements of crustal motion in Southwest China. *Geology*, 25:179–182.

27. Shen, Z., D. Dong, T. Herring, K. Hudnut, D. Jackson, R. King, S. McClusky, and L. Sung, 1997. Crustal deformation measured in Southern California. *EOS: Transactions, American Geophysical Union*, 78(43):477, 482.

28. Shevchenki, V. I., T. V. Guseva, A. A. Lukk, A. V. Mishin, M. T. Prilepin, R. E. Reilinger, M. W. Hamburger, A. G. Shempelev, and S. L. Yunga, 1999. Recent geodynamics of the Caucuses Mountains from GPS and seismological evidence. *Izvestiya, Physics of the Solid Earth*, 35:692–704.

29. Walpersdorf, A., C. Vigny, J. C. Ruegg, P. Huchon, L. M. Asfaw, and K. S. Al, 1999. 5 years of GPS observations of the Afar triple junction area. *Journal of Geodynamics*, 28:225–236.

30. Bevis, M., 1991. GPS networks: the practical side. *EOS: Transactions, American Geophysical Union*, 72:49, 55–56.

31. Segall, P., and J. L. Davis, 1997. GPS applications for geodynamics and earthquake studies. *Annual Reviews of Earth and Planetary Sciences*, 25:301–336.

32. Lu, Z., D. Mann, and J. Freymuller, 1998. Satellite Radar Interferometry measures deformation at Okmok Volcano. *EOS: Transactions, American Geophysical Union*, 79(39):461–468.

33. Massonnet, D., M. Rossi, C. Carmona, F. Adragna, G. Peltzer, K. Feigl, and T. Rabaute, 1993. The displacement field of the Landers earthquake mapped by radar interferometry. *Nature*, 364:138–142.

34. Peltzer, G., F. Crampé, and G. King, 1999. Evidence of nonlinear elasticity of the crust from the M_W 7.6 Manyi (Tibet) earthquake. *Science*, 286:272–276.

35. Joughin, I., L. Gray, R. Bindschadler, S. Price, D. Morse, C. Hulbe, K. Mattar, and C. Werner, 1999. Tributaries of West Antarctica ice streams revealed by RADARSAT interferometry. *Science*, 286:283–286.

36. Stein, S., 1993. Space geodesy and plate motions. In D. E. Smith and D. L. Turcotte (eds.), *Contributions of Space Geodesy to Geodynamics: Crustal Dynamics*. American Geophysical Union, Geodynamics Series, vol. 23. 5–20.

37. DeMets, C., R. G. Gordon, D. F. Argus, and S. Stein, 1990. Current plate motions. *Geophysical Journal International*, 101:425–478.

38. DeMets, C., R. G. Gordon, D. F. Argus, and S. Stein, 1994. Effect of recent revisions to the geomagnetic reversal time scale on estimates of current plate motions. *Geophysical Research Letters*, 21:2191–2194.

39. Gordon, R. G, and S. Stein, 1992. Global tectonics and space geodesy. *Science*, 256:333–342.

40. Larson, K. M., J. T. Freymueller, and S. Philipsen, 1997. Global plate velocities from the Global Positioning System. *Journal of Geophysical Research*, 102:9961–9981.

41. Argus, D. F., and M. B. Heflin, 1995. Plate motion and crustal deformation estimated with geodetic data from the Global Positioning System. *Geophysical Research Letters*, 22:1973–1976.

42. Dixon, T. H., F. Farina, C. DeMets, P. Jansma, P. Mann, and E. Calais, 1998. Relative motion between the Caribbean and North American plates and related boundary zone deformation from a decade of GPS observations. *Journal of Geophysical Research*, 103:115, 157–15, 182.

43. Pollitz, F. F., and T. H. Dixon, 1998. GPS measurements across the northern Caribbean plate boundary zone: impact of postseismic relaxation following historical earthquakes. *Geophysical Research Letters*, 25:2233–2236.

44. Antonelis, K., D. J. Johnson, M. M. Miller, and R. Palmer, 1999. GPS determination of current Pacific-North American plate motion. *Geology*, 27:299–302.

45. Jordan, T. H., and J. B. Minster, 1988. Measuring crustal deformation in the American West. *Scientific American*, 259(2):48–56.

46. DeMets, C., and T. H. Dixon, 1999. New kinematic models for Pacific-North America motion from 3 Ma to present; I, Evidence for steady motion and biases in the NUVEL-1A model. *Geophysical Research Letters*, 26:1921–1924.

47. Sieh, K. E., and R. H. Jahns, 1984. Holocene activity of the San Andreas Fault at Wallace Creek, California. *Geological Society of America Bulletin*, 95:883–896.

48. Argus, D. F., and R. G. Gordon, 1991. Current Sierra Nevada-North America motion from very long baseline interferometry: Implications for the kinematics of the western United States. *Geology*, 19:1085–1088.

49. Bennett, R. A., J. L. Davis, and B. P. Wernicke, 1999. Present-day pattern of Cordilleran deformation in the western United States. *Geology*, 27:371–374.

50. Dixon, T. H., M. Miller, F. Farina, H. Wang, and D. Johnson, 2000. Present-day motion of the Sierra Nevada block and some tectonic implications for the Basin and Range Province, North American Cordillera. *Tectonics*, 19:1–24.

51. Norabuena, E., L. Leffler-Griffin, A. Mao, T. Dixon, S. Stein, I. S. Sacks, L. Ocolo, and M. Ellis, 1998. Space geodetic observations of Nazca-South America convergence across the central Andes. *Science*, 279:358–362.

52. Abdrakhmatov, K. Y., et al., 1996. Relatively recent construction of the Tien Shan inferred from GPS measurements of present-day crustal deformation rates. *Nature*, 384:450–453.

53. Newman, A., S. Stein, J. Weber, J. Engeln, A. Mao, and T. Dixon, 1999. Slow deformation and lower seismic hazard at the New Madrid Seismic Zone. *Science*, 284:619–621.

54. Thatcher, W., 1986. Geodetic measurements of active-tectonic processes. In *Active Tectonics*. Washington, DC: National Academy Press. 155–163.

55. Savage, J. C., R. O. Burford, and W. T. Kinoshita, 1975. Earth movements from geodetic measurements. *California Division of Mines and Geology Bulletin*, 196:175–186.

56. Lin, J., and R. S. Stein, 1989. Coseismic folding, earthquake recurrence, and the 1987 source mechanism at Whittier Narrows, Los Angeles Basin, California. *Journal of Geophysical Research*, 94:9614–9632.

57. Miller, M. M., F. H. Webb, D. Townsend, M. P. Golombek, and R. K. Dokka, 1993. Regional co-seismic deformation from the June 28, 1992 Landers, California, earthquake: results from the Mojave GPS network. *Geology*, 21:868–872.

58. Blewitt, G., M. B. Heflin, K. J. Hurst, D. C. Jefferson, F. H. Webb, and J. F. Zumberge, 1993. Absolute far-field displacements from the 28 June 1992 Landers earthquake sequence. *Nature*, 361:340–342.

59. Hudnut, K. W., Z. Shen, M. Murray, S. McClusky, R. King, T. Herring, B. Hager, Y. Feng, A. Donnellan, and Y. Bock, 1996. Co-seismic displacements of the 1994 Northridge, California, earthquake. *Bulletin of the Seismological Society of America*, 86(Suppl. B):19–36.

60. Stramondo, S., M. Tesauro, P. Briole, E. Sansosti, S. Salvi, R. Lanari, M. Anzidei, P. Baldi, G. Fornaro, A. Avallone, G. Buongiorno, G. Franceschetti, and E. Boschi, 1999. The September 26, 1997 Colfiorito, Italy, earthquakes: modeled coseismic surface displacement for SAR interferometry and GPS. *Geophysical Research Letters*, 26:883–886.

61. Khazaradze, G., A. Qamar, and H. Dragert, 1999. Tectonic deformation in western Washington from continuous GPS measurements. *Geophysical Research Letters*, 26:3153–3156.

62. Dvorak, J., and D. Dzurisin, 1997. Volcano geodesy: the search for magma reservoirs and the formation of eruptive vents. *Reviews in Geophysics*, 35:343–384.

63. Mattioli, G. S., T. H. Dixon, F. Farina, E. S. Howell, P. E. Jansma, and A. L. Smith, 1998. GPS measurement of surface deformation around Soufriere Hills Volcano, Montserrat from October 1995 to July 1996. *Geophysical Research Letters*, 25:3417–3420.

64. Dixon, T. H., A. Mao, M. Bursik, M. Heflin, J. Langbein, R. Stein, and F. Webb, 1997. Continuous monitoring of surface deformation at Long Valley Caldera, California, with GPS. *Journal of Geophysical Research*, 102:12,017–12,034.

65. Owen, S., et al., 1995. Rapid deformation of the south flank of Kilauea volcano, Hawaii. *Science*, 267:1328–1332.

66. Thatcher, W., 1990. Precursors to eruption. *Nature*, 343:590–591.

67. Shimada, S., Y. Fujinawa, S. Sekiguchi, S. Ohmi, T. Eguchi, and Y. Okada, 1990. Detection of a volcanic fracture opening in Japan using Global Positioning System measurements. *Nature*, 343:631–633.

68. Yabuki, T., T. Kanazawa, and H. Wakita, 1991. Anomalous movements in Oshima Volcano associated the Ito submarine eruption revealed from GPS measurements. In Y. Ida and M. Mizoue (eds.), Seismic and volcanic activity in and around the Izu Peninsula and its tectonic implications. *Journal of Physics of the Earth*, 39:155–164.

4

Geomorphic Indices of Active Tectonics

INTRODUCTION

Morphometry is defined as quantitative measurement of landscape shape. At the simplest level, landforms can be characterized in terms of their size, elevation (maximum, minimum, or average), and slope. Quantitative measurements allow geomorphologists objectively to compare different landforms and to calculate less straightforward parameters (**geomorphic indices**) that may be useful for identifying a particular characteristic of an area (for example, its level of tectonic activity).

Some geomorphic indices have been developed as basic reconnaissance tools to identify areas experiencing rapid tectonic deformation [1]. This information is used for planning research to obtain detailed information about active tectonics. Other indices were developed to quantify description of landscape [2]. Geomorphic indices are particularly useful in tectonic studies because they can be used for rapid evaluation of large areas, and the necessary data often can be obtained easily from topographic maps and aerial photographs [1]. Some of the geomorphic indices most useful in studies of active tectonic include:

- The hypsometric integral [2]
- Drainage basin asymmetry [3, 4]
- Stream length–gradient index [5]
- Mountain-front sinuosity (S_{mf} index) [6, 7]

• Ratio of valley floor width to valley height (V_f index) [6, 7]

The results of several of the indices may also be combined, along with other information such as uplift rates, to produce **tectonic activity classes** [7], which are broad-based assessments of the relative degree of activity in an area.

HYPSOMETRIC CURVE AND HYPSOMETRIC INTEGRAL

The **hypsometric curve** describes the distribution of elevations across an area of land, ranging in scale from one drainage basin to the entire planet. The curve is created by plotting the proportion of total basin height (h/H = relative height) against the proportion of total basin area (a/A = relative area) (Figure 4.1) [2]. The total height (H) is the relief within the basin (the maximum elevation minus the minimum elevation). The total surface area of the basin (A) is the sum of the areas between each pair of adjacent contour lines. The area (a) is the surface area within the basin above a given line of elevation (h). The value of relative area (a/A) always varies from 1.0 at the lowest point in the basin (where $h/H = 0.0$) to 0.0 at the highest point in the basin (where $h/H = 1.0$).

A useful attribute of the hypsometric curve is that drainage basins of different sizes can be compared with each other because area and elevation are plotted as functions of total area and total elevation. That is, the hypsometric curve is independent of

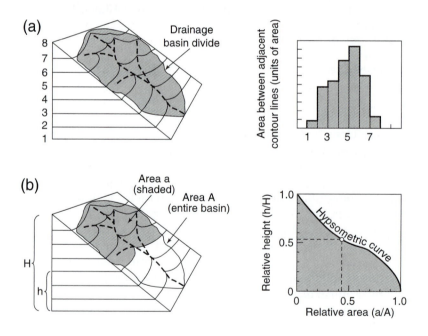

Figure 4.1 Hypothetical drainage basin showing how one point on the hypsometric curve is derived. Plotting several other values (for different contours) of a/A and h/H allows the curve to be constructed. (After Strahler, 1952 [2] and Mayer, 1990 [9].)

differences in basin size and relief [2]. As long as the topographic maps being used are of a sufficiently large scale to accurately characterize the basins being measured, there should be no effect of different scales.

A simple way to characterize the shape of the hypsometric curve for a given drainage basin is to calculate its **hypsometric integral** (H_i). The integral is defined as the area under the hypsometric curve. One way to calculate the integral for a given curve is as follows [8, 9]:

$$H_i = \frac{\text{mean elevation} - \text{minimum elevation}}{\text{maximum elevation} - \text{minimum elevation}}. \tag{4.1}$$

Thus only three values, two of them easily obtained from a topographic map, are necessary to calculate the integral. Maximum and minimum elevations are read directly from the map. Mean elevation can be obtained by point sampling (on a grid) of at least 50 values of elevation in the basin and calculating the mean [8], or by analysis of Digital Elevation Models (DEMs). High values of the hypsometric integral indicate that most of the topography is high relative to the mean, such as a smooth upland surface cut by deeply incised streams. Intermediate to low values of the integral are associated with more evenly dissected drainage basins.

The relationship between the hypsometric integral and degree of dissection permits its use as an indicator of a landscape's stage in the Cycle of Erosion (see Chapters 2 and 9). The Cycle of Erosion describes the theoretical evolution of a landscape through several stages: a "youthful" stage characterized by deep incision and rugged relief, a "mature" stage where many geomorphic processes operate in approximate equilibrium, and an "old age" stage characterized by a landscape near base level with very subdued relief. A high hypsometric integral indicates a youthful topography (Figure 4.2a). An intermediate value of the hypsometric integral and a sigmoidal-shaped hypsometric curve indicate a mature stage of development (Figure 4.2b). Further development to the old-age stage will not change the value of the integral, unless high-standing erosional remnants are preserved (Figure 4.2c). However, more sophisticated numerical descriptions of the hypsometric curve that are sensitive to continued evolution of the topography are available [2, 9, 10]. In summary, hypsometric analysis remains a powerful tool for differentiating tectonically active from inactive regions. The calculation of hypsometric curves and integrals has become almost trivial with the advent of DEMs [11]. Hypsometry at continental and planetary scales is discussed in Chapter 9.

DRAINAGE BASIN ASYMMETRY

The geometry of stream networks can be described in several ways, both qualitatively and quantitatively. Where drainage develops in the presence of active tectonic deformation, the network often has a distinct pattern and geometry (see Chapter 5). The **asymmetry factor** was developed to detect tectonic tilting transverse to flow at drainage-basin or larger scales [3]. The asymmetry factor (AF) is defined as

$$AF = 100(A_r/A_t) \tag{4.2}$$

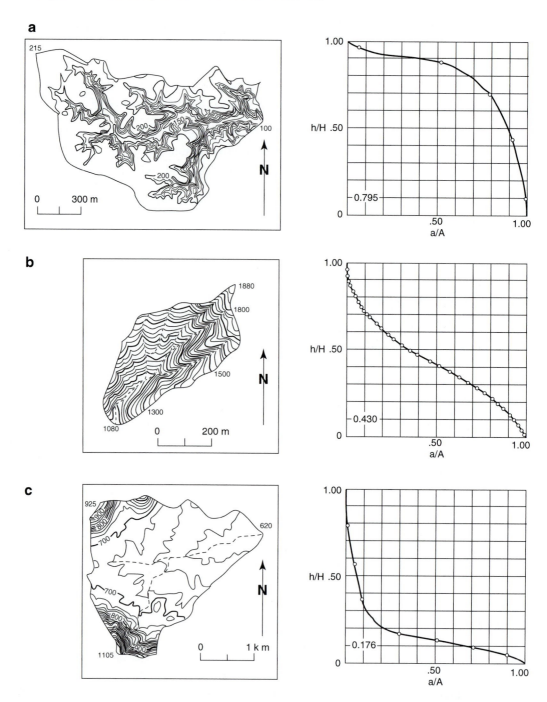

Figure 4.2   Three examples of different values of the hypsometric integral. See text for explanation. (After Strahler, 1952 [2].)

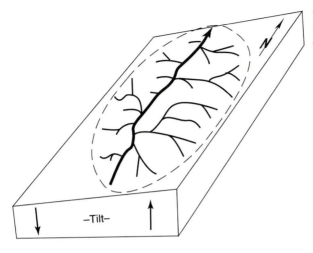

Figure 4.3   Block diagram showing how the asymmetry factor is calculated.

$$AF = 100 \left(\frac{A_r}{A_t} \right)$$

$$= 100 \left(\frac{3.2 \text{ km}^2}{4.9 \text{ km}^2} \right) = 65$$

AF > 50 implies tilt down to the
left of basin (looking downstream)

where A_r is the area of the basin to the right (facing downstream) of the trunk stream, and A_t is the total area of the drainage basin. For most stream networks that formed and continue to flow in stable settings, AF should equal about 50. The AF is sensitive to tilting perpendicular to the trend of the trunk stream. Values of AF significantly greater or less than 50 may suggest tilt. For example, in a drainage basin where the trunk stream flows north and tectonic rotation is down to the west (Figure 4.3), tributaries on the east (right) side of the main stream are long compared to tributaries on the west side, and AF is greater than 50. If the tilting was in the opposite direction, then the larger streams would be on the left side of the main stream and the AF would be less than 50.

Like most geomorphic indices, the AF works best where each drainage basin is underlain by the same rock type. The method also assumes that neither lithologic controls (such as dipping sedimentary layers) nor localized climate (such as vegetation differences between north- and south-facing slopes) causes the asymmetry [12]. Finally, tributaries to streams that flow down steep regional slopes may be asymmetrical without active tilting, with longer distances between the channel and the drainage divide on the "high" sides of the basin [13].

An example of application of the drainage basin asymmetry factor comes from the Pacific coast of Costa Rica [3]. The Nicoya Peninsula is a broad, emergent area of the outer arc of the Middle America Subduction Zone. The Nicoya Peninsula has been the site of active uplift probably since at least Oligocene-Miocene time (about 25 m.a.). Measurements of the drainage-basin asymmetry of large southwest-draining rivers show that deformation coincides with the major faults on the peninsula (Figure 4.4). In particular, the area southeast of the Montaña Lineament Zone is marked by drainage basins

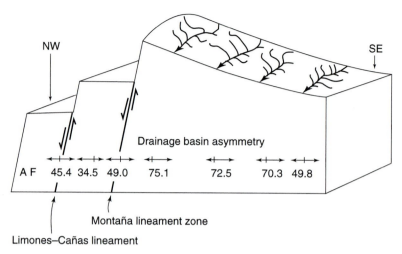

Figure 4.4   Asymmetry factors for the Nicoya Peninsula indicate tilting down to the southeast, south of the Montaña Lineament. (After Hare and Gardner, 1985 [3].)

tilted down to the southeast. Together, the morphometric and the geologic data support a faulted half-dome model for the active deformation of the Nicoya Peninsula [3].

Another quantitative index to evaluate basin asymmetry is the **Transverse Topographic Symmetry Factor** (T) [4].

$$T = D_a/D_d \qquad (4.3)$$

where D_a is the distance from the midline of the drainage basin to the midline of the active meander belt, and D_d is the distance from the basin midline to the basin divide (Figure 4.5). For a perfectly symmetric basin, T = 0. As asymmetry increases, T increases and approaches a value of 1. Assuming that the dip of the bedrock can be shown to have negligible influence on the migration of stream channels, then the direction of regional migration is an indication of the ground tilting in that direction [4]. Thus, T is a vector with a bearing (direction) and magnitude from 0 to 1. Values of T are calculated for different segments of valleys (Figure 4.5) and indicate preferred migration of streams perpendicular to the drainage-basin axis. This analysis is most appropriate to dendritic drainage patterns, where evaluation of tributary valleys as well as the main or trunk valley allows for a larger range of T. Analysis of a number of drainage basins in an area results in a field of T vectors, and this field can be spatially averaged and analyzed to define anomalous zones of basin asymmetry [14]. Calculation of T, as with AF described earlier, does not provide direct evidence of ground tilting, but like AF, it is a method for rapidly identifying possible tilt. This method was used to suggest the direction of possible Holocene tilting of the ground in the Mississippi Embayment [4, 14].

STREAM LENGTH–GRADIENT (SL) INDEX

The **stream length–gradient index** (or **SL index**) is calculated for a particular reach of interest and defined as

$$SL = (\Delta H/\Delta L)L \qquad (4.4)$$

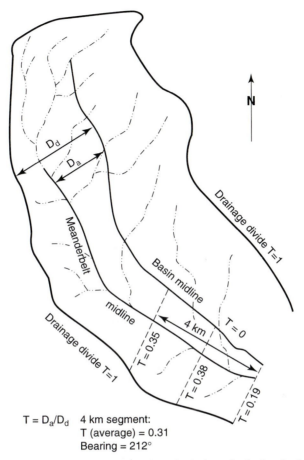

Figure 4.5   Diagram of a portion of a drainage basin showing how
the transverse topographic symmetry factor (T) is calculated for one
stream segment within the basin. (After Cox, 1994 [4].)

where SL is the stream length–gradient (SL) index, $\Delta H / \Delta L$ is the channel slope or gradient of the reach (ΔH is the change in elevation of the reach and ΔL is the length of the
reach), and L is the total channel length from the midpoint of the reach of interest upstream to the highest point on the channel. In most cases, these parameters are measured
from topographic maps. Figure 4.6 shows how the SL index is calculated for a hypothetical example.

The SL index correlates to stream power. Total stream power available at a particular reach of channel is an important hydrologic variable because it is related to the
ability of a stream to erode its bed and transport sediment. Total or available stream
power is proportional to the product of the slope of the water surface and discharge. The
slope of the water surface generally is approximated by the slope of the channel bed, and
there is a good correlation between total channel length upstream and bankful discharge
(the discharge with return period of about 1.5 yr), which is thought to be important in
forming and maintaining rivers.

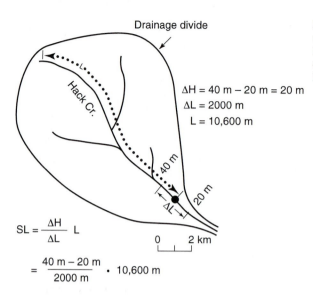

Drainage divide

Hack Cr.

$\Delta H = 40\ m - 20\ m = 20\ m$

$\Delta L = 2000\ m$

$L = 10{,}600\ m$

$SL = \dfrac{\Delta H}{\Delta L}\ L$

$= \dfrac{40\ m - 20\ m}{2000\ m}\cdot 10{,}600\ m$

$= 106\ \text{gradient meters}$

0   2 km

The SL index is sensitive to changes in channel slope, and this sensitivity allows the evaluation of relationships among possible tectonic activity, rock resistance and topography. The sensitivity of the SL index to rock resistance is illustrated in Figure 4.7, which shows the longitudinal profile of the Potomac River upstream from Washington, D.C. SL index values are relatively low in the Valley and Ridge province and in the Appalachian Valley, where the rock types are shale, siltstone, some sandstone, and carbonate rocks. The index increases significantly where the river crosses the relatively hard rocks of the Blue Ridge, then decreases on relatively soft rocks of the Triassic basin and the Piedmont. Finally, the index increases significantly again at the resistant rocks of the Great Falls in the lower reaches of the river [5]. Studies in the Appalachian Mountains of the eastern United States suggest that a good correlation exists between rock resistance and stream length–gradient index. In other words, the form of the land is well adjusted to rock resistance.

In landscape evolution, the adjustment of stream profiles to rock resistance is assumed to occur fairly quickly. Therefore, the SL index is used to identify recent tectonic activity by identifying anomalously high index values on a particular rock type. For example, an area of high SL indices on soft rock, such as shale, may indicate recent tectonic activity. Anomalously low values of the index also may represent tectonic activity. For example, along linear valleys produced by strike-slip faulting, low indices are expected because the rocks in the valleys are often crushed by fault movement, and the streams flowing through those valleys should have a lesser slope.

A map of the SL indices is produced by the following method:

1. Obtain a topographic map of the area of interest. The method is particularly useful for large areas, so 1 : 250,000 scale maps often are useful. Using a trans-

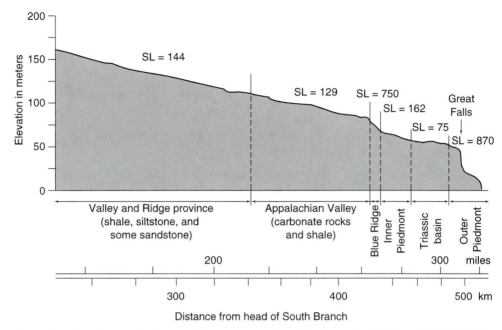

Figure 4.7   SL indices for the Potomac River upstream from Washington, D.C. (After Hack, 1973 [5].)

parent overlay, outline the major streams and rivers. Extend streams to where contour lines no longer "V" in the upstream direction.

2. Select a convenient contour interval (say 100 m for a map scale of 1 : 250,000). Where these contours cross the mapped streams, mark the locations on the overlay.

3. Connect points of equal elevation between the streams. This will create a map of the level to which the streams have eroded, called a **subenvelope map**. Figure 4.8 shows a hypothetical map of the level to which streams have eroded Bull Mountain.

4. Along each stream, measure the distance, ΔL, between successive contours along the stream as well as the total upstream stream length. The calculations are simple because the contour interval is constant, and thus ΔH is constant. Calculate the SL index (Equation 4.4) for each stream segment between successive contours, and record that value on the stream at the midpoint between contours on the subenvelope map. These locations are noted as dots in Figure 4.8. The values of the index farthest upstream, near the drainage divide, may be spurious, so it may be necessary to start calculating the index a standard distance downstream from the divide.

5. Finally, construct the SL index map by contouring the SL values. The subenvelope map for Bull Mountain (Figure 4.8) suggests that a belt of high indices lies between the contour intervals of 200 and 400 m for streams 1, 2, and 3. This zone may reflect a resistant rock unit, or perhaps an active fault zone along the southern flank of the range. Field work to examine evidence for presence of

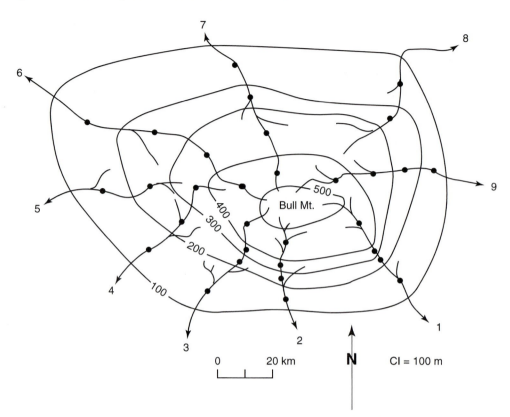

Figure 4.8   Hypothetical mountain showing construction of map of level to which streams have eroded. Dots are points where the SL index should be calculated.

resistant rocks or active tectonics will help differentiate between various interpretations concerning the pattern of values of the SL index. As discussed later, the SL index has proven useful as a reconnaissance tool in active tectonics work in several studies [15–17].

STREAM LENGTH-GRADIENT INDICES IN THE SAN GABRIEL MOUNTAINS OF SOUTHERN CALIFORNIA

SL indices can be calculated for areas of several thousand square kilometers or more, utilizing small-scale topographic maps (1:50,000 to 1:250,000). For example, Figure 4.9 shows SL indices for the San Gabriel Mountains of Southern California. The mountain range is part of the Transverse Ranges, which have been uplifted from below sea level to several thousand meters elevation in the past few million years. The range is bounded on the north and northeast by the San Andreas fault zone and on the southwest and south by a system of active reverse faults, including the San Fernando fault and the Sierra Madre fault zone. The San Gabriel fault zone, also located near the southern and

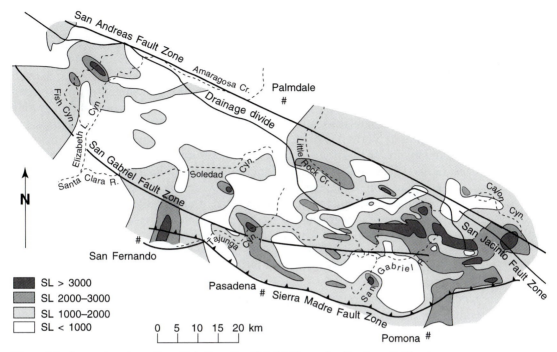

Figure 4.9   Stream length-gradient SL indices, San Gabriel Mountains, Southern California.

southwestern boundary, is a right-lateral strike-slip fault that is thought to have been more active in the past than at present.

Anomalously high values of the SL index occur along the southern and eastern front of the San Gabriel Mountains. Although it has been known for some years that uplift rates are relatively high there, the SL index map of the San Gabriel Mountains demonstrates the utility of the index for identifying zones of tectonic activity. Of particular interest is the zone of high indices near San Fernando. This is the site of the 1971 (M_W 6.6) San Fernando earthquake, which caused widespread damage in the Los Angeles area. Thus, if nothing were known about the San Gabriel Mountains, but good topographic maps and other elevation control were available, the SL index would have indicated that the southern and southeastern flanks of the range should be studied in more detail. There the active reverse faults that generated the 1971 earthquake would likely be found.

Other interesting aspects of the SL index map for the San Gabriel Mountains are the belts of relatively low indices along the San Andreas and San Gabriel fault zones. These index values are most likely related to soft rocks produced by crushing caused by long-term fault movement. Low indices are also found in a belt of rocks with low resistance between Elizabeth Canyon and Soledad Canyon (western-central portion of Figure 4.9). In the central part of the ranges (at the southeastern end of the map, Figure 4.9), resistant metamorphic rocks are present and high indices are found.

STREAM LENGTH–GRADIENT INDICES AT THE MENDOCINO TRIPLE JUNCTION, NORTHERN CALIFORNIA

A study of tectonic geomorphology at the Mendocino Triple Junction (Figure 4.10) sought to identify zones of active tectonics [17]. This triple junction is the location where the North American, Pacific, and Juan de Fuca Plates come together. The research evaluated the response of coastal streams to uplift related to the triple junction (also see discussion in Chapter 5). The position of the triple junction is not fixed in time; it is migrating northward at approximately 56 mm/yr—the same rate as the relative movement between the Pacific and North American Plates. North of the triple junction, the tectonics are dominated by northeast-directed compression, producing northwest-trending folds and thrust faults [18]. South of the triple junction, the tectonics are related to the right-lateral strike-slip motion of the San Andreas fault system, with localized extension and compression. At the triple junction itself, high rates of uplift have been observed.

As the triple junction migrates northward, it acts similarly to a ship plowing through calm lake water (Figure 4.11). In front of the metaphorical ship, uplift of the water surface increases to a maximum near the ship's bow. In the wake of the ship, uplift decreases back to the calm water level. Similarly, uplift rates due to the passage of the triple junction increase from about 1 mm/yr near Eureka, to 2.8 mm/yr at Singley Flat, to a maximum of about 4 mm/yr at Big Flat, and then uplift rates decline gradually to the south. At Fort Bragg, rates are approximately 0.3 mm/yr (Figure 4.10) [17]. The rates of uplift along the coast were established by careful measurement and dating of uplifted marine platforms.

Both climate and rock resistance are relatively uniform along the coast of the study area (Figure 4.10). Detailed study of the coastal streams suggests that the first-order streams (those farthest headward in the drainage basin) are most sensitive to tectonics and thus are the best indicators of areas with high rates of uplift. Figure 4.12 shows the uplift rate and channel gradients of the first-order stream channels. In contrast, the channel gradients of higher-order stream channels (not illustrated) do not correlate well with uplift rates [17]. Evidently, larger streams have sufficient stream power to overwhelm the effects of tectonics. For a more detailed explanation of the relationship between stream power that flow over bedrock and tectonics, see Chapter 5.

SL indices were also evaluated for the Mendocino Triple Junction area. The index allowed discrimination between streams characterized by high to intermediate uplift rates and streams characterized by low uplift rates. Figure 4.13 shows the profiles of three streams (approximate locations shown in Figure 4.10) along with average SL indices for stream reaches. DeHaven Creek is an area with a low uplift rate, less than 1 mm/yr, and SL indices are also relatively low, except in the headward part of the stream profile. Telegraph Creek is an area of intermediate uplift rates, and the SL indices are of intermediate value along the entire profile. Finally, Fourmile Creek is in an area of rapid uplift and has relatively high SL indices along its entire length, as well as a convex profile that is characteristic of streams undergoing rapid uplift [17].

A conclusion of this study is that the SL index is clearly able to distinguish between low, intermediate, and high rates of uplift. In particular, first-order stream channels are most sensitive to recent tectonic activity. This study confirmed the usefulness of

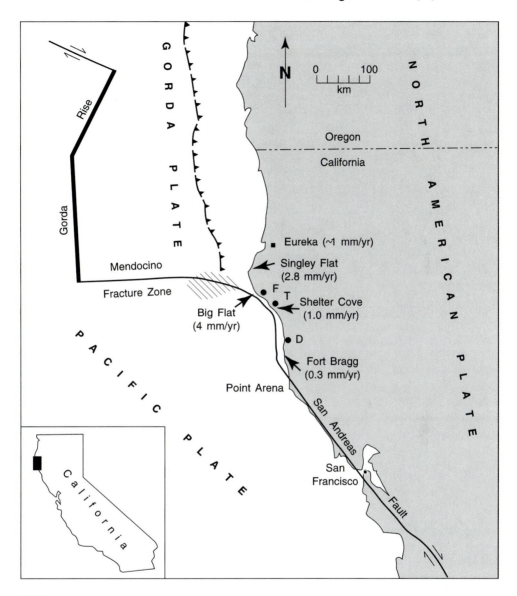

Figure 4.10 Tectonic framework of the Mendocino Triple Junction area (F is Fourmile Creek, T is Telegraph Creek, and D is DeHaven Creek). Velocity values indicate uplift rates at those locations. (After Merritts and Vincent, 1989 [17].)

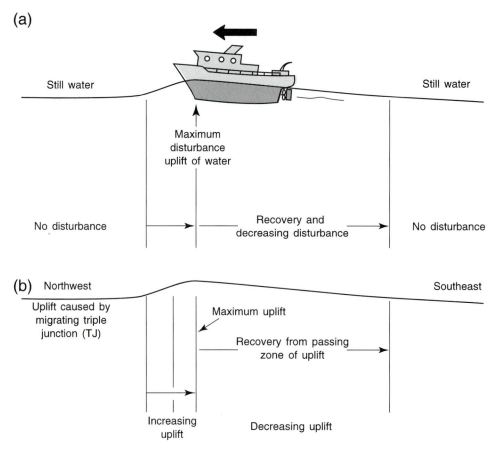

Figure 4.11 Idealized diagram to illustrate uplift process related to migrating Mendocino Triple Junction. (a) The moving ship represents the migrating junction disturbing the water surface. (b) The migrating triple junction produces a similar pattern of uplift.

the SL index as a reconnaissance tool in categorizing relative magnitudes of uplift in an area.

MOUNTAIN-FRONT SINUOSITY

Mountain-front sinuosity [6, 7] is defined as

$$S_{mf} = L_{mf}/L_s \qquad (4.5)$$

where S_{mf} is the mountain-front sinuosity; L_{mf} is the length of the mountain front along the foot of the mountain, at the pronounced break in slope; and L_s is the straight-line length of the mountain front (Figure 4.14). Mountain-front sinuosity is an index that reflects the balance between erosional forces that tend to cut embayments into a mountain front and tectonic forces that tend to produce a straight mountain front coincident with an active range-bounding fault. Those mountain fronts associated with active tec-

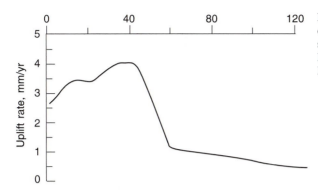

Figure 4.12   Relationship between channel gradient of first-order streams and uplift rate near the Mendocino Triple Junction. (After Merritts and Vincent, 1989 [17].)

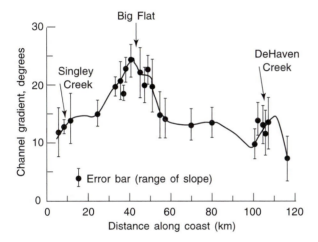

tonics and uplift are relatively straight, with low values of S_{mf}. If the rate of uplift is reduced or ceases, then erosional processes will carve a more irregular mountain front, and S_{mf} will increase.

In practice, the values of S_{mf} may be calculated easily from topographic maps or aerial photographs. However, values of S_{mf} depend on image scale [19], and small-scale topographic maps (1 : 250,000) produce only a rough estimate of mountain-front sinuosity. Aerial photographs and larger-scale maps, with resolution greater than the irregularity of the mountain front, are more useful when calculating S_{mf}.

MOUNTAIN-FRONT SINUOSITY NEAR THE GARLOCK FAULT, CALIFORNIA

One of the first studies that used S_{mf} evaluated relative tectonic activity north and south of the Garlock fault in California [19]. In Figure 4.15, the dashed lines represent the straight-line lengths of the mountain fronts that were evaluated. The solid lines are the outlines of the actual mountain fronts, reflecting their sinuosity. The values of sinuosity seem to define three groups of activity [19]:

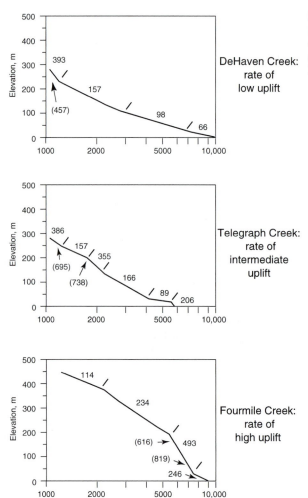

Figure 4.13 Stream length gradient (SL) indices for streams in areas of contrasting rates of uplift. Locations shown in Figure 4.10. (After Merritts and Vincent, 1989 [17].)

- The area north of the Garlock fault, with relatively low values of S_{mf}
- A transitional area in the central part of the map, north of and adjacent to the Garlock fault, with higher values of S_{mf}
- An area south of the Garlock fault, with relatively high values of S_{mf}

- This study concluded that the most active mountain fronts—those associated with active, range-bounding faults—generally have an S_{mf} between 1.0 and 1.6. Mountain fronts with lesser activity, but that still reflect active tectonics, have sinuosities between approximately 1.4 and 3. Inactive mountain fronts have sinuosities from about 1.8 to greater than 5. In general, sinuosity values greater than 3 are associated with fronts that are so eroded and embayed that the topographic range fronts, which at one time were coincident with active range-bounding geologic structures, have now eroded back into the range 1 km or more [19].

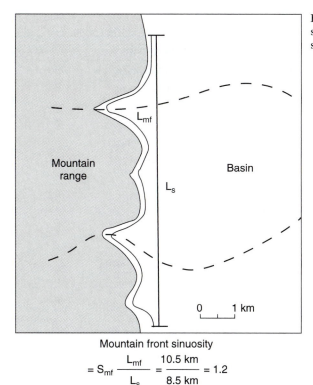

Figure 4.14   Idealized diagram showing how mountain-front sinuosity (S_{mf}) is calculated.

Mountain front sinuosity

$$= S_{mf} \frac{L_{mf}}{L_s} = \frac{10.5 \text{ km}}{8.5 \text{ km}} = 1.2$$

Like the stream length–gradient index, mountain-front sinuosity is a potentially valuable reconnaissance tool used to identify areas of relative tectonic activity.

RATIO OF VALLEY FLOOR WIDTH TO VALLEY HEIGHT

The ratio of valley floor width to valley height (V_f) may be expressed as

$$V_f = 2V_{fw}/[(E_{ld} - E_{sc}) + (E_{rd} - E_{sc})] \qquad (4.6)$$

where V_f is the valley floor width-to-height ratio; V_{fw} is the width of the valley floor; E_{ld} and E_{rd} are elevations of the left and right valley divides, respectively; and E_{sc} is the elevation of the valley floor [6, 7] (Figure 4.16). When calculating V_f, these parameters are measured at a set distance from the mountain front for every valley studied. This index differentiates between broad-floored canyons, with relatively high values of V_f, and V-shaped valleys, with relatively low values. High values of V_f are associated with low uplift rates, so that streams cut broad valley floors. Low values of V_f reflect deep valleys with streams that are actively incising, commonly associated with uplift.

The study that evaluated the mountain-front sinuosity north and south of the Garlock fault also calculated values of the ratio of valley floor width to valley height [19]. The V_f values in that area ranged from 0.05 to 47.0. Lower values were associated with valleys north of the Garlock fault, where tectonic activity is assumed to be more vigorous.

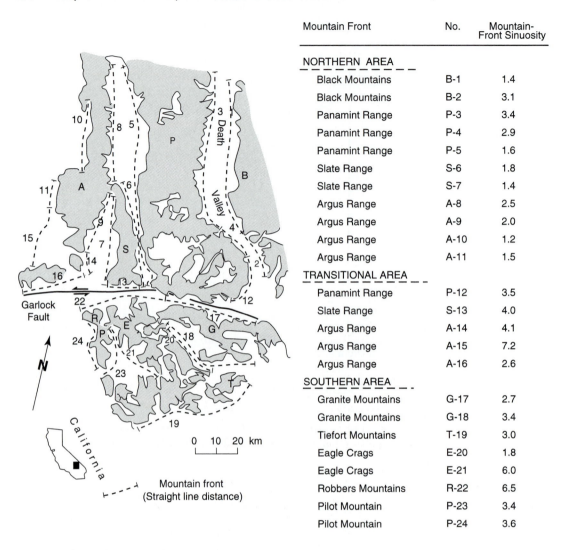

Mountain Front	No.	Mountain-Front Sinuosity
NORTHERN AREA		
Black Mountains	B-1	1.4
Black Mountains	B-2	3.1
Panamint Range	P-3	3.4
Panamint Range	P-4	2.9
Panamint Range	P-5	1.6
Slate Range	S-6	1.8
Slate Range	S-7	1.4
Argus Range	A-8	2.5
Argus Range	A-9	2.0
Argus Range	A-10	1.2
Argus Range	A-11	1.5
TRANSITIONAL AREA		
Panamint Range	P-12	3.5
Slate Range	S-13	4.0
Argus Range	A-14	4.1
Argus Range	A-15	7.2
Argus Range	A-16	2.6
SOUTHERN AREA		
Granite Mountains	G-17	2.7
Granite Mountains	G-18	3.4
Tiefort Mountains	T-19	3.0
Eagle Crags	E-20	1.8
Eagle Crags	E-21	6.0
Robbers Mountains	R-22	6.5
Pilot Mountain	P-23	3.4
Pilot Mountain	P-24	3.6

Figure 4.15   Mountain-front sinuosity (S_{mf}) north and south of the Garlock fault. (After Bull and McFadden, 1977 [19].)

ALLUVIAL FANS AND TECTONIC ACTIVITY AT MOUNTAIN FRONTS

As earthquakes are concentrated on discrete fault zones, mountain building can be concentrated on distinct mountain fronts (a steep escarpment that may reflect long-term uplift of one side of the front relative to the other side; Figure 4.17). The geomorphology of mountain fronts reveals a great deal about the tectonic activity occurring there (see the previous discussion of mountain-front sinuosity). For example, mountain fronts in arid and semiarid regions are often characterized by alluvial fans (see Figure 4.17). An alluvial fan may be thought of as the endpoint of an erosional-depositional system in

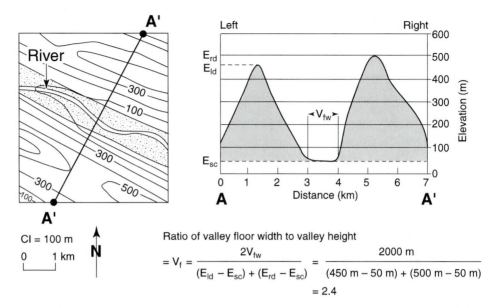

$$\text{Ratio of valley floor width to valley height}$$

$$= V_f = \frac{2V_{fw}}{(E_{ld} - E_{sc}) + (E_{rd} - E_{sc})} = \frac{2000 \text{ m}}{(450 \text{ m} - 50 \text{ m}) + (500 \text{ m} - 50 \text{ m})}$$

$$= 2.4$$

Figure 4.16   Idealized diagram illustrating how the ratio of valley floor width to valley height (V_f) is calculated. Note: Left and right are determined by looking downstream.

Figure 4.17   Front of the White Mountains, California. A single escarpment, about 2500 m high, separates the mountain range from the adjacent valley. The prominent alluvial fan forms where a stream exits the confines of a canyon within the range. Alluvial fans are a common feature at mountain fronts in arid and semiarid settings.

which sediment eroded from a mountain source area is transported to the mountain front. At the front, this material is deposited as a fan-shaped body (segment of a cone) of fluvial and/or debris-flow deposits [20]. The connecting link between the erosional and depositional parts of the system is the stream. The morphology of an alluvial fan is a function of variables including size of drainage basin contributing sediment to the fan, geology of the source area, relief of the source area, climate and vegetation of the source area, and tectonic activity. Profiles of alluvial fans in cross section are generally concave but often contain significant breaks in slope that mark boundaries between relatively straight-line fan segments. In fact, most alluvial fans are segmented, and younger fan segments may be identified from older segments based on relative soil-profile development, weathering of surficial clasts, erosion of the surface, and development of desert varnish [20, 21]. They may also be numerically dated using exposure dating techniques (see Chapter 2).

The morphology of a segmented alluvial fan may be used as an indicator of active tectonics because the fan form may reflect varying rates of tectonic processes, such as faulting, uplift, tilting, and folding along and adjacent to the mountain front. In the simplest case, if the rate of uplift along a mountain front is high relative to the rate of downcutting of the stream channel in the mountain and to deposition on the fan, then deposition tends to occur in the fanhead area. The youngest fan segment is found near the apex of the fan. Such mountain fronts would also tend to have low values of mountain-front indices S_{mf} and V_f. On the other hand, if the rate of uplift of the mountain front is less than or equal to the rate of downcutting of the stream in the mountain source area, then fanhead incision occurs, and deposition is shifted downfan [20]. As a result, younger fan segments are located well away from the mountain front (Figure 4.18 illustrates these two conditions). Such mountain fronts would have relatively high values of S_{mf} and V_f, suggesting relatively low rates of mountain-front activity. The mountain block is wearing down with time; the front is more eroded and sinuous, and mountain-front valleys are wider.

For example, work on alluvial fans in Death Valley found that tilting produced segmented fans [22]. Alluvial fans on the west side of the valley are tilted down to the east, and this shifted the locus of fan deposition downfan. That is, fanhead incision has occurred, and younger fan segments are located well away from the mountain front and fan apex. On the east side of Death Valley, the basin is being downdropped relative to the mountain along normal faults, and segments of alluvial fans are often (but not always) located near the apex area at the mountain front.

The overall shape of an alluvial fan also can reveal the pattern of tectonic activity at and near a mountain front. Because alluvial fans are roughly conical in shape, topographic contours across simple fans are approximately circular (the intersection of a cone with a horizontal plane is a circle). Where alluvial fans are not simple, however, and have undergone tectonic tilt, contour lines across the fans form segments of ellipses, not circles (the intersection of a tilted cone and a horizontal plane is an ellipse; Figure 4.19) [23, 24]. Where conditions are appropriate, the amount of tilt can be calculated by fitting ideal ellipses to the contours across an alluvial fan and measuring the length of the long axes (a) and the short axes (b) of the ellipses [25]. It turns out that the amount of tilt (β) equals

$$\beta = \arccos(((b/a)^2 \sin^2 \alpha + \cos^2 \alpha)^{0.5}) \qquad \textbf{(4.7)}$$

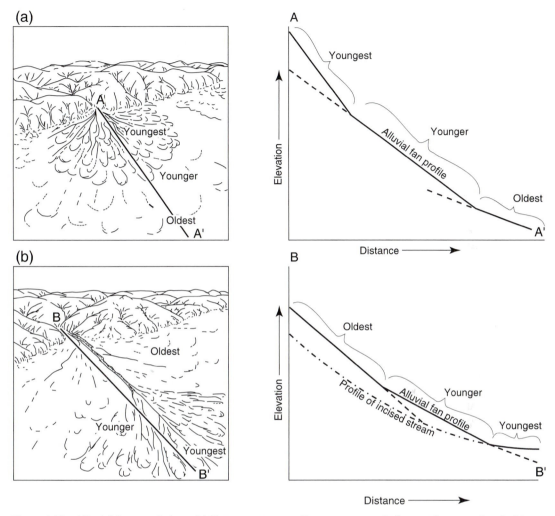

Figure 4.18   Alluvial-fan morphology: (a) Fan segments are adjacent to mountain front and are associated with active uplift. (b) Youngest fan segments are away from the mountain front and are associated with erosion of the mountain block rather than uplift. (After Bull, 1977 [20].)

where α is the slope of the fan along the short axes of the ellipses (see Figure 4.19) [25]. Ideal conditions for this measurement include a relict fan or fan-segment surface that has been depositionally inactive during tilting.

SAN EMIGDIO CANYON: AN EXAMPLE OF A DEFORMED ALLUVIAL FAN

Study of the tectonic geomorphology of active folding over the buried Wheeler Ridge and Los Lobos reverse faults at the San Emigdio Mountain front, southern San Joaquin Valley, California, provides information concerning the tectonic and geomorphic development of mountain fronts produced by active folding and faulting. Monoclinally

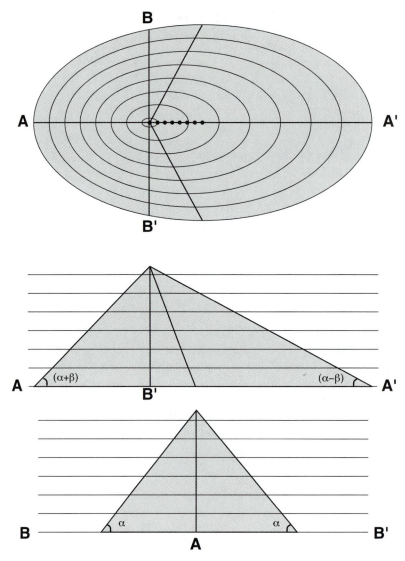

Figure 4.19 Geometric model of a tilted cone, in map view (top diagram) and in cross section (middle and lower diagrams). Fine lines represent topographic contours (lines of equal elevation across the surface of the cone). Note that the contour lines form ellipses in map view. A tilted alluvial fan consists of some fraction of the total cone shape, typically up to 180° of the total.

flexed gravels with a minimum age of 82 to 125 ka and with dips as great as 50° provide evidence of late Pleistocene deformation at the present active range front above the Wheeler Ridge fault (Figure 4.20). Studies of the surface folding of alluvial fans and fluvial terraces suggest a Holocene vertical deformation rate of 1.9 to 3.0 m/ky at the active range front and 0.8 to 1.3 m/ky ~2 km basinward at the Los Lobos folds. Geomorphic evidence also suggests that the locus of active folding and vertical deformation along the northern flank of the San Emigdio Mountains has migrated basinward with time. This evidence includes a relic intermontane front, now within the uplifted block, 5 km from the present active mountain front, and the existence of recently initiated folds in the alluvial fan 2 km basinward from the mountain front at the Los Lobos folds. Northward migration of tectonic activity results in progressive widening of the uplifted block as the location of active folding moves basinward. This migration of tectonic activity appears to occur through increase of vertical deformation along more northerly folds and faults accompanied by reduction of activity along the older, more southerly structures [26].

Alluvial fans have been accumulating since at least the late Pleistocene on the north flank of the San Emigdio Mountains. Because alluvial fans are sensitive to changes within the erosional-depositional system, geomorphic studies of alluvial fans can be used to evaluate tectonic perturbations that occurred during and after fan deposition [20]. Several segments occur on the San Emigdio fan, the largest and best-formed fan (Figure 4.21). Study of the fan suggests that the upper (folded) fan segments have a different fanhead apex than the fan-toe segment. Vertical deformation along the active mountain front has produced the younger fanhead segments of Q_2 (0.5 to 1.0 ka) and Q_3 (4.7 to 7.5 ka) age, the apex of which is at the active front (Figure 4.20). The segment found at the fan toe is believed to be a morphologic remnant of an older Q_4 (82 to 125 ka) fan, the apex of which is 5 km south of the active mountain front at an intermontain front (Figure 4.20). Although most of the Q_4 soil profiles of the fan-toe deposits are thinly buried by very recent San Joaquin Valley fill or disturbed by farming, the Q_4 fan-toe morphology remains [26].

The San Emigdio alluvial fan may be reconstructed based on the assumption that contour lines are concentric arcs having their centers of curvature at the apex of the fan. Thus, by fitting alluvial-fan contours with circular arcs, one can reconstruct the radii and thus determine the location of the fan apex at the time of deposition of that portion of the fan. Using this methodology, the fan-toe contours project back to a different apex than mid-fan and fan-head contours. We interpret this as evidence for two alluvial fans, each with a different apex (Figure 4.22) and a different time of deposition [26]. Reconstruction of the fan apex indicates that the apex of the older, larger fan (Q_4) is located south (mountainward) of the active front at a prominent topographic break (Figure 4.23). This topographic break hypothetically is an older, intermontane (relic) mountain front [27], marking the former northern boundary of the San Emigdio Mountains [26, 27].

In order to reconstruct the morphology of the Q_4 fan, it is assumed that the fan-toe segment of the San Emigdio fan consists of nearly undeformed Q_4 gravels, which is reasonable because the toe segment is several kilometers north of any identified deformation. Using topographic measurements from the fan-toe segment, the depositional morphology of the Q_4 fan is reconstructed using a series of equations developed by

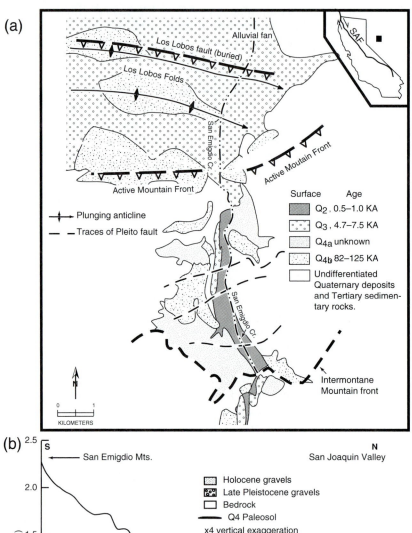

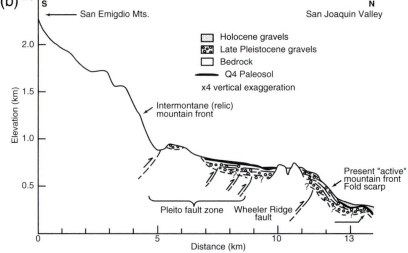

Figure 4.20 Generalized map of Holocene and late Pleistocene deposits (Q_2–Q_4) of the San Emigdio Canyon area, the buried trace of the Wheeler Ridge fault, and Los Lobos folds (a); and idealized topographic profile constructed approximately along the crest of the west side of San Emigdio Canyon (b). (Modified after Keller et al., 2000 [26].)

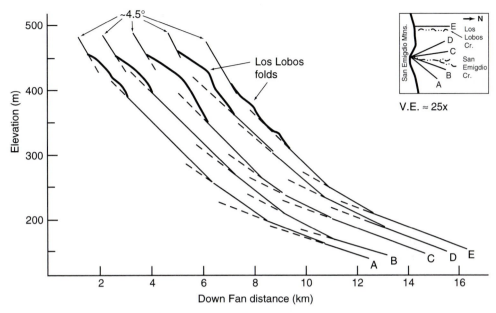

Figure 4.21   Radial profiles of the San Emigdio alluvial fan downstream from the present mountain front (buried Wheeler Ridge fault). Notice the deformation of the Los Lobos folds and the several segments of fans that are present. (Modified after Keller et al., 2000 [26].)

Troeh [28]. These equations utilize three points of known elevation and of known radial distance from the fan apex to predict the elevation of any other point on the fan. The equations are as follows

$$S = \frac{(E_a - E_c)(R_b^2 - R_c^2) - (E_b - E_c)(R_a^2 - R_c^2)}{(R_a - R_c)(R_b^2 - R_c^2) - (R_b - R_c)(R_a^2 - R_c^2)} \tag{4.8}$$

$$L = \frac{E_a - E_c - S(R_a - R_c)}{R_a^2 - R_c^2} \tag{4.9}$$

$$P = E_a - SR_a - LR_a^2 \tag{4.10}$$

$$E_z = P + SR_z + LR_z^2 \tag{4.11}$$

where

$$E_a, E_b, E_c = \text{elevation of any three known points A, B, C on the fan}$$

$$R_a, R_b, R_c = \text{radial distance from apex to points A, B, C}$$

$$L = \frac{1}{2} \text{ the rate of change of slope}$$

$$P = \text{elevation of fan apex}$$

$$S = \text{slope of fan at P}$$

$$E_z = \text{Elevation of any point Z at known radial distance } R_z$$

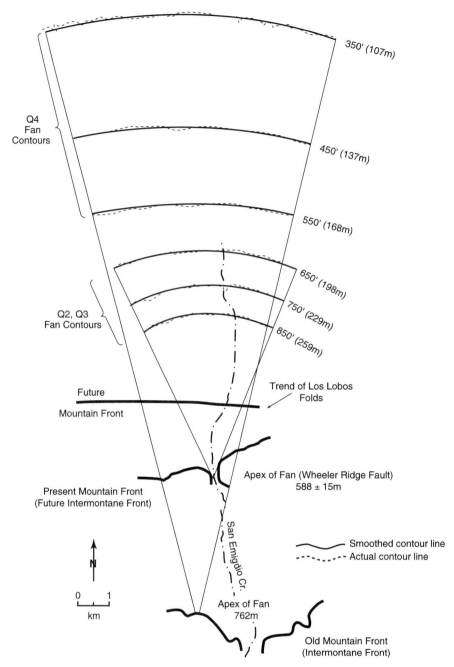

Figure 4.22   Selected, slightly smoothed contours of the San Emigdio alluvial fan. Two different fans and associated radial profiles are clearly present, one with an apex at the active mountain front, which is underlain by the buried Wheeler Ridge fault, and the other suggesting an apex at an intermontane front several kilometers to the south underlain by the Los Lobos fault. (Modified after Keller et al., 2000 [26].)

Intermontane (relic) front

Active mountain front

(a)

0   ~2km

(b)

Figure 4.23   (a) High-altitude view of San Emigdio Canyon showing the intermontane and active mountain fronts. (b) Apex area of Q_4 deposits associated with the intermontane mountain front in the San Emigdio Mountains. The flat, mesalike landform is the Pleistocene alluvial fan. (Courtesy of NASA.)

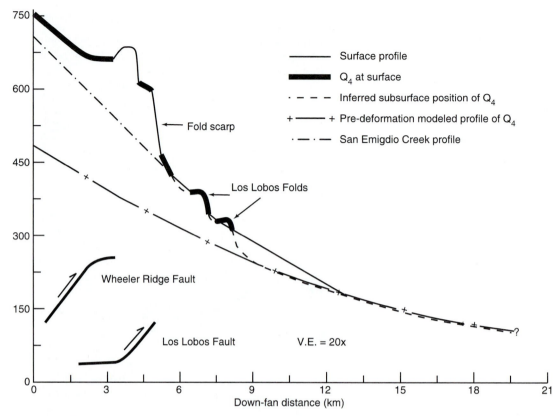

Figure 4.24   Reconstruction of the San Emigdio Canyon Q_{4b} alluvial fan using Troeh's equations [27]. Also shown is the topographic profile of the Q_4 surface along the western crest of the canyon, and the profile of San Emigdio Creek. The buried Wheeler Ridge and Los Lobos faults are shown for illustrative purposes. (Modified after Keller et al., 2000 [26].)

The reconstructed morphology of the Q_4 alluvial fan produced by Troeh's equations can then be applied to estimate vertical displacement since Q_4 time. Using present-day data from the fan-toe segment (contours at 107 m, 137 m, and 168 m) (Figure 4.22), it is estimated that the average elevation of the fan apex at the time of Q_4 deposition was about 490 m. The present elevation of the apex is about 760 m. Troeh's equations can be used to reconstruct an entire radial profile of an alluvial fan. Figure 4.24 shows the reconstructed Q_4 profile, as well as the present-day surface profile of Q_4 and the present-day profile of San Emigdio Creek [26]. Also shown are the approximate positions of the Wheeler Ridge and Los Lobos faults. Using the reconstructed profile, the amount of deformation that has occurred since Q_4 gravels were deposited can be estimated. The amount of vertical deformation of the Q_4 fan above the Wheeler Ridge fault is 760 m - 490 m or about 270 m, resulting from the combined deformation on both the Wheeler Ridge and Los Lobos faults. Vertical deformation over the Los Lobos folds is approximately 90 m. Thus, 180 m of the vertical deformation is attributed to the Wheeler Ridge fault [26].

RELIC AND INTERMONTANE FRONTS

Mountain-front sinuosity and the ratio of valley floor width to height and our discussion of the San Emigdio alluvial fan established that some mountain fronts are more tectonically active than others. Older mountain fronts, located within ranges (**intermontane fronts**), may remain active or may be no longer active. Similarly, a mountain front at the piedmont may no longer be tectonically active or have much reduced activity. Such fronts are **relic landforms**, landforms produced by processes no longer active or experiencing greatly reduced rates. Hypothetically, mountain fronts formed early in a range's development are active for a period of time; then deformation migrates toward and beyond the edges of the range, and new mountain fronts form. In the western Transverse Ranges of California, we see a central highlands with several relic intermontane fronts, flanked by fold-and-thrust belts that are presently active (Figure 4.25). The interior, now relic intermontane fronts at one time had the same basic morphology as the present active, outer fronts. Streams that emerged from the intermontane fronts commonly fed a series of alluvial fans on what was then the margins of the ranges. When tectonic activity shifted, the old alluvial fans were sometimes consumed by the active mountain-building process as new fronts developed (Figure 4.26). As the mountain ranges grow outward (wider with time), they consume their own alluvial fans [26].

CLASSIFICATION OF RELATIVE TECTONIC ACTIVITY

Each of the indices discussed in this chapter provides a relative classification of tectonic activity useful in reconnaissance studies. When more than one index is applied to a particular region, the results are more meaningful than those of any single analysis. This concept was tested by evaluating mountain fronts on the southern flank of the central Ventura Basin in California (Figure 4.27). The basin has a complex geologic history characterized by extension and rotation during the Miocene and strong compression from the Pliocene to the present [29]. Eight mountain fronts in the area were evaluated in terms of S_{mf} and V_f. In addition, SL indices were calculated for the same area. All

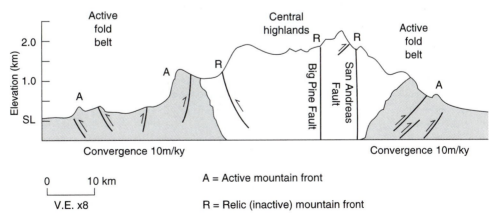

Figure 4.25   Active and relic mountain fronts, western Transverse Ranges. The San Andreas and Big Pine faults are right-lateral and left-lateral strike-slip faults, respectively.

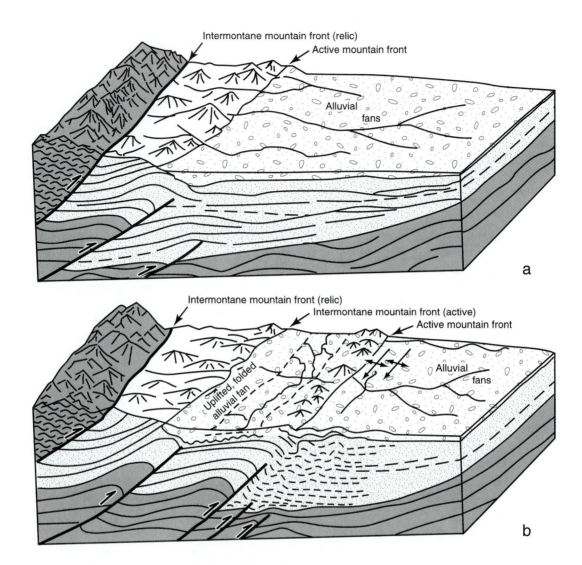

Figure 4.26 Idealized model illustrating the concept of mountain-front migration. At (a) a relic intermontane mountain front is present, as is an active mountain front from which alluvial fans are being shed. At (b) the upper parts of the fans have been faulted, folded, and incorporated into the mountain block and a new active mountain front is forming. (Modified with permission from D. Burbank, 1987, written correspondence.)

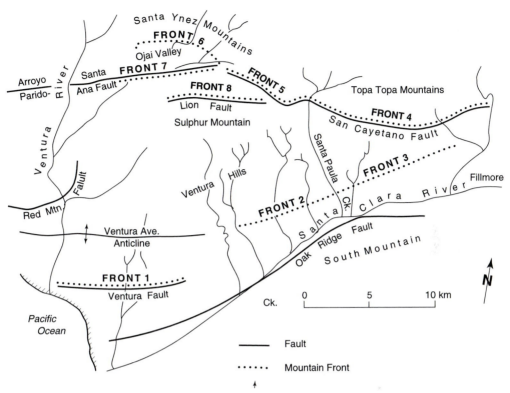

Figure 4.27   Mountain fronts and major structures used for the tectonic geomorphic analysis in the central Ventura Basin. (Values of S_{mf}, V_f, and tectonic activity class for these mountain fronts are listed in Table 4.1.) (After Rockwell et al., 1985. In *Tectonic Geomorphology*, Morisawa and Hack (eds.). Boston: Allen & Unwin, 183–207.)

of the mountain fronts are bounded by active folds or faults. Mountain fronts associated with the San Cayetano fault (fronts 4–8 in [Figure 4.27; Figure 4.28]) generally have high SL indices, whereas the fronts associated with the Ventura Avenue anticline and the buried Ventura fault (fronts 1–3) have relatively low indices. The S_{mf} of the young mountain front at the San Cayetano fault (front 5) is 1.14, and the V_f is 0.47, reflecting a relatively high tectonic activity. Low SL indices are present along front 1 (Figure 4.27) because the rocks are weak shales, and the gradients of streams that cross these rocks do not reflect the tectonic activity. However, other indices of relative tectonic activity do indicate active deformation; front 1 has a very low mountain-front sinuosity and low valley floor width-to-height ratios (see Table 4.1, which summarizes selected geomorphic parameters for the eight mountain fronts).

When all the information from the mountain fronts and SL indices is combined, it is possible to produce a **relative tectonic activity class** designation [6, 7]. Mountain fronts that are suggestive of the highest tectonic activity are designated as class 1 fronts. These fronts typically have low values of S_{mf}, low V_f, and high SL indices. In the central

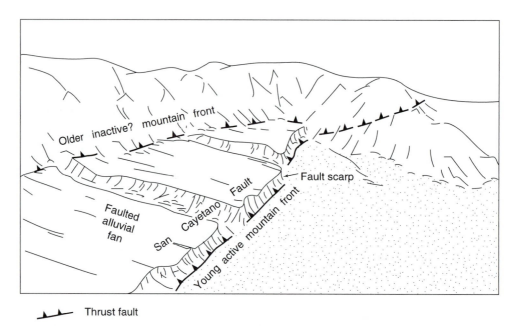

Figure 4.28 (a) Illustration of a truncated alluvial fan at the San Cayetano fault along mountain front 5 (Figure 4.27, Table 4.1). The nearly straight fault scarp at this location is approximately 60 m high. Upstream of this location (in upper left part of the illustration) is an older mountain front (intermontane mountain front) that may no longer be active.

Ventura Basin, these fronts are usually associated with an uplift rate of greater than 1 mm/yr. Class 2 mountain fronts are associated with less tectonic activity, reflected in higher S_{mf} and V_f values and lower SL values. Class 3 mountain fronts are still associated with active tectonics, but geomorphic indices suggest less activity than for class 2 fronts. An example of a class 2 front in the central Ventura Basin is mountain front 6, which has an S_{mf} of 2.7 and a V_f of approximately 1.9 (Table 4.1). Mountain fronts characterized by minimal tectonic activity or those that are now inactive may be classified as class 4 or class 5 [6, 7].

Attempting to classify mountain fronts in terms of relative tectonic activity is a fairly new endeavor [6]. The classification and boundaries between classes are arbitrary and only indicate relative differences. No attempts have been made to put specific bounds on values of geomorphic indices, because these indices reflect local conditions of rock type, structure, and climate. Their real usefulness is for differentiating mountain fronts as very active, moderately active, or inactive. This information is valuable in areas where detailed field studies have not yet been conducted.

Table 4.1

LOCATIONS OF MOUNTAIN FRONTS OF THE WESTERN TRANSVERSE RANGES SHOWN IN FIGURE 4.27.

Front Number	Location of Tectonic Mountain Front	Type of Front	Orientation of Front	Front Length (km)	Maximum Relief (m)	Mountain Front Sinuosity S_{mf}	Valley-Floor Width to Valley Heights V_f	Tectonic Activity Class	Uplift Rate
1	Ventura River to Harmon Cyn.	Bounding folded (VAA)	E–W	6.45	438	1.09	0.7	1	4 mm/yr
2	Aliso Cyn. to Fagon Cyn.	Bounding folded	N60E	8.88	670	1.57	1.8	2	unknown
3	Orcutt Cyn. to Snow Cyn.	Bounding folded	N60E	10.1	1280	1.83	1.91	2	unknown
4	S.P. Creek to Sespe Creek	Internal faulted (SCF)	E–W	6.45	915	1.14	0.43	1	2–8 mm/yr
5	Sisar to Santa Paula Creek	Internal/bounding faulted (SCF)	N60W	5.65	1639	1.14	0.47	1	0.5–1.5 mm/yr
6	Wilsie to West Gridley Canyon	Bounding folded	Curving front NE–W	5.25	1363	2.72	1.89	2–3	unknown
7	Wilsie to San Antonio Creek	Bounding (APF) faulted	N75E	5.85	305	1.01	<0.73	1	0.4 mm/yr
8	South side Upper Ojai Valley	Bounding faulted (Lion F.)	N80E to EW	4.23	451	1.34	0.80	1	unknown

(After Rockwell et al., 1985. In M. Morisawa and J. T. Hack (eds.), *Tectonic Geomorphology.* Boston: Allen & Unwin.)

SUMMARY

Quantitative measurements have allowed geomorphologists to objectively compare landforms and calculate geomorphic indices that are useful for identifying a particular characteristic of the area (for example, its level of tectonic activity). The hypsometric curve and hypsometric integral are related to the degree of dissection of a landscape. Hypsometric analysis is a useful tool for differentiating tectonically active from tectonically inactive regions. Drainage-basin asymmetry is defined in terms of the Asymmetry Factor as well as the Transverse Topographic Symmetry Factor. Both of these are valuable in rapid evaluations of drainage basins to determine if tectonic tilt may have occurred. The stream length–gradient index (SL) is a useful tool for studying tectonic geomorphology; high values commonly are found where streams cross resistant rocks or where streams cross active structures. Values of the stream length–gradient index may be calculated easily from topographic maps to provide information where more detailed studies would be fruitful. Mountain-front sinuosity (S_{mf}) is an index that reflects the balance between erosional forces and tectonic forces at a mountain front. In general, active mountain fronts have relatively low values of sinuosity. The ratio of valley-floor width to valley height (V_f) differentiates between broad-floored canyons and V-shaped valleys. Low values of V_f are associated with active tectonics.

Alluvial fans are potential indicators of active tectonics. Most alluvial fans are segmented, and age and location of segments is related to tectonic activity of a mountain front. Cross-sectional profiles and fan contours may be useful in identifying the position of a fan apex, identifying relic mountain fronts, and calculating of tectonic tilt. Modeling of alluvial-fan morphology may be useful in evaluating the amount of uplift a fan apex has experienced. The presence of relic, intermontane fronts is evidence that with time a mountain range may grow wider. As a result, older alluvial fans may become incorporated into the growing range.

When several indices of relative tectonic activity are evaluated for a particular region, it is possible to develop a system of relative tectonic-activity classes. Commonly, it is useful to classify areas as being very active, moderately active, or inactive. Such basic classification is useful in delineating areas where more detailed field studies will identify active structures and calculate rates of active tectonic processes.

REFERENCES CITED

1. Keller, E. A., 1986. Investigations of active tectonics: use of surficial earth processes. In R. E. Wallace (chairperson), *Active Tectonics*. Washington, DC: National Academy Press, 136–147.

2. Strahler, A. N., 1952. Hypsometric (area-altitude curve) analysis of erosional topography. *Geological Society of America Bulletin*, 63:1117–1141.

3. Hare, P. W., and T. W. Gardner, 1985. Geomorphic indicators of vertical neotectonism along converging plate margins, Nicoya Peninsula, Costa Rica. In M. Morisawa and J. T. Hack (eds.), *Tectonic Geomorphology: Proceedings of the 15th Annual Binghamton Geomorphology Symposium*, September 1984. Boston: Allen & Unwin, 75–104.

4. Cox, R. T., 1994. Analysis of drainage basin symmetry as a rapid technique to identify areas of possible Quaternary tilt-block tectonics: an example from the Mississippi Embayment. *Geological Society of America Bulletin*, 106:571–581.

5. Hack, J. T., 1973. Stream-profile analysis and stream-gradient index. *U. S. Geolological Survey Journal of Research*, 1:421–429.

6. Bull, W. B., 1977. Tectonic geomorphology of the Mojave Desert. U.S. Geological Survey Contract Report 14-08-001-G-394. Menlo Park, CA: Office of Earthquakes, Volcanoes, and Engineering.

7. Bull, W. B., 1978. Geomorphic tectonic classes of the south front of the San Gabriel Mountains, California. U.S. Geological Survey Contract Report 14-08-001-G-394. Menlo Park, CA: Office of Earthquakes, Volcanoes, and Engineering.

8. Pike, R. J., and S. E. Wilson, 1971. Elevation-relief ratio, hypsometric integral and geomorphic area-altitude analysis. *Geological Society of America Bulletin*, 82:1079–1083.

9. Mayer, L., 1990. *Introduction to Quantitative Geomorphology: An Exercise Manual.* Englewood Cliffs, NJ: Prentice Hall.

10. Harlin, J. M., 1978. Statistical moments of the hypsometric curve and its density function. *Mathematical Geology*, 10:59–72.

11. Gardner, T. W., K. C. Sasowski, and R. L. Day, 1990. Automated extraction of geomorphometric properties from digital elevation data. *Zeitshrift fur Geomorphologie Supplement*, 80:57–68.

12. Gardner, T. W., W. Back, T. F. Bullard, P. W. Hare, R. H. Kesel, D. R. Lowe, C. M. Menges, S. C. Mora, F. J. Pazzaglia, I. D. Sasowski, J. W. Troester, and S. G. Wells, 1987. Central America and the Caribbean. In W. L. Graf (ed.), *Geomorphic Systems of North America*, Centennial Special Volume 2. Boulder, CO: Geological Society of America, 343–402.

13. Cox, R. T., personal communication.

14. Cox, R. T., R. B. Van Arsdale, and J. B. Harris, 2001. Identification of possible Quaternary deformation in the northeastern Mississippi embayment using quantitative geomorphic analysis of drainage-basin asymmetry. *GSA Bulletin*, 113:615–624.

15. Keller, E. A., 1977. Adjustments of drainage to bedrock in regions of contrasting tectonic framework. *Geological Society of America Abstracts with Programs*, 9:1046.

16. Zhao, X., 1990. Tectonic geomorphology and soil chronology of the Frazier Mountain area, western Transverse Ranges, California. Ph.D. dissertation. University of California, Santa Barbara, CA.

17. Merritts, D., and K. R. Vincent, 1989. Geomorphic response of coastal streams to low, intermediate, and high rates of uplift, Mendocino Triple Junction region, Northern California. *Geological Society of America Bulletin*, 101:1373–1388.

18. Clarke, S., and G. Carver, 1992. Late Holocene tectonics and paleoseismicity, southern Cascadia subduction zone. *Science*, 255:188–192.

19. Bull, W. B., and L. D. McFadden, 1977. Tectonic geomorphology north and south of the Garlock fault, California. In D. O. Doehring (ed.), *Geomorphology in Arid Regions*. Proceedings of the Eighth Annual Geomorphology Symposium. Binghamton, NY: State University of New York at Binghamton, 115–138.

20. Bull, W. B., 1977. The alluvial fan environment. *Progress in Physical Geography*, 1:222–270.

21. Bull, W. B., 1964. Geomorphology of segmented alluvial fans in western Fresno County, California. *U.S. Geological Survey Professional Paper*, 352-E:89–128.

22. Hooke, R. L., 1972. Geomorphic evidence of late-Wisconsin and Holocene tectonic deformation, Death Valley, California. *Geological Society of America Bulletin*, 83:2073–2098.

23. Verhoogen, J., F. J. Turner, L. E. Weiss, C. Wahrhaftig, and W. S. Fyfe, 1970. *The Earth: An Introduction to Physical Geology*. New York: Holt, Rinehart and Winston.

24. West, R. B., 1991. Tectonic geomorphology, landform modeling, and soil chronology of alluvial fans of the Tejon embayment, southernmost San Joaquin Valley, California. Master's thesis, University of California: Santa Barbara, CA.

25. Pinter, N., and E. A. Keller, 1995. Geomorphic analysis of neotectonic deformation, northern Owens Valley, California. *Geologische Rundschau*, 84:200–212.

26. Keller, E. A., D. B. Seaver, D. L. Laduzinsky, D. L. Johnson, and T. L. Ku, 2000. Tectonic geomorphology of active folding over buried reverse faults: San Emigdio Mountain front, southern San Joaquin Valley, California, 2000. *Geological Society of America Bulletin*, 112:86–97.

27. Davis, T. L., 1983. Late Cenozoic structure and tectonic history of the western "Big Bend" of the San Andreas fault and adjacent of San Emigdio Mountains. Ph. D. dissertation, University of California, Santa Barbara, CA.

28. Troeh, F. R., 1965. Landform equations fitted to contour maps. *American Journal of Science*, 263:616–627.

29. Keller, E. A., and T. K. Rockwell, 1984. Tectonic geomorphology, Quaternary chronology and paleoseismicity. In *Developments and Applications of Geomorphology*, J. E. Costa and P. J. Fleisher (eds.). Berlin: Springer-Verlag, 203–238.

The face of [the river], in time, became a wonderful book ... which told its mind to me without reserve, delivering its most cherished secrets as clearly as if it uttered them with a voice.... In truth, the passenger who could not read this book saw nothing but all manner of pretty pictures in it, painted by the sun and shaded by the clouds, whereas to the trained eye these were not pictures at all, but the grimmest and most dead-earnest of reading matter.

Mark Twain
Life on the Mississippi

INTRODUCTION

The study of active river systems is known as fluvial geomorphology (from the Latin *fluvius*, meaning "river"). The forms of rivers or streams and the processes that occur within them are described by a large number of parameters: channel width and depth, dissolved sediment load, suspended load, bed load, channel slope and sinuosity, flow velocity, channel roughness, and many others. The delicate balance between these parameters in a river system means that rivers are very sensitive to any kind of change. Climatic changes that have repeatedly swept over the Earth during the Quaternary Period have had profound effects on most geomorphic systems, including rivers [1].

Changes in global sea level over the same interval (see Chapter 6) have caused large-magnitude cycles of aggradation (accumulation of sediment in river valleys) and degradation (removal of material) in the downstream reaches of many rivers. Perhaps the single most effective agent of geomorphic change has been humans, who in their brief tenure on the Earth have nearly doubled the sediment supply to the world's rivers [2] and otherwise monkeyed with fluvial systems worldwide. Finally, with the growth of the science of tectonic geomorphology has come the growing realization that active tectonic process also can influence river form and process [3–6].

BEDROCK RIVERS, ALLUVIAL RIVERS, AND RIVER GRADE

One of the most fundamental subdivisions of river systems is between those that flow over bedrock channels and those that flow on a bed of alluvium (river sediments). Alluvial rivers are those that "flow between banks and on a bed composed of sediment that is transported by the river" [7]. The form of a given river depends on the balance between driving forces (gravity, amount of discharge, channel slope) and resisting forces (viscosity, channel roughness, friction). In alluvial rivers, long-term resisting forces are greater than or equal to the driving forces, so that the system can barely transport all of the available sediment—the result is that the river flows in a bed of its own detritus. In bedrock rivers, driving forces tend to be greater than resisting forces, and all sediment supplied can be transported away, with the result that the river flows over a channel of exposed bedrock over much or all of its length. Bedrock rivers generally are associated with smaller drainage basins, higher relief, and/or stronger bedrock. As discussed later in this chapter, some studies have documented a systematic tectonic signature in the longitudinal profiles of bedrock rivers, but more subtle adjustments to tectonics typically are masked by local variations in rock strength and structure.

In general, alluvial rivers obey a stricter set of rules than do bedrock rivers because the sedimentary medium can act quickly and effectively to balance driving forces and resisting forces. A river that maintains itself in such a state of dynamic equilibrium (see Chapter 2) is called a **graded river**. The concept of graded rivers was developed by Mackin [8] and refined by Leopold and Maddock [9]:

> A graded river is one in which, over a period of years, slope and channel characteristics are delicately adjusted to provide, with available discharge, just the velocity required for the transportation of the load supplied from the drainage basin. The graded stream is a system in equilibrium; its diagnostic characteristic is that any change in any of the controlling factors will cause a displacement of the equilibrium in a direction that will tend to absorb the effect of the change.

The key to understanding river grade is the last sentence in the preceding definition—any change in one variable in a graded river system will cause change in other variables in the system in order to reestablish equilibrium. When looking at alluvial rivers in a tectonic context, this adjustment implies that anomalous fluvial characteristics, in the absence of other plausible explanations, may record active tectonic deformation or deformation in the recent past.

COSEISMIC MODIFICATION OF RIVER SYSTEMS

The clearest and most direct impacts of tectonics on river systems occur when a large earthquake either ruptures or deforms the surface, or when ground shaking modifies fluvial processes or forms. Earthquakes may cause sudden shifts in river flow, or they may even permanently alter the course of the river. The Owens River of eastern California brings rainfall and snow melt from the peaks of the Sierra Nevada down to the desert of the Owens Valley. In 1872, a M 8 earthquake struck the Owens Valley and, according to eyewitness reports, "the disturbance of the water in the river ... was so severe that fish were thrown out upon the bank; and men stopping there, who were engaged in building a boat, did not hesitate to capture them, and served them up for breakfast" [10]. In 1872, the Owens River system included Owens Lake, a large lake that was later drained to supply water to the Los Angeles region. During the great earthquake, it was reported that "the water had receded from the shore, and that it stood in a perpendicular wall.... The wave, however, returned to shore in the course of two or three minutes, breaking and flowing some two hundred feet beyond the former edge of the shore" [10].

The New Madrid earthquake sequence of 1811–1812 was introduced in Chapter 1. These earthquakes are noteworthy because they may have been the largest historical events in the continental United States (although this has been disputed [11]) and because these earthquakes occurred in the interior of the North American Plate. The New Madrid sequence is also noteworthy because its epicentral region coincided with the Mississippi River, and near-surface faulting and strong ground shaking had dramatic effects on the fluvial system. Eyewitness accounts of the river "running backwards" in 1811 and 1812 [12] may record coseismic uplift of an area known as the Lake County Uplift [13]. Other witnesses reported the formation of "rapids" on the Mississippi during one or more of the New Madrid earthquakes, the roar of which could be heard several miles away [14, 15]. Other areas were dropped downward during the earthquakes, creating what are known as the "sunklands". Strong seismic shaking caused pervasive bank failures along the Mississippi [15], and this enormous input of sediment apparently affected the river downstream for many years [16]. Other coseismic effects in the vicinity of the river included landsliding, ground fissuring, and eruptions of liquefied sand ("sand blows") over a broad region [13, 17, 18].

FLUVIAL RESPONSES TO TECTONIC ACTIVITY

Of the many effects of active tectonics discussed in this chapter, the majority are neither so sudden nor so dramatic as the coseismic events described previously. More typically, tectonic deformation of the Earth's surface takes place slowly over thousands of years or longer. Recall from Chapter 3 that, although deformation is imperceptible to human eyes, it often can be measured by the most sensitive scientific instruments. Geomorphologists have used the characteristics of river systems to confirm or refute geodetic measurements that suggest areas of active deformation [3, 19]. In some cases, rivers may record active tectonic movement even more precisely than the best space-based geodetic techniques. With the best available vertical precisions of about ±1 cm and the longest period of monitoring of most areas now about 10 years, space-based geodesy can identify vertical tectonic strain rates as slow as ~1 mm/yr; however, if a river is sensitive to perturbations of only 1 m (a conservative estimate), then a Holocene floodplain (10,000 years old) could record strain rates as subtle as 0.1 mm/yr, ten times slower than detectable using geodesy. Furthermore, geodetic measurements can detect only deformation that is ongoing up to and including the present, whereas fluvial landforms and other geomorphic indicators may also record episodic or cyclic activity that is no longer occuring.

AGGRADATION AND DEGRADATION

Tectonic activity exerts its most fundamental control over fluvial and other geomorphic and sedimentary systems by determining the spatial distribution of deposition and erosion. Over both long periods of time and short, and over broad regions as well as local areas, deposition occurs—or at least occurs most rapidly—in areas of subsidence. In contrast, tectonic uplift favors erosion or at least reduced deposition. For example, in areas of crustal extension, which are characterized by upthrown blocks (horsts) and downthrown blocks (grabens), the drainage pattern quickly assumes the structural pattern of the extension. Most throughgoing rivers flow within the grabens. Major tributary streams tend to find their way into the grabens at irregularities in the bounding faults, such as fault step-overs [20].

Tectonic activity also exerts large-scale control on fluvial systems by influencing the patterns and rates of erosion and therefore sediment supply. Sediment yields tend to be greatest, other factors being equal, in areas where tectonic activity has created the highest and most rugged topography [21–23]. In depositional settings, thinning of fluvial sediments in a sequence can indicate uplift of local tectonic structures [e.g., 24–26].

CHANGES IN DRAINAGE AND STREAM PATTERN

It has long been noted that rock structure and/or ongoing tectonic activity in an area often will influence the geometry of the fluvial system [27–29]. This influence is reflected in the overall arrangement of streams forming regional drainage networks ("drainage pattern") as well as in the location and character of individual stream channels ("stream pattern" [27]). The structural/tectonic signature in drainage pattern has been used for much of this century as a reconnaissance tool in various applications, including hydrocarbon exploration [27, 30–32]. A drainage network formed on a stable, homogeneous landscape will develop a characteristic pattern known as dendritic (Figure 5.1a). Other drainage patterns can indicate distinct bedrock structures. For example, a trellis pattern (Figure 5.1c) can form as a result of strong jointing in the rock. This kind of structural control over the drainage pattern does not necessarily imply any active deformation, only heterogeneity in the bedrock. In contrast, some drainage patterns can result from active deformation. For example, radial drainage (Figure 5.1d) typically is formed by ongoing upwarp of a small dome, such as caused by salt diapirs (columns of salt that rise through denser surrounding sediment). The first great petroleum reservoirs of the

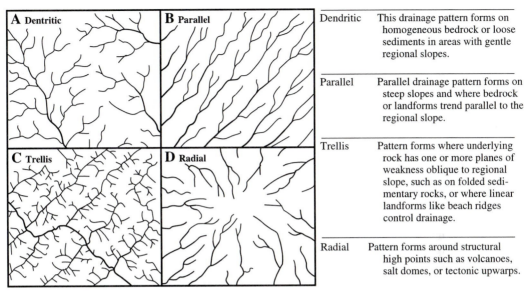

Figure 5.1   Classification of the basic drainage patterns. (After Howard, 1967 [27].)

Gulf of Mexico region and the Middle East were found around salt diapirs (oil and gas are trapped against the margins of the salt) using surficial information such as drainage patterns.

Where and when streams encounter actively uplifting structures, there can be three possible consequences: either (1) the stream will continue to flow in spite of the deformation, but its path will be **deflected** by the deformation; or (2) the fluvial system will have enough power to maintain itself through downward incision; or (3) uplift will **defeat** the stream, forcing drainage to flow by some other route. Deflection of streams by tectonic deformation has been noted by several early researchers [30, 31]. Even subtle warps may divert streams sufficiently to leave a geomorphic or sedimentological fingerprint. In the Mississippi Embayment of the southern United States, many small tributary streams appear to have shifted their positions in response to subtle and slow tectonic tilting [33, 34]. This subtle deflection of streams is the basis of the drainage-basis asymmetry indices discussed in Chapter 4. In syntectonic sedimentary deposits, fluvial diversions may result in local thinning, intraformational unconformities, or paleocurrent trends away from the structural highs [5]. In tectonically active areas like the Tien Shan [35] and New Zealand [36], the evolution of active structures can be documented in the drainage patterns and the Quaternary fluvial histories of those areas.

Maintenance of the path of an **antecedent river**—meaning one that predates the structure through which it cuts—is favored by high stream power, a weak geologic substrate, a slow rate of uplift, and a low sediment flux [35, 37]. As a river incises through a growing structure, relief between the channel and the surrounding topography will increase, forming a **water gap** (Figure 5.2, left side). A water gap is a notch cut through an active structure—or indeed any topographic barrier—through which a river flows. (For additional discussion about the formation of water gaps, see Chapter 7 and Figure 7.24).

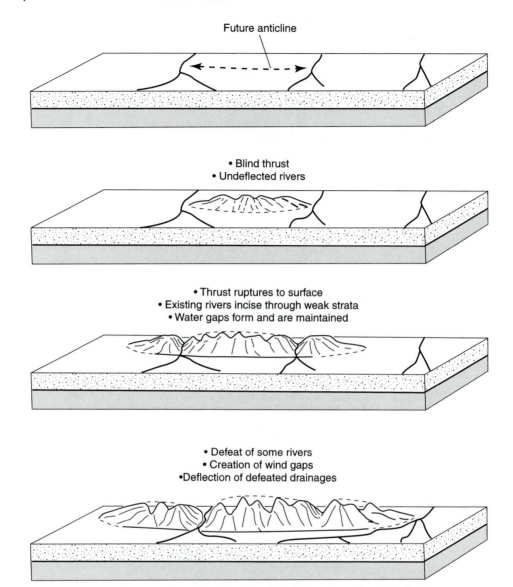

Future anticline

• Blind thrust
• Undeflected rivers

• Thrust ruptures to surface
• Existing rivers incise through weak strata
• Water gaps form and are maintained

• Defeat of some rivers
• Creation of wind gaps
•Deflection of defeated drainages

Figure 5.2   Formation of wind and water gaps across a growing anticline. The larger stream can incise as quickly as the surface is rising up, forming a deep notch through the structure, which is a water gap. The smaller stream keeps up for a while, but eventually is defeated by continuing uplift, forming a wind gap. (After Burbank et al., 1999 [35].)

Continuing uplift of a structure through which a river flows may exceed the ability of the river to downcut, forcing it to find an entirely different path around the structure. In such a case, the abandoned notch in the ridge that began as a water gap is called a **wind gap** (Figure 5.2, right side). Wheeler Ridge (Figures 7.26 and 7.27) is an active anticline in the southern San Joaquin Valley of California that is rising upward at a rate of ~3 mm/yr and is propagating eastward at ~30 mm/yr [38]. Propagation of the anticline out into the valley has pushed the north-flowing streams in the area steadily eastward during the Quaternary. Two water gaps are being maintained, but a large wind gap to the west records an earlier path of one of the streams that was defeated by the growing anticline.

Offset Streams One of the clearest tectonic signatures occurs where one or more streams cross an active strike-slip fault (see discussion in Chapter 2). Lateral translation of the stream reaches below the fault relative to the reaches upstream results in systematic offsets of the channels across the fault, in either a right-lateral or left-lateral sense (Figure 5.3) [e.g., 39–43]. Offset streams reflect movement on the fault during one or more earthquakes and/or by gradual aseismic motion. One problem with measuring displacement from offset streams is that continuing fluvial activity attempts to maintain the stream's course across the fault, smoothing the channel path across the fault into a rounded S-shaped pattern (see Figure 8.7) [41, 44]. Progressive slip can capture upstream channels, leading to underestimates of the total magnitude

Figure 5.3   Oblique aerial photograph of streams offset across the San Andreas fault at Wallace Creek. (From Wallace, 1990. U.S. Geological Survey Professional Paper 1515.)

of displacement [40]. Other factors can obscure even the sense of displacement, so systematic examination of a large number of fault-crossing streams is recommended [43].

Where the problems discussed previously can be overcome and where offset channels can be dated, they can be used to calculate fault slip rates. Wallace Creek is offset nearly 400 m across the San Andreas fault at a site approximately 150 km northwest of the city of Los Angeles. Wallace Creek (Figure 5.3) is an ephemeral stream (a stream that flows only during times of precipitation) that flows at nearly right angles to the San Andreas fault [45]. Figure 5.4 shows a series of sketch maps depicting the recent history of faulting at Wallace Creek, where trenches were excavated to determine the sequence of deposition and erosion. This sequence allowed determination of slip rates for this section of the San Andreas fault. Age control was provided by carbon-14 analysis of organic material in the stream alluvium. Figure 5.4 shows five stages in the development of Wallace Creek [45]:

- Stage 1: deposition of alluvium dated at approximately 19 ka
- Stage 2: channel incision sometime before about 5.9 ka
- Stage 3: offset of Wallace Creek by approximately 250 m between 5.9 and 3.9 ka
- Stage 4: the offset stream was abandoned in favor of a more direct route across the fault sometime during the last 3900 yr
- Stage 5: continued offset along the fault produced 380 m of offset on the Stage 2 channel and 130 m of offset on the Stage 4 channel between 3900 ka and present

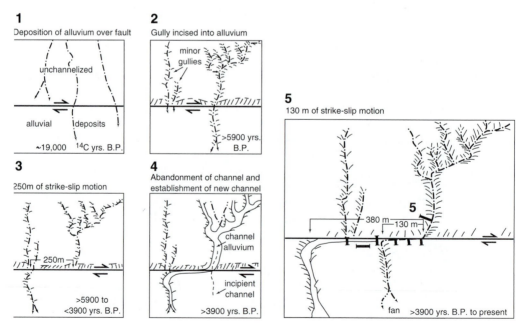

Figure 5.4   The history of earthquake surface rupture at Wallace Creek revealed by subsurface trenches. (From Sieh, 1981. In D.W. Simpson and P.G. Richards (eds.), *Earthquake Prediction: An International Review*. Washington, DC: American Geophysical Union.)

Reasoning that the total offset of 380 m must be older than or approximately equal to the oldest materials deposited after the first stream incision (>5900 ka), the 380 m of offset and the 5900-yr interval provide a maximum slip rate of 64 mm/yr. The 130-m offset occurred in <3900 yr, providing a minimum slip rate of 33 mm/yr. Taken together, these two conclusions suggest that the long-term slip rate at Wallace Creek is between 33 and 64 mm/yr [45]. Offset streams along strike-slip faults are often difficult to work with because, by themselves, there is no way to date the offset. At Wallace Creek, it was possible to bracket the slip rate with minimum and maximum age estimates because of the datable stratigraphy associated with the two stream offsets found there.

RESPONSES OF BEDROCK CHANNELS

As outlined previously, bedrock channels lack an appreciable cover of alluvium, and water flows over bare rock along much or all of the channel length during high discharge conditions [46]. Because bedrock rivers do not achieve "grade" in the same sense that alluvial rivers do, they are less sensitive indicators of local tectonic deformation. At more regional scales, however, forms and processes in bedrock channels appear to be linked to the rates and patterns of tectonic activity, and a flurry of recent studies has attacked the problem of describing this complex relationship [47, various articles in 48, 49].

In a bedrock channel, driving forces exceed resisting forces so that the channel incises over time. Several numerical models have created realistic looking longitudinal profiles of bedrock rivers and simulated the evolution of those profiles by assuming that incision rate is proportional either to shear stress (the force of moving water acting per unit area of channel bed) or to stream power (power per unit length of channel) [47, 50, 51]. Channel incision is a complex process—including bed abrasion, plucking, dissolution, cavitation, and other mechanisms—and other studies have pointed out that additional refinements in the models must be made [46, 52, 53]. Incision of bedrock channels is considered important because this process may drive the character and evolution of the entire landscape in many mountainous areas by controlling the local base level of erosion [e.g., 54].

Bedrock channel incision is also important because it appears to be the principal mechanism by which these rivers respond to, and in some cases apparently equilibrate with, regional tectonic activity. This link has been investigated by several studies along the Pacific Northwest coast of the United States, near the Mendocino Triple Junction. The Mendocino Triple Junction, where the Pacific, North American, and Juan de Fuca Plates meet (see Figure 7.5), is a zone of high uplift rates—4 mm/yr, versus 0.3 mm/yr away from the triple junction [55]. Through time, northward migration of the Mendocino Triple Junction has caused a wave of uplift to move along the coast. As discussed in Chapter 4, the gradients (steepness) of the bedrock channels are greatest near the region of greatest uplift rate (Figure 5.5b). Although this pattern is seen in all the types of channels plotted in Figure 5.5b, the smallest, most upstream tributaries in each basin seem to "feel" the uplift the most, and the downstream trunk channels "feel" the uplift the least [56]. Figure 5.6 illustrates why the effect of varying uplift rates is not evenly distributed through each drainage basin. In a setting undergoing uplift, the rate (R) at which a point on a river increases in altitude through time is equal to

$$R = \text{uplift rate} - \text{rate of incision.} \tag{5.1}$$

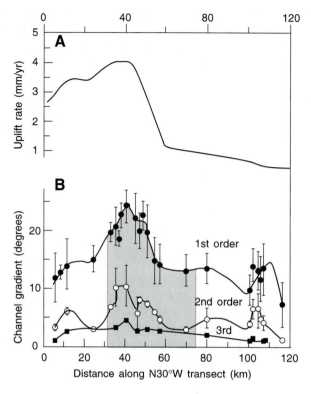

Figure 5.5 Plots of (a) uplift rate and (b) channel gradients of first-, second-, and third-order channels along a transect along the coast of northern California near the Mendocino Triple Junction. Note that the channels are steepest where the uplift rate is greatest, and that the first-order gradients show the most pronounced effect. (From Merritts and Vincent, 1989 [56].)

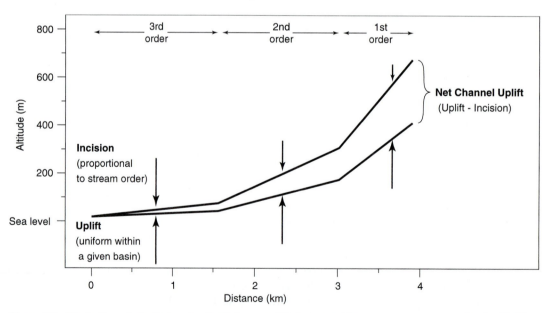

Figure 5.6 Illustration of why first-order channels are uplifted most and become steepest as a result of uplift. The change in elevation of a point on the channel is the net effect of uplift (uniform across the drainage basin) and erosion of the channel (greater farther downstream, where the channel carries more water).

Even though entire drainage basins are being uplifted, the largest streams have the most energy for erosion and are best able to maintain equilibrium profiles. The largest streams will not increase in altitude through time as much as the small tributaries, nor will their gradients increase as much if the rate of uplift increases.

Another conclusion from river studies near the Mendocino Triple Junction is that streams seem to respond surprisingly quickly to tectonic stimuli [51, 53]. Because the triple junction and its associated pulse of uplift are moving northward at a fixed rate—5.6 cm/yr [57]—distance along the coast can be translated into time. The streams with the steepest gradients are about 7.3 km away from the zone of the greatest uplift rates—suggesting that the drainage systems require about 130 ky (at 5.6 cm/yr) to reequilibrate to accelerated uplift. This estimated response time is specific to rivers in this study area; basins of a different size, in a different climate, or with a different assemblage of rocks will respond differently. However, time is a very difficult variable to quantify, and the 130-ky lag time remains one of the firmest estimates available. The issues of lag times and regional balances between uplift and degradation are discussed in detail in Chapter 9.

CHANGES IN LONGITUDINAL PROFILE

The longitudinal profile of a stream is a cross-sectional plot of its elevation from its headwaters downstream to its mouth (Figure 5.7). The longitudinal profiles of most streams form a relatively smooth curve, with low gradients in downstream reaches that gradually increase upstream. Local perturbations in a stream's longitudinal profile, such as a local steepening, can be caused by variations in the geologic substrate, water and sediment input from tributaries, or other variations along the length of the channel. In the absence of such variations, however, local anomalies in a longitudinal profile—particularly anomalies that can be traced across several streams—may indicate zones of tectonic activity. The stream length–gradient index outlined in Chapter 4 is one method for mapping such anomalies.

Longitudinal profiles can be plotted for the channel, the floodplain, or for any terraces along the length of a stream. Large, recent tectonic deformation may be recorded

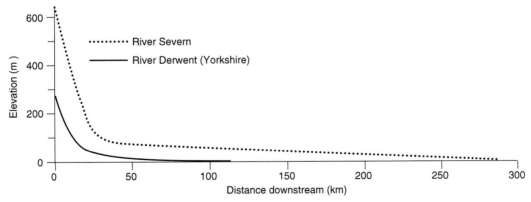

Figure 5.7 Equilibrium longitudinal profiles of two rivers in Great Britain. The concave-upward shapes of these profiles are characteristic of most graded rivers. (After Richards, 1982. *Rivers: Form and Process in Alluvial Channels*. New York: Methuen & Co.)

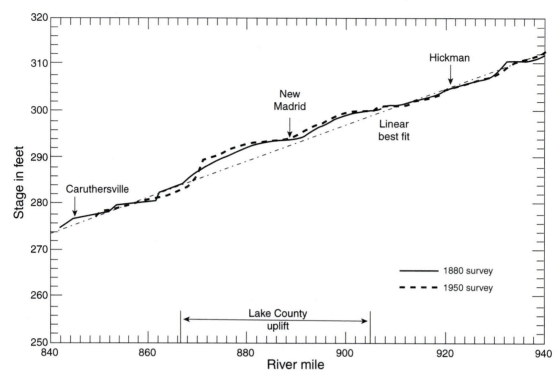

Figure 5.8   Water-surface profile at bankfull conditions of the Mississippi River where it crosses the Lake County Uplift in the New Madrid seismic zone. The active channel appears to still be perturbed across the crest of the warp at the times of these surveys. (After Schumm et al., 2000 [6].)

in the profile of the channel itself, usually measured as the water surface at bankfull discharge. For example, the 1811–1812 earthquakes, discussed previously, uplifted a broad area known as the Lake County Uplift. Where the Mississippi River crosses this uplifted area (Figure 5.8), the longitudinal profile of the channel still appears to be warped upward. That the channel of the mighty Mississippi still records this uplift is testament to the magnitude of the coseismic deformation as well as the relative recency of the earthquake sequence. Figure 5.8 also shows that the profile anomaly appears to be wearing down over time. Earthquakes that occurred farther in the past or that resulted in less deformation are less likely to be preserved in channel profiles.

Alternatively, longitudinal plots of floodplain elevations may preserve a tectonic signature longer than channel profiles. Whereas the active channel of a stream may quickly reestablish equilibrium across some past coseismic uplift, floodplains typically are formed over time scales of decades to a few centuries and therefore may preserve a longer duration record. For example, the 1886 Charleston, South Carolina, earthquake (M_b 6.6–6.9) was the largest on the eastern coast of the United States in history, but the exact location of the fault that caused the earthquake is unknown because it did not rupture the surface. However, longitudinal profiles of rivers across the epicentral area are convex upward across a northeast-trending zone (Figure 5.9) [58–60]. The implica-

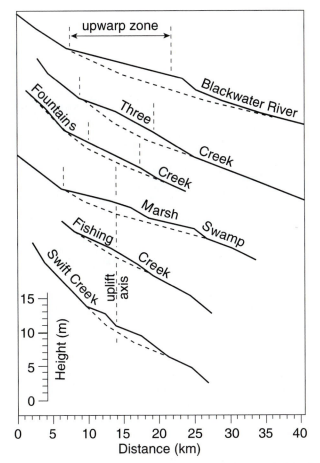

Figure 5.9   A series of streams in the U.S. Atlantic coastal plain show apparent upwarping. The axis of this warp can be traced for several hundred kilometers and may represent a regionally continuous fault zone. Rupture on this fault may have caused the 1886 Charleston, South Carolina earthquake. (After Marple and Talwani, 1999 [60].)

tion of this and other evidence is that rupture on this fault system, recently called the East Coast fault system, was responsible for the 1886 earthquake, but rupture did not penetrate the thick sedimentary cover of the Atlantic coastal plain [60].

FLUVIAL TERRACES

Where uplift occurs, base level falls, or other factors disrupt the equilibrium of the system, a river may incise through its floodplain in order to reach a new graded profile and begin cutting a new floodplain. The old floodplain becomes a river terrace—an inactive bench stranded above the new level of the river (Figure 5.10). Repeated episodes of

Figure 5.10   Terraces of the Minjar River, Kyrgyzstan, looking to the south. The Trans Alai Range is in the background. The modern river flows through a deep gorge cut into the main terrace surface, with one prominent inset terrace visible on the left wall of the gorge. (Photo courtesy of J.R. Arrowsmith, Arizona State University.)

downcutting may preserve several terraces above a river. River terraces can be classified in terms of the three processes that lead to their formation (Figure 5.11) [61]:

- Tectonically induced downcutting
- Climatically induced aggradation
- Complex response

Terraces in a setting undergoing long-term uplift typically consist of a thin layer of river sediment overlying a bedrock-cut platform (these terraces are called **strath terraces**). Downcutting is characterized by brief periods of equilibrium and floodplain construction separated by periods of incision. In contrast, terraces with a thick sedimentary cover (called **fill terraces**) reflect extended periods when there is an abundant supply of sediment, such as during major climatic shifts [61]. Complex response—changes in river systems caused directly by internal variables and thresholds, without tectonic or other external stimuli (see Chapter 2)—forms terraces that are typically small and unpaired (isolated, not found on both sides of a river).

Each fluvial terrace preserves the longitudinal profile of the river at the time that floodplain was active. Depending on the river and the local setting, fluvial terraces may be preserved for centuries up to many millennia. As a result, terraces often are the best tools for measuring long-term deformation. When terrace surfaces become inactive and incised, they do so over short periods of time. As a result, the surfaces become good tools for studying both the distribution and the timing of tectonic deformation. Three basic types of deformation can be traced across terrace surfaces (Figure 5.12):

- Surface faulting

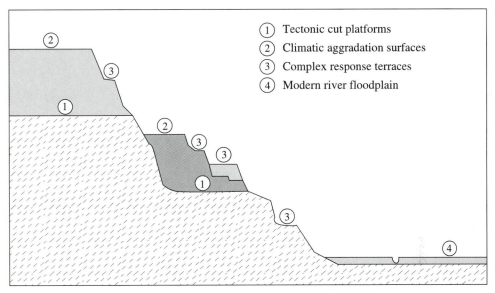

Figure 5.11   Classification of tectonic, climatic, and complex-response terraces. In this model, platforms cut onto bedrock reflect periods of tectonic uplift. Aggradation of thick fluvial deposits reflects episodes of climate change. Small unpaired benches reflect critical thresholds and adjustment of internal variables. (After Bull, 1990 [61].)

- Warping
- Tilting (convergence or divergence of terraces)

Surface Faulting on Terraces   Faults are as likely to cut the surface of a fluvial terrace as any other landform, but several aspects of terraces make them useful for studying faulting. Terraces are quite flat, so fault scarps are preserved longer than they would be on a steep slope. Furthermore, terrace surfaces represent approximate geomorphic timelines. If a fault cuts a given terrace, then faulting is known to postdate the age of that surface; if a fault crosses but does not cut a given terrace, then faulting must predate that surface. Finally, the age of fault scarps cut entirely in loose sediment sometimes can be estimated directly by diffusion-equation modeling, a process that is outlined in Chapter 8.

Many studies have recognized faulting on river terraces [e.g., 62–64]. In one case, initial field mapping recognized a large number of different surfaces, but careful remapping and correlation of surfaces using soil ages revealed that there were only four terrace levels [62]. The four terraces were cut and offset by a number of fault strands. By knowing the ages of the different terraces and measuring the surface displacements, it was possible to calculate the rate of slip on the fault system [62].

Warping of Terraces   Terrace surfaces are formed as river floodplains. Deformation will alter the longitudinal profiles of terraces compared with the profile of the

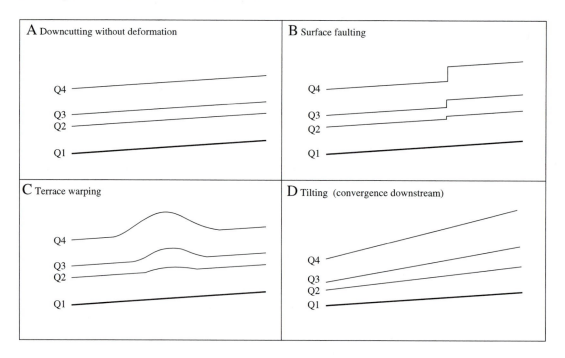

Figure 5.12   Four types of tectonic deformation of fluvial terraces: (a) uplift and incision without differential deformation, (b) surface faulting, (c) terrace warping, and (d) tilting. Note that cases (b), (c), and (d) illustrate progressive deformation through time—the oldest terraces are most deformed.

modern channel, and the shapes of deformed terraces reveal the character of deformation (Figure 5.12c). An assumption implicit in this comparison is that the original terrace profile was the same shape as the profile of the modern river. Changes in discharge over time, sediment load, or bedrock substrate strength may invalidate this assumption. An important test of tectonic deformation is whether higher and older terraces are more deformed than younger terraces (as in Figure 5.12). Climatically or lithologically controlled variations in profiles are unlikely to progress systematically through time.

   Warping of terraces has been used to recognize recent deformation in many studies [e.g., 64–66]. In the heart of the Los Angeles metropolitan area, Pleistocene and Holocene terraces and broad geomorphic surfaces are warped over two active anticlines, revealing as much as 150 m of uplift and up to 30° of tilting [64]. These anticlines probably represent the surface manifestations of buried thrust faults similar to the one that caused the 1994 Northridge earthquake (see Chapter 7).

   Tilting of Terraces   Simple downcutting on a river system without any contemporaneous deformation will lead to a sequence of parallel terraces. Where downcutting is accompanied by regional-scale tilting, however, longitudinal terrace profiles will diverge or converge (Figure 5.12d). As with terrace warping, other processes can cause divergence or convergence; progressive displacement of a series of terraces is an important test before tectonic deformation can be inferred.

OTHER RESPONSES TO LONGITUDINAL DEFORMATION

As discussed previously, achievement of "grade" means that a tectonically induced change in an alluvial river can induce changes in other characteristics of the system. Like with longitudinal profiles, local anomalies in *channel character* can result from variations in geology, hydrology, and so on, but such anomalies can also result from tectonic activity. Three sets of channel parameters have been identified as potential indicators of deformation where other causes can be ruled out: (1) channel pattern, (2) sinuosity, and (3) other hydrologic variables.

Channel Pattern   Alluvial channels are traditionally classified as either straight, meandering, or braided (Figure 5.13). The occurrence of a particular channel pattern depends on fluvial conditions such as discharge and sediment supply and, as illustrated in Figure 5.13, valley slope. Where streams flow across tectonic upwarps, erosion across the crest of the warp can generate sufficient sediment to change the local channel pattern from meandering to braided downstream. In addition, because of the dependence on slope, longitudinal tilting of just a few tenths of a percent also can change the pattern of a river (a 0.1% change in slope is equivalent to differential uplift of 1 m over a distance of 1 km). The Sefid Rud River in northwest Iran has both straight and meandering reaches. The largest meandering reach coincides with the epicentral region of a M_W 7.3 earthquake in 1990 that killed 40,000 people. It is interpreted that uplift during and between previous earthquakes has caused upwarping, and higher gradients on the downstream flank of the warp have altered the river's course from straight to meandering [67]. This evidence suggests that upwarping in the region has been occurring for

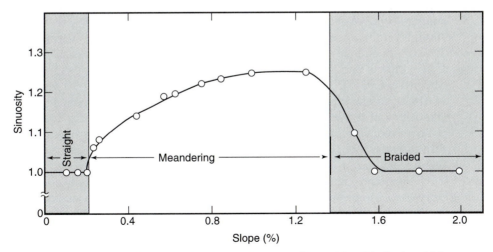

Figure 5.13   Relationship between valley slope and sinuosity (channel length/valley length) for a given discharge. Channels are classified as either straight, meandering, or braided. (After Schumm and Kahn, 1972. *Geological Society of American Bulletin*, 83:1755–1770.)

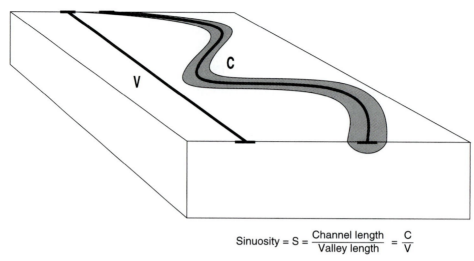

$$\text{Sinuosity} = S = \frac{\text{Channel length}}{\text{Valley length}} = \frac{C}{V}$$

Figure 5.14   Definition and calculation of the sinuosity of a meandering river channel.

a long time and probably is caused by the same process that caused the 1990 earth-quake [67].

Sinuosity Rivers meander in order to maintain a channel slope in equilibrium with discharge and sediment load. On a meandering river, sinuosity is the ratio of chan-nel length to valley length (Figure 5.14). A river meanders when the straight-line slope of the valley is too steep for equilibrium—the sinuous path of the meanders reduces the slope of the channel. Any tectonic deformation that changes the slope of a river valley may result in a corresponding change in sinuosity to maintain the equilibrium channel slope. Where meandering rivers cross tectonic upwarps, they tend to be less sinuous on the upstream flank and more sinuous on the downstream flank [5, 64, 68]. Repeated geodetic surveys since the late 1800s along the Mississippi River between Cairo, Illinois, and St. Louis, Missouri, show relative elevation changes of several tens of centimeters [3]. This amount of deformation is large for the stable interior of the North American Plate, and the pattern of the Mississippi reveals a correlation between sinuosity and the geodetic results (Figure 5.15). The river is most sinuous in zones of purported downtilt and least sinuous in zones of uptilt, just as expected. The river pat-tern provides independent evidence that the geodetic measurements are valid and the tectonic deformation is real [3].

Figure 5.13 illustrates that the proportional relationship between slope and sinu-osity should hold true within the meandering regime. But as discussed in the previous section, tilting can cause a channel to cross hydraulic thresholds, switching from one channel pattern to another. Within the braided regime, sinuosity is a difficult parameter to measure, but the relative density of individual channels and bars seems to be affect-ed by tectonic warping. Experimental results suggest that aggradation in a braided chan-nel, which would tend to occur upstream of an upwarp, is associated with increased

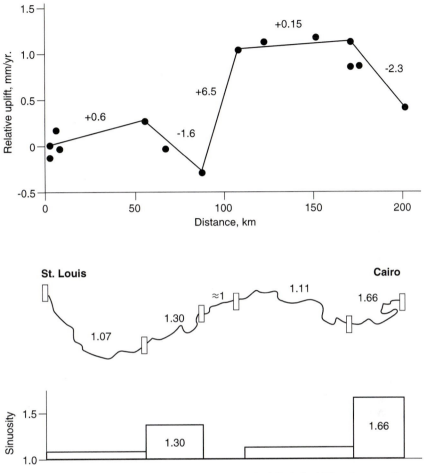

Figure 5.15   (a) Relative uplift and subsidence along the Mississippi River between St. Louis, Missouri, and Cairo, Illinois, as determined by geodetic releveling. (b) Average sinuosity of the Mississippi for the same intervals measured in (a). Note that where the geodetic measurements indicate uptilt, the sinuosities are relatively low; where there is downtilt, sinuosities are relatively high. (After Adams, 1980 [3].)

density of channels and bars; degradation, which tends to occur across the crest of up-warps and on their downstream flanks, is associated with decreased channel and bar density [6, 69].

Additional Hydrologic Variables Changes in channel pattern and sinuosity have been shown to be potentially good indicators of tectonic deformation, but there is at least preliminary evidence that other characteristics of streams may also be useful. Jorgensen [69] measured a wide range of hydrologic variables on the Jefferson River, Montana, where the river crosses a zone of uplift. While many parameters showed no

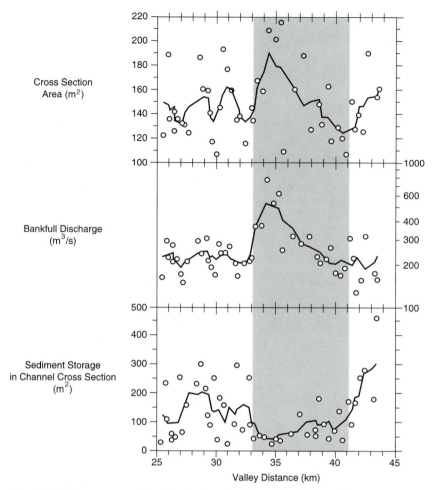

Figure 5.16 Measurement of hydrologic characteristics of the Jefferson River, Montana, across a zone of tectonic activity (shaded) suggests that anomalous values of several parameters may correlate with the deformation. Channel cross-sectional area and bankfull discharge are both anomalously high just upstream of the warp, whereas stored sediment in the channel appears to be extremely low across the entire active zone. (After Jorgensen, 1990 [69], cited in Holbrook and Schumm, 1999 [5].)

systematic change across the active zone, channel cross-sectional area, bankfull discharge, and sediment storage showed some promising correlations (Figure 5.16). Similarly, it has been suggested that overbank flooding can be anomalously frequent upstream of upwarps or across and downstream of subsided zones [69, 70]. Bed-load grain size may also be influenced by tectonically steepened or lowered slopes, with somewhat coarser sediments associated with steepening and finer sediments in back-tilted stream reaches [6]. The parameters above seem to show systematic tectonic influence in certain settings, but additional studies are needed to confirm the links and explore over what range of conditions they are valid.

RESPONSES TO LATERAL DEFORMATION

Most of the fluvial responses discussed so far are associated with longitudinal tectonic deformation, which is to say deformation parallel to the channel. Most of these responses are mechanistic responses to tectonic increases or decreases in local stream gradient. An entirely different class of responses results from deformation oriented in the perpendicular direction, resulting in cross-channel tilting of some kind. For example, lateral tilting is the dominant form of deformation in extensional half grabens—downthrown blocks in which one of the bounding faults is dominant and the other is subsidiary or inactive altogether (Figure 5.17). Through time, offset on the dominant fault will tilt the floor of a half-graben block down in that direction. Streams in a half graben typically respond to active tilting by shifting toward the fault [71, 72]. This phenomenon has been recognized in the Hebgen Lake area of Montana, just west of Yellowstone National Park, the site of a magnitude 7.5 earthquake in 1959. The earthquake was accompanied by up to 6.7 m of fault offset and a corresponding amount of down-to-the-northeast tilt on the Hebgen Lake tilt block [73]. Where the Madison River crosses the tilt block, it

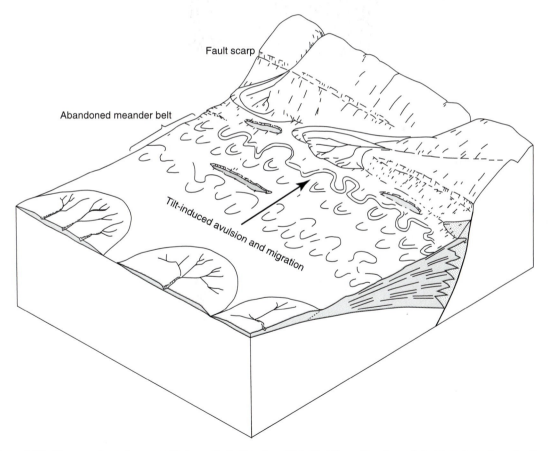

Figure 5.17   Topography and geomorphology of a half graben. Note that the active channel draining down the axis of the half graben has shifted toward the active fault scarp by ongoing tilting. (From Leeder and Gawthorpe, 1987 [71].)

flows in a broad meander belt, 225 to 1600 m wide. The active channel, however, is located at the very northeast margin of the belt, suggesting that tilt has shifted the channel [74–76].

In the preceding example, the asymmetry of the Madison River alone does not reveal the mechanism by which the river shifted. If it suddenly jumped (**avulsed**) to one side of the meander belt, such as in a single earthquake, old meander scars would be randomly oriented within the belt (Figure 5.18a). The meander scars along the Madison River, in contrast, are preferentially oriented, concave to the northeast. This pattern is consistent with gradual migration of the channel (called **combing**) as a result of steady tectonic tilt (Figure 5.18b) that must have predated the 1959 Hebgen Lake earthquake [74, 75]. On the U.S. Atlantic coastal plain, long-term lateral tilting—coincident with stream incision and terrace cutting—has led to formation of asymmetrical terrace sequences as the stream channels have migrated down-tilt (Figure 5.19) [60].

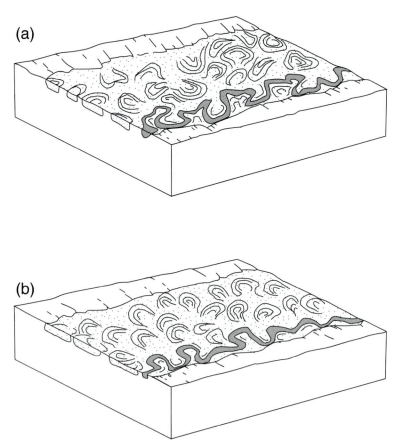

Figure 5.18   Models illustrating the effects of tilting perpendicular to a meandering channel. (a) In the case of a sudden tilting event, the channel shifts with a sudden avulsion and leaves randomly oriented meander scars. (b) In the case of gradual tilting, the channel shifts gradually, leaving meander scars preferentially oriented toward the direction of relative subsidence. (From Leeder and Alexander, 1987 [74].)

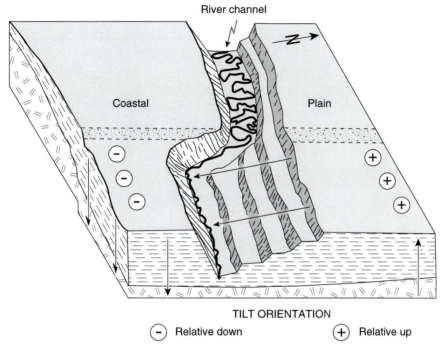

Figure 5.19   Active tilting transverse to an incising river can result in the formation of unpaired terrace sequences (that is, terraces on one side of the valley only). The example illustrated here is from the Atlantic coastal plain of South Carolina. (After Marple and Talwani, 1999 [60].)

Avulsion of streams in response to tectonic deformation is also well documented. In Long Valley caldera in eastern California, the Owens River (mentioned earlier in this chapter, but about 150 km upstream) crosses a zone of resurgent magmatism in which periodic magma injection causes pulses of uplift [77], most recently 0.5 m of uplift in 1980. The Owens River runs in one of two parallel meander belts, one closer to the zone of uplift and one farther away (Figure 5.20) [78]. Historical and geologic evidence indicate that the two belts are maintained by the periodic nature of uplift in the caldera—the downslope channel is active when the magma chamber is inflating, and the upslope channel is active when the chamber is deflating [78]. In another example, where the Carson River, Nevada, runs near the Genoa fault, it occupies one of six parallel channel belts. Dating of deposits in the different channel belts suggests that the river jumps closer to the fault during large earthquake ruptures and migrates gradually away during the interseismic periods [79].

The asymmetry imparted by lateral tectonic deformation is also seen in fluvial sedimentary deposits [71, 72]. For example, asymmetrical fluvial deposits in northern Wales reveal tilting at the time of accumulation in the Ordovician (~450 Ma) [80]. The asymmetry of fluvial deposits also can be used to recognize the nature and timing of tectonic and other geologic processes in the past. In the southern Rio Grande Rift of New Mexico, slip on the rift-bounding fault and the resulting tilting force the Rio Grande River against the alluvial fans on the east side of the valley [81]. Pleistocene deposits in

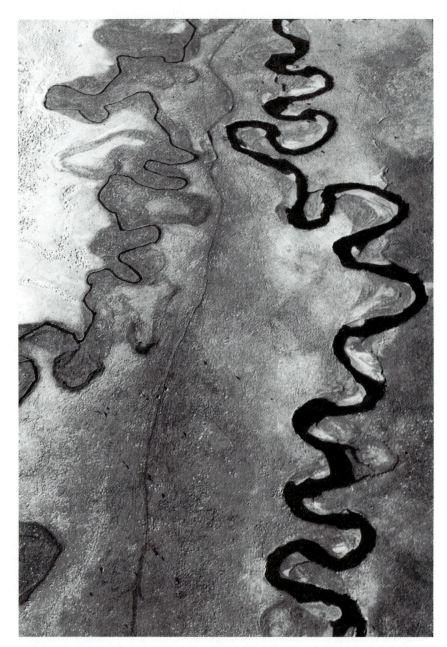

Figure 5.20 Parallel meandering channels of the Owens River in Long Valley caldera, looking down toward the northwest. Recent uplift of the resurgent dome of the caldera (left of photograph) has been shifting river flow from the upslope channel to the downslope one. (From Reid, 1992 [78].)

this area record repeated migration of the fluvial-alluvial fan boundary closer to the fault during periods of intense tectonic tilting and then farther away when tilt wanes and fan deposition predominates. These cycles and similar patterns elsewhere are interpreted as earthquake clusters on the bounding fault, with periods of intense activity separated by periods of relative tectonic quiescence [81, 82].

MODELS OF TECTONIC ADJUSTMENT

Researchers have used a variety of approaches to understand the full range of fluvial responses to active tectonic deformation. Integrated models help to explore fluvial behavior in various tectonic settings and predict the range of responses. Such models include mathematical simulations of river response and large sandbox models that recreate fluvial systems in the laboratory. In addition, some field locations provide a broad variety of evidence that helps to demonstrate the broad spectrum of fluvial landforms and features that may result from active tectonic stimuli.

NUMERICAL MODELS

The dynamics of fluvial systems are sufficiently well known to be modeled computationally. Computer models sometimes provide a system-wide perspective that may be difficult in field-based studies, where limited data and extraneous variability may obscure crucial relationships. In addition, these models typically integrate changes in the system over time and over a broader range of conditions than may be possible to observe in present-day streams or even recent fluvial deposits. For example, in one simulation, the effects of several different types of tectonic deformation were modeled for a large river system (Figure 5.21) [83]. The figure illustrates responses along a 200-km-long river reach across which there is uplift, centered at the 100 km mark, beginning at time = 0. The zone of channel incision widens, and the point of maximum channel-elevation increase (the net effect of uplift and incision) shifts upstream through time. The results of this simulation also emphasize the complexity of such systems. Even though the model was greatly simplified, simple tectonic deformations led to complex river responses, including the following:

1. Reversals of deposition and erosion through time at some points on the channel
2. Significant lag times between stimuli and responses
3. Critical transition points, such as between zones of aggradation and degradation, that do not coincide with the axes of deformation
4. In general, a mismatch between the location of tectonic deformation and the distribution of fluvial response

Note that the complex responses in this study, sometimes called "nonlinear" responses, are different from the concept of "complex response" introduced in Chapter 2.

Additional modeling efforts have focused on the nature and geometry of fluvial deposits in tilting half grabens. As discussed previously in this chapter, observations of present-day streams in actively tilting areas demonstrate a strong tectonic influence on fluvial forms and processes. Together with descriptions of fluvial sedimentary deposits,

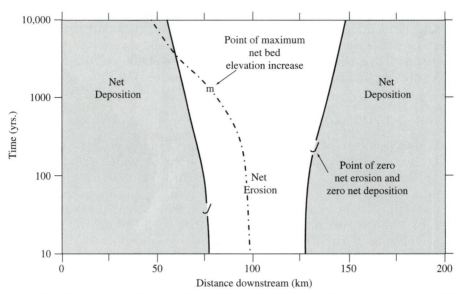

Figure 5.21 Zones of aggradation and degradation and critical transition points on a simulated river channel undergoing anticlinal uplift across the channel at the 100-km mark (distance = 0 is upslope). Note that the zones and the transition points gradually shift locations after uplift begins at time = 0. (After Snow and Slingerland, 1990 [83].)

this information has been used to develop a class of qualitative and quantitative models that are sometimes called "LAB models" (after the contributions of Leeder [84], Allen [85], and Bridge [86]). These models predict different stratigraphic relationships in response to different conditions of tectonic forcing, such as tilt rate and slip per faulting event, and other factors, such as discharge and sediment supply (Figure 5.22). Various field-based studies have confirmed some of the model predictions [79] whereas other studies have suggested some refinements in the assumptions on which these models are based [e.g., 24].

EXPERIMENTAL MODELS

An alternative to numerical modeling is to observe the behavior of small-scale channels in laboratory flumes (a flume is a sediment-filled box, inclined and provided with a continuous flow of water from high end to low end). Such experiments also provide systemwide insights into fluvial responses to tectonic activity. In some of the first and most sweeping experiments, Ouchi [70] created braided and meandering fluvial patterns in a 9.1-m-long by 2.4-m-wide by 0.6-m-deep flume. The central portion of the flume was fitted with a flexible base that raised or lowered the channel to mimic uplift and subsidence. Four experiments were run using this apparatus:

1. Uplift across a braided channel
2. Subsidence across a braided channel
3. Uplift across a meandering channel
4. Subsidence across a meandering channel

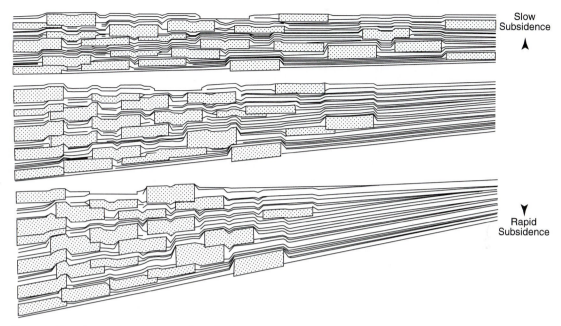

Figure 5.22   Numerical simulation of fluvial sedimentation in a subsiding half graben under conditions ranging from slow tilt to fast. This model suggests that the channel bodies will be thicker, more continuous, and concentrated near the fault when tectonic tilting is rapid. (After Alexander and Leeder, 1987 [72].)

The results are summarized in Table 5.1. In experiment 1, the braided channel incised through the axis of uplift, and it aggraded both upstream and downstream of the upwarp. Aggradation downstream was caused by increased sediment load supplied by the incision; aggradation upstream was caused by the reduction in slope. In experiment 2, the effects were opposite those of experiment 1—aggradation across the axis of deformation and degradation upstream and downstream. In experiment 3, the meandering channel adjusted to uplift by changes in sinuosity and flow velocity, with little or no aggradation or degradation. Downstream of the upwarp, sinuosity increased and, in the process, the outer banks of the meanders eroded and point bars inside the meanders grew; upstream of the uplift, flow velocity was reduced and water flooded over the point bars. In experiment 4, the effects were opposite those of experiment 3—sinuosity increased upstream and the current slowed, and flooding occurred downstream [70].

Schumm et al. [6] outline additional flume runs on the same apparatus that confirm many of the same phenomena and highlight other aspects of the fluvial response to deformation. In these experiments, axial uplift across the flume's channel created permanent increases in sinuosity and changes in meander geometry downstream of the warp, as observed previously. Upstream of the uplift axis, sinuosity decreased and meander geometry also changed as predicted, but the changes disappeared soon after the uplift ceased [6, 87]. Erosion and incision downstream of the warp may lock the fluvial channel into place and preserve the anomalous tectonic signatures, but it is suggested that changes on the up-tilting flank of such a warp may be more ephemeral.

Table 5.1

RESPONSE OF EXPERIMENTAL BRAIDED AND MEANDERING CHANNELS TO UPLIFT AND SUBSIDENCE ACROSS THE CHANNEL. (AFTER OUCHI, 1985 [70].)

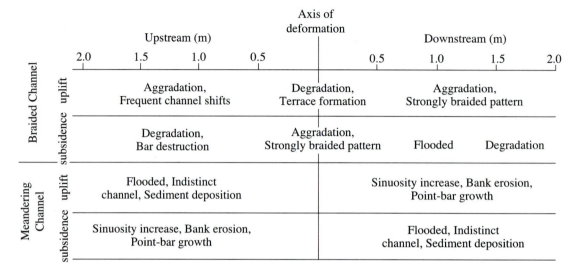

		Upstream (m)			Axis of deformation		Downstream (m)		
		2.0 1.5 1.0 0.5				0.5 1.0 1.5 2.0			
Braided Channel	uplift	Aggradation, Frequent channel shifts			Degradation, Terrace formation		Aggradation, Strongly braided pattern		
	subsidence	Degradation, Bar destruction			Aggradation, Strongly braided pattern		Flooded	Degradation	
Meandering Channel	uplift	Flooded, Indistinct channel, Sediment deposition					Sinuosity increase, Bank erosion, Point-bar growth		
	subsidence	Sinuosity increase, Bank erosion, Point-bar growth					Flooded, Indistinct channel, Sediment deposition		

EMPIRICAL MODELS

It has been argued here that the main benefit of creating numerical or laboratory simulations of fluvial behavior is that they provide overarching perspective into the functioning and sensitivity of stream systems to tectonic stimuli. Where a broad range of mutually supporting and overlapping evidence is available, some field studies can provide the same insights. Deformation across the Monroe and Wiggins uplifts of Louisiana and Mississippi (Figure 5.23) is amply documented by warped Tertiary and Quaternary sedimentary strata, deformed Pleistocene and Holocene terraces, and geodetic measurements of ongoing motion [88]. The rivers that cross the zones of uplift show many of the same modifications seen in the flume experiments. Above the axes of uplift, channel sinuosity as well as channel depth are reduced, and overbank flooding is increased. Below the uplifts, sinuosity and depth are increased, bank erosion is accelerated, and some reaches are locally braided in response to increased sediment loads [88]. In addition, the longitudinal profiles of both the river-valley floors and the channels are convex upward across the Monroe and Wiggins uplifts on all but the largest rivers. The largest rivers apparently have sufficient energy for the channels to cut through the deformation without measurable perturbation [88].

SUMMARY

Rivers are profoundly influenced by active tectonics. Fluvial systems are sensitive to both faulting and regional surface deformation. Fluvial features may record small-scale tectonic deformation accumulated over many thousand years, in some cases making

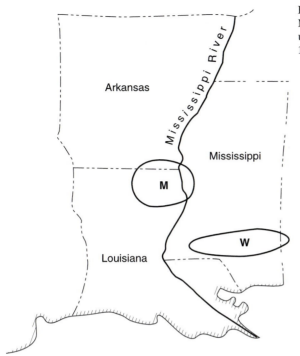

Figure 5.23   Locations of the Monroe (M) and Wiggins (W) uplifts. (After Burnett and Schumm, 1983 [88].)

them sensitive to subtle tectonic processes at the limits of even the most precise geodetic measurement tools. As a result, it is possible to use perturbations in fluvial systems to locate, characterize, and measure recent tectonic activity. This type of research is a relatively recent development in geomorphology, and the study of rivers and active tectonics continues to evolve. Several avenues of research are now being developed that may lead to great advances in the future. Digital elevation models and a variety of satellite images make recognition of fluvial patterns increasingly straightforward [4]; many parameters can even be measured by automatic computer analysis [89]. In addition, advances in dating techniques [90, 91] should increase opportunities for dating terraces and other landforms, widening the range of settings from which tectonic rates can be calculated. Finally, we look forward to the full integration of fluvial geomorphology and tectonic geomorphology, with the recognition of tectonic effects in many fluvial systems and the recognition that rivers present valuable tools for use in studies of active tectonics.

REFERENCES CITED

1. Bull, W. B., 1991. *Geomorphic Response to Climate Change*. New York: Oxford University Press.

2. McLennan, S. M., 1993. Weathering and global denudation. *Journal of Geology*, 101:295–303.

3. Adams, J., 1980. Active tilting of the United States midcontinent: geodetic and geomorphic evidence. *Geology*, 8:442–446.

4. Deffontaines, B., and J. Chorowicz, 1991. Principles of drainage basin analysis from multisource data: application to the structural analysis of the Zaire Basin. *Tectonophysics*, 194:237–263.

5. Holbrook, J., and S. A. Schumm, 1999. Geomorphic and sedimentary response of rivers to tectonic deformation: a brief review and critique of a tool for recognizing subtle epeirogenic deformation in modern and ancient settings. *Tectonophysics*, 305:287–306.

6. Schumm, S. A., J. F. Dumont, and J. M. Holbrook, 2000. *Active Tectonics and Alluvial Rivers*. Cambridge, U.K.: Cambridge University Press.

7. Schumm, S. A., 1986. Alluvial river response to active tectonics. In R. E. Wallace (ed.), *Active Tectonics: Studies in Geophysics*. Washington, DC: National Academy Press. 80–94.

8. Mackin, J. H., 1948. Concept of the graded river. *Geological Society of America Bulletin*, 59:463–512.

9. Leopold, L. B., and T. Maddock, 1953. Hydraulic geometry of stream channels and some physiographic implications. *U.S. Geological Survey Professional Paper* 252.

10. Whitney, J. D., 1872. The Owens Valley earthquake. *The Overland Monthly*, 9:130–140, 266–278.

11. Newman, A., S. Stein, J. Weber, J. Engeln, A. Mao, and T. Dixon, 1999. Slow deformation and lower seismic hazard at the New Madrid Seismic Zone. *Science*, 284:619–621.

12. Hamilton, R. M., 1980. Quakes along the Mississippi. *Natural History*, 89:70–75.

13. Russ, D. R., 1982. Style and significance of surface deformation in the vicinity of New Madrid, MO. *U.S. Geological Survey Professional Paper*, 1236-H:95–114.

14. Fuller, M. L., 1912. The New Madrid earthquake. *U.S. Geological Survey Bulletin*, 494.

15. Penick, J., Jr., 1981. *The New Madrid Earthquakes of 1811–12*. Columbia, MO: The University of Missouri Press.

16. Walters, W. H., Jr., and D. B. Simons, 1984. Long-term changes of lower Mississippi River meander geometry. In C. M. Elliot (ed.), *River Meandering*. New York: American Society of Civil Engineers, 318–329.

17. Johnston, A. C., and E. S. Schweig, 1996. The enigma of the New Madrid earthquakes of 1811–1812. *Annual Reviews of Earth and Planetary Sciences*, 24:339–384.

18. Van Arsdale, R., 1997. Hazard in the heartland: The New Madrid seismic zone. *Geotimes*, 42(5):16–19.

19. Saucier, R. T., 1987. Geomorphological interpretation of late Quaternary terraces in western Tennessee and their regional tectonic implications. In D. P. Russ and A. J. Crone (eds.), The New Madrid, Missouri, Earthquake Region—Geological, Seismological, and Geotechnical Studies. *U.S. Geological Survey Professional Paper* 1336-A.

20. Paton, S., 1992. Active normal faulting, drainage patterns, and sedimentation in southwestern Turkey. *Journal of the Geological Society, London*, 149:1031–1044.

21. Milliman, J. D., 1997. Fluvial sediment discharge to the sea and the importance of regional tectonics. In W. F. Ruddiman (ed.), *Tectonic Uplift and Climate Change*. New York: Plenum Press. 239–257.

22. Milliman, J. D., and J. P. M. Syvitski, 1992. Geomorphic/tectonic control of sediment discharge to the ocean; the importance of small mountainous rivers. *Journal of Geology*, 100:525–544.

23. Ahnert, F., 1970. Functional relationships between denudation, relief, and uplift in large mid-latitude drainage basins. *American Journal of Science*, 268:243–263.

24. Guiseppe, A. C., and P. L. Heller, 1998. Long-term river response to regional doming in the Price River Formation, central Utah. *Geology*, 26:239–242.

25. Holbrook, J. M., and R. Wright-Dunbar, 1992. Depositional history of Lower Cretaceous strata in northeastern New Mexico: implications for regional tectonics and depositional sequences. *Geological Society of America Bulletin*, 104:802–813.

26. Pivnik, D. A., and G. D. Johnson, 1995. Depositional responses to Pliocene-Pleistocene foreland partitioning in northwest Pakistan. *Geological Society of America Bulletin*, 107:895–922.

27. Howard, A. D., 1967. Drainage analysis in geologic interpretation: a summation. *American Association of Petroleum Geologists Bulletin*, 51:2246–2259.

28. Gilbert, G. K., 1877. *Geology of the Henry Mountains, U.S. Geological Survey of the Rocky Mountains*. Washington, DC: U.S. Government Printing Office.

29. Gilbert, G. K., 1884. A theory of the earthquakes of the Great Basin, with a practical application. *American Journal of Science*, 3rd Series, 27:49–53.

30. Zernitz, E. R., 1932. Drainage patterns and their significance. *Journal of Geology*, 40:498–521.

31. DeBlieux, C., 1949. Photogeology of Gulf Coast exploration. *American Association of Petroleum Geologists Bulletin*, 33:1251–1259.

32. Lattman, L. H., 1959. Geomorphology: New tool for finding oil. *Oil and Gas Journal*, 57:231–236.

33. Cox, R. T., 1994. Analysis of drainage-basin symmetry as a rapid technique to identify areas of possible Quaternary tilt-block tectonics: an example from the Mississippi Embayment. *Geological Society of America Bulletin*, 106:911–931.

34. Cox, R. T., R. B. Van Arsdale, and J. B. Harris, 2001 Identification of possible Quaternary deformation in the northeastern Mississippi Embayment using quantitative geomorphic analysis of drainage-basin asymmetry. *Geological Society of America Bulletin*, 113:615–624.

35. Burbank, D. W., J. K. McLean, M. Bullen, K. Y. Abdrakhmatov, and M. M. Miller, 1999. Partitioning of intermontane basins by thrust-related folding, Tien Shan, Kyrgyzstan. *Basin Research*, 11:75–92.

36. Jackson, J., R. Norris, and J. Youngson, 1996. The structural evolution of active fault and fold systems in central Otago, New Zealand: evidence revealed by drainage patterns. *Journal of Structural Geology*, 18:217–234.

37. Humphrey, N. F., and S. K. Konrad, 2000. River incision or diversion in response to bedrock uplift. *Geology*, 28:43–46.

38. Keller, E. A., R. L. Zepeda, T. K. Rockwell, T. L. Ku, and W. S. Dinklage, 1998. Active tectonics at Wheeler Ridge, southern San Joaquin Valley, California. *Geological Society of America Bulletin*, 110:298–310.

39. Grant, L. B., and K. Sieh, 1994. Paleoseismic evidence of clustered earthquakes on the San Andreas fault in the Carrizo Plain, California. *Journal of Geophysical Research*, 99:6819–6841.

40. Gaudemer, Y., P. Taponnier, D. Turcotte, 1989. River offsets across active strike-slip faults. *Annales Tectonicae*, III:55–76.

41. McGill, S. F., and K. Sieh, 1991. Surficial offsets on the central and eastern Garlock fault associated with prehistoric earthquakes. *Journal of Geophysical Research*, 96:21,597–21,621.

42. Zhang, P., M. Ellis, D. B. Slemmons, and F. Mao, 1990. Right-lateral displacements and the Holocene slip rate associated with prehistoric earthquakes along the southern Panamint Valley fault zone: implication for southern Basin and Range tectonics and coastal California deformation: *Journal of Geophysical Research*, 95:4857–4872.

43. Huang, W., 1993. Morphological patterns of stream channels on the active Yishi Fault, southern Shandong Province, Eastern China: implications for repeated great earthquakes in the Holocene. *Tectonophysics*, 219:283–304.

44. Pinter, N., S. B. Lueddecke, E. A. Keller, and K. Simmons, 1998. Late Quaternary slip on the Santa Cruz Island fault, California. *Geological Society of America Bulletin*, 110:711–722.

45. Sieh, K., and R. H. Jahns, 1984. Holocene activity of the San Andreas fault at Wallace Creek, California. *Geological Society of America Bulletin*, 95:883–896.

46. Howard, A. D., 1998. Long profile development of bedrock channels: Interaction of weathering, mass wasting, bed erosion, and sediment transport. In K. J. Tinkler and E. E. Wohl (eds.),

Rivers over Rocks: Fluvial Processes in Bedrock Channels. American Geophysical Union, Geophysical Monograph, 107:297–319.

47. Howard, A. D., W. E. Dietrich, and M. A. Seidl, 1994. Modeling fluvial erosion on regional to continental scales. *Journal of Geophysical Research,* 99:13,971–13,986.

48. Tinkler, K. J., and E. E. Wohl, 1998. *Rivers over Rocks: Fluvial Processes in Bedrock Channels.* American Geophysical Union, Geophysical Monograph, Washington, D.C. 107.

49. Whipple, K. X., G. S. Hancock, and R. S. Anderson, 2000. River incision in bedrock: mechanics and relative efficacy of plucking, abrasion, and cavitation. *Geological Society of America Bulletin,* 112:490–503.

50. Howard, A. D., and G. Kerby, 1983. Channel changes in badlands. *Geological Society of America Bulletin,* 94:739–752.

51. Whipple, K. X., and G. E. Tucker, 1999. Dynamics of the stream-power river incision model: implications for height limits of mountain ranges, landscape response timescales, and research needs. *Journal of Geophysical Research,* 104:17,661–17,674.

52. Sklar, L., and W. E. Dietrich, 1998. River longitudinal profiles and bedrock incision models: Stream power and the influence of sediment supply. In K. J. Tinkler and E. E. Wohl (eds.), *Rivers over Rocks: Fluvial Processes in Bedrock Channels.* American Geophysical Union, Geophysical Monograph, 107:237–260.

53. Snyder, N. P., K. X. Whipple, G. E. Tucker, and D. J. Merritts, 2000. Landscape response to tectonic forcing: digital elevation model analysis of stream profiles in the Mendocino Triple Junction region, northern California. *Geological Society of America Bulletin,* 112:1250–1263.

54. Burbank, D. W., J. Leland, E. Fielding, R. S. Anderson, N. Brozovic, M. R. Reid, and C. Duncan, 1996. Bedrock incision, rock uplift and threshold hillslopes in the northwestern Himalayas. *Nature,* 379:505–510.

55. Merritts, D., and W. B. Bull, 1989. Interpreting Quaternary uplift rates at the Mendocino Triple Junction, northern California, from uplifted marine terraces. *Geology,* 17:1020–1024.

56. Merritts, D., and K. R. Vincent, 1989. Geomorphic response of coastal streams to low, intermediate and high rates of uplift, Mendocino triple junction region, northern California. *Geological Society of America Bulletin,* 101:1373–1388.

57. Engebretson, D. C., A. Cox, and R. A. Gordon, 1985. Relative motion between oceanic and continental plates in the Pacific basin: *Geological Society of America Special Paper* 205.

58. Rhea, S., 1989. Evidence for uplift near Charleston, South Carolina. *Geology,* 17:311–315.

59. Marple, R. T., and P. Talwani, 1993. Evidence of possible tectonic upwarping along the South Carolina coastal plain from an examination of river morphology and elevation data. *Geology,* 21:651–654.

60. Marple, R. T., and P. Talwani, 2000. Evidence for a buried fault system in the Coastal Plain of the Carolinas and Virginia—Implications for neotectonics in the southeastern United States. *Geological Society of America Bulletin,* 112:200–220.

61. Bull, W. B., 1990. Stream-terrace genesis: implications for soil development. *Geomorphology,* 3:351–367.

62. Rockwell, T. K., E. A. Keller, M. N. Clark, and D. L. Johnson, 1984. Chronology and rates of faulting of Ventura River terraces, California. *Geological Society of America Bulletin,* 95:1466–1474.

63. Bishop, P., and J. C. Bousquet, 1989. The Quaternary terraces of the Lergue River and activity of the Cévennes fault in the lower Hérault Valley (Languedoc), southern France. *Zeitscrift fur Geomorphologie,* 33:405–415.

64. Bullard, T. F., and W. R. Lettis, 1993. Quaternary fold deformation associated with blind thrust faulting, Los Angeles Basin, California. *Journal of Geophysical Research*, 98:8349–8369.

65. Reheis, M. C., 1985. Evidence for Quaternary tectonism in the northern Bighorn Basin, Wyoming and Montana. *Geology*, 13:364–367.

66. Pinter, N., and E. A. Keller, 1995. Geomorphic analysis of neotectonic deformation, northern Owens Valley, California. *Geologische Rundschau*, 84:200–212.

67. Berberian, M., M. Qorashi, J. A. Jackson, K. Priestley, and T. Wallace, 1992. The Rudbar-Tarom earthquake of 20 June in NW Persia: preliminary field and seismological observations, and its tectonic significance. *Bulletin of the Seismological Society of America*, 82:1726–1755.

68. Boyd, K. F., and S. A. Schumm, 1995. Geomorphic evidence of deformation in the northern part of the New Madrid seismic zone. *U.S. Geological Survey Professional Paper*, 1538-R.

69. Jorgensen, D. W., 1990. Adjustment of Alluvial River Morphology and Process to Localized Active Tectonics. Unpublished Ph.D. dissertation, Colorado State University, Fort Collins.

70. Ouchi, S., 1985. Response of alluvial rivers to slow active tectonic movement. *Geological Society of America Bulletin*, 96:504–515.

71. Leeder, M. R., and R. L. Gawthorpe, 1987. Sedimentary models for extensional tilt-block/half-graben basins. In M. P. Coward, J. F. Dewey, and P. L. Hancock (eds.), *Continental Extensional Tectonics*. Geological Society Special Publication, London, England, 28:139–152.

72. Alexander, J., and M. R. Leeder, 1987. Active tectonic control on alluvial architecture. In F. G. Etheridge, R. M. Flores, and M. D. Harvey (eds.), *Recent Developments in Fluvial Sedimentology*. Society of Economic Paleontologists and Minerologists Special Publication, 39:243–252.

73. Myers, W. B., and W. Hamilton, 1964. Deformation associated with the Hebgen Lake earthquake of August 17, 1959. *U.S. Geological Survey Professional Paper*, 435:55–98.

74. Leeder, M. R., and J. Alexander, 1987. The origin and tectonic significance of asymmetrical meander-belts. *Sedimentology*, 34:217–226.

75. Alexander, J., and M. R. Leeder, 1990. Geomorphology and surface tilting in an active extensional basin, SW Montana, U.S.A. *Journal of the Geological Society, London*, 147:461–467.

76. Alexander, J., J. S. Bridge, M. R. Leeder, R. E. L. Collier, and R. L. Gawthorpe, 1994. Holocene meander-belt evolution in an active extensional basin, southwestern Montana. *Journal of Sedimentary Research*, B64:542–559.

77. Bailey, R. A., G. B. Dalrymple, and M. A. Lanphere, 1976. Volcanism, structure, and geochronology of Long Valley Caldera, Mono County, California. *Journal of Geophysical Research*, 81:725–744.

78. Reid Jr., J. B., 1992. The Owens River as a tiltmeter for Long Valley caldera, California. *Journal of Geology*, 100:353–363.

79. Peakall, J., 1998. Axial river evolution in response to half-graben faulting: Carson River, Nevada, U.S.A. *Journal of Sedimentary Research*, 68:788–799.

80. Orton, G. J., 1991. Emergence of subaqueous depositional environments in advance of a major ignimbrite eruption, Cape Curig Volcanic Formation, Ordovician, North Wales—an example of regional volcanotectonic uplift? *Sedimentary Geology*, 74:251–286.

81. Mack, G. H., and M. R. Leeder, 1999. Climatic and tectonic controls on alluvial-fan and axial-fluvial sedimentation in the Plio-Pleistocene Palomas half graben, southern Rio Grande Rift. *Journal of Sedimentary Research*, 69:635–652.

82. Dorsey, R. J., P. J. Umhoefer, and P. D. Falk, 1997. Earthquake clustering inferred from Pliocene Gilbert-type deltas in the Loreto basin, Baja California Sur, Mexico. *Geology*, 25:679–682.

83. Snow, R. S., and R. L. Slingerland, 1990. Stream profile adjustment to crustal warping: nonlinear results from a simple model. *Journal of Geology*, 98:699–708.

84. Leeder, M. R., 1978. A quantitative stratigraphic model for alluvium, with special reference to channel deposit density and interconnectedness. In A. D. Miall (ed.), *Fluvial Sedimentology*. Canadian Society of Petroleum Geologists Memoir, 5:578–596.

85. Allen, J. R. L., 1978. Studies in fluviatile sedimentation: An exploratory quantitative model for the architecture of avulsion-controlled alluvial suites. *Sedimentary Geology*, 21:129–147.

86. Bridge, J. S., and M. R. Leeder, 1979. A simulation model of alluvial stratigraphy. *Sedimentology*, 26:617–644.

87. Jin, D., and S. A. Schumm, 1987. A new technique for modeling river morphology. In V. Gardner (ed.), *International Geomorphology*. Chichester, U.K.: John Wiley & Sons, 681–690.

88. Burnett, A. W., and S. A. Schumm, 1983. Alluvial-river response to neotectonic deformation in Louisiana and Mississippi. *Science*, 222:49–50.

89. Gardner, T. W., K. C. Sasowsky, and R. L. Day, 1990. Automated extraction of geomorphometric properties from digital elevation data. *Zeitscrift fur Geomorphologie Supplement*, 80:57–68.

90. Dunai, T. J., and J. R. Wijbrans, 2000. Long-term cosmogenic ^{3}He production rates (152 ka–1.35 Ma) from ^{40}Ar/^{39}Ar dated basalt flows at 29 degrees N latitude. *Earth and Planetary Science Letters*, 176:147–156.

91. Fleming, A., M. A. Summerfield, J. O. Stone, L. K. Fifield, and R. G. Cresswell, 1999. Denudation rates for the southern Drakensberg Escarpment, SE Africa, derived from in-situ-produced cosmogenic ^{36}Cl; initial results. *Journal of the Geological Society of London*, 156, Part 2:209–212.

6

Active Tectonics and Coastlines

INTRODUCTION

On April 25, 1992 a magnitude 7.1 earthquake struck the northern California coast, causing moderate damage to small towns in the area. The earthquake occurred along the Cascadia Subduction Zone, where the Juan de Fuca Plate subducts beneath North America (see Figure 6.10 and discussion of **coseismic subsidence** later in this chapter). In the days and weeks after the earthquake, subtle changes to the coastline of northern California became apparent. The first accounts came from fishermen and residents of the coast near the epicenter, who reported changes in familiar offshore rocks and reefs. Within one to two weeks, attached marine organisms began to die at the upper margin of the intertidal zone, forming a stripe of bleached-out death along 23 km of the shoreline (Figure 6.1). Scientists who came to investigate the natural destruction determined that the earthquake had uplifted the coast by up to 1.4 m, carrying a whole intertidal ecosystem up and out of the ocean [1]. While such events are rare in the time scale of human events, they are common in geologic time.

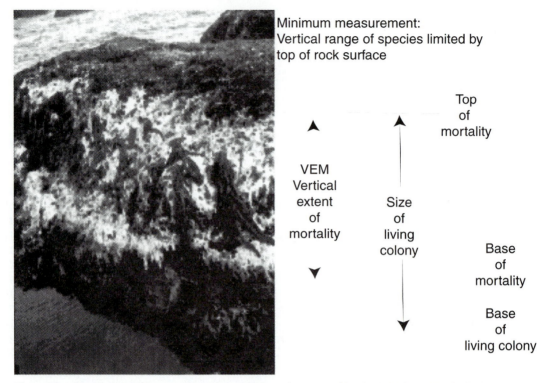

Figure 6.1.   Photograph of the mortality of intertidal organisms caused by the 1992 Cape Mendocino earthquake. (From Carver et al., 1994. *Geology*, 22:195–198.)

Tectonic activity occurs in all settings of the planet, from the deepest ocean trench to the highest mountain peak. But activity that coincides with ocean or lake coastlines has unique effects, and the record of tectonic activity at coasts contains data of great value in determining past earthquake history. The ocean surface is Earth's best approximation of a perfect horizontal datum (see discussion of the **geoid** in Chapter 3). Apart from tides and variations in atmospheric pressure, which average out over time, sea level precisely follows a surface of equal-gravity potential everywhere around the Earth. Any change of position relative to sea level disrupts the delicate balance of biological, chemical, and physical forces concentrated there.

Coastal landforms can be used to decode several aspects of active tectonics. After the April, 1992 Cape Mendocino earthquake discussed previously, scientists used the height of the "kill zone" of intertidal organisms to determine the amount of coseismic uplift (up to 1.4 m) and the pattern of the uplift (an east-west trending arch) [1]. Although these kill zones are useful for measuring coseismic uplift, they are not as useful for measuring gradual, long-term uplift or deformation of coastlines, simply because the biological evidence is not preserved longer than a few months to years. Long-term uplift is recorded by more durable features, such as coastal terraces, beach ridges, and coral reefs. Coastlines are further useful for studies of active tectonics because coastal landforms typically form over short periods of geologic time, so that inactive coastal features can act as effective isochrons (timelines) for estimating the ages of subsequent deformation events. Finally, an additional advantage of coastal landforms is that they are more likely than other features to include material suitable for numerical dating.

The character of a coastline is a function of many variables, but four stand out:

1. The local tectonic setting
2. The supply of sediment
3. The amount of erosional energy
4. The nature and amount of biological activity

The rugged, cliff-lined coasts of California, Oregon, and Washington are distinct from the sandy, barrier-island coastlines of the East Coast of the United States because, in much of the West, the land is steadily rising relative to sea level. In addition, the Pacific coast is starved of sand compared with the Atlantic, and the energy of breaking waves is focused onto the cliffs. Figure 6.2 illustrates a classification of coastlines along two axes, from **emergent** (rising relative to sea level) to **submergent** (falling relative to sea level),

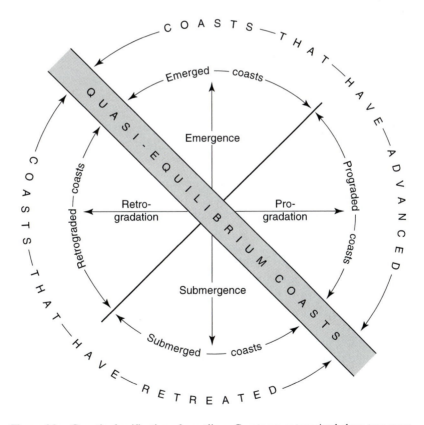

Figure 6.2.   Genetic classification of coastlines. Coasts are categorized along two axes: from emerged to submerged, and from prograded to retrograded. (After Valentin, 1970. Paper read at the Symposium of the IGU Commission on Coastal Geomorphology: Moscow, cited in Bloom, 1998 [4].)

and from **retrograded** (retreating landward through time) to **prograded** (advancing oceanward through time). Consider for example, the effects of changes in sea level on coastal landscapes. Since the maximum of the last glaciation about 18 ka (18,000 years before present), melting glaciers and warming oceans have raised global sea level by about 125 m, drowning continental shelves worldwide and creating coastal landforms intimately associated with rising sea levels.

Yet in spite of the worldwide and seemingly paramount importance of global sea level, we find modern examples of coastlines exhibiting all of the behaviors just listed: There are emerging coasts as well as submerging coasts, and advancing as well as retreating. For example, the Pacific coast of the United States is emergent, the Gulf of Mexico near the Mississippi River is submergent, and the southern Atlantic coast of the United States is close to neutral. Clearly, the local tectonic and geomorphic processes at work in different locations play a major role in determining the overall character of a coastline. In this chapter, we will discuss three broad classes of coasts, each characterized by a distinct set of dominant geomorphic/sedimentary processes:

- **Erosional** ("Pacific type"): high erosional energy relative to sediment supply
- **Elastic** ("Atlantic type"): high sediment supply relative to erosional energy
- **Carbonate**: coasts characterized by the formation of coral reefs

These three classes are associated with different assemblages of landforms, each recording the occurrence, the character, and the rate of active tectonic deformation in a given region.

COASTAL LANDFORMS

EROSIONAL COASTS

Bold erosional coasts have a characteristic morphology (Figure 6.3). At low tide, you can walk along the **wave-cut platform**, a wide and flat surface cut by the abrasive action of wave-churned sand, chemical weathering of the rock by seawater, and the action of intertidal organisms. Where organisms are the principal agent of erosion, such as tropical shorelines protected from wave action, the cut platform is called a "bioerosion platform" [2]. At the landward margin of a wave-cut or bioerosional platform is the **seacliff**. The seacliff is a geomorphic anomaly—the steep face is maintained only by the perpetual battering of high-tide and storm waves at its base. Long-term seacliff retreat makes cliff-top homes scenic but ephemeral real estate [3]. Wave-cut platforms typically have a natural upward slope away from the ocean, so that a given platform covers a fairly broad range of elevations (depths, actually) relative to sea level. The best uniform sea-level datum on a wave-cut platform is the **shoreline angle** (sometimes called the "inner edge" of the wave-cut platform), which is the line of intersection between the old platform and the associated seacliff (see Figure 6.3). In other settings, shoreline erosion of a durable rock may form a **wave-cut notch** at the base of the cliff, an overhung depression that is also a good indicator of sea level.

When a platform is uplifted above sea level, the landform becomes known as a **coastal terrace**. These features are also commonly referred to as "marine terraces", but because the processes that form them are truly coastal and not marine, the former des-

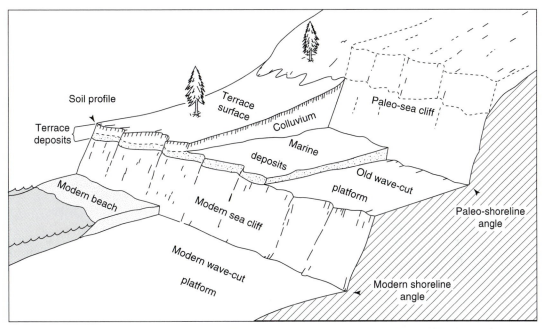

Figure 6.3.   General morphology of an erosional coastline. Active features include the modern wave-cut platform, beach, and seacliff. An older, uplifted, and now-inactive shoreline is recorded by the old wave-cut platform, its cover of marine sediment, and the eroded remnant of the old seacliff. Note that the shoreline angles, both modern and uplifted, are the only uniform reference elevations to sea level. (From Hanson et al., 1990. *Guidebook, Friends of the Pleistocene Fall Field Trip*.)

ignation may be considered preferable. The flat platform of an uplifted coastal terrace may, with luck, still contain fossils of its original inhabitants, and it generally will be covered with a thin veneer of loose nearshore sediments. The uplifted seacliff of a coastal terrace is no longer maintained by wave erosion, so it will degrade. Through time, the terrace becomes covered by colluvium from the old seacliff and by wash from farther upslope. This colluvial cover thickens inland and in places may range up to many tens of meters in thickness. Alternatively, old terraces may become stripped, with the terrace surface representing an erosional remnant degraded by some unknown amount. More typically, however, as a terrace ages, it is most aggressively erased by the incision and widening of stream valleys draining across the terrace surface.

Experience demonstrates that uplifted coastal terraces are powerful tools in the study of active tectonics, but the same experience also shows that considerable care and rigor must be exercised in the recognition, measurement, and utilization of coastal terraces. First, it is important to note that not all that is flat near a coastline is a coastal terrace. Various constructional and erosional processes can create horizontal or subhorizontal surfaces. In addition, any quantitative application of terrace elevations requires that a valid geomorphic datum be measured. On wave-cut coastal terraces, the best datum is the shoreline angle, the next best is the uplifted platform (with appropriate corrections), and the worst is the topographic surface of the terrace. Sadly, a review of the literature still finds studies utilizing terrace surface elevations with no adequate correction for the substantial and variable thickness of the marine and colluvial cover. The authors here have found that precise measurements must be made either *in outcrop*, such as where transverse stream valleys cut the terraces, or with appropriate subsurface control. Finally, the precision of terrace elevations varies substantially depending on the measurement technique employed in different studies and the datum to which those measurements were tied. Common approaches have included the use of EDMs, GPS positioning, corrected and uncorrected altimeter measurements, optical transits and levels, hand levels, and estimation from topographic maps. Equally important, measurements may have been tied to global frames of reference (such as with GPS), local geodetic benchmarks, or to "eyeball" estimates of present sea level—the last sometimes corrected for tides and/or wave run-up.

CLASTIC COASTS

Sandy barrier islands from Fire Island, New York, to Chincoteague, Maryland, to the Outer Banks of North Carolina, to Padre Island, Texas, are all examples of clastic coastlines. All are associated with ample sediment supply and shore-parallel transport (**longshore transport**) of sand along the beach. The particular nature of barrier islands, separated from the mainland by broad lagoons, is attributed to rising postglacial sea level [4]. Clastic coasts are associated with a variety of landforms, but the shore has a generalized profile (Figure 6.4) that reflects the fundamental processes of wave action and sediment movement both perpendicular and parallel to the shore.

Of the features of the shoreface profile, the **winter berm** (or "storm berm") is the "high-water mark" of a given shoreline. The berm is a topographically high, laterally continuous feature that may be preserved if the active shoreline shifts oceanward in position. Such a shift can be caused by: (1) an influx of sediment and progradation of the coast (Figure 6.5), (2) a relative fall in sea level (see discussion of sea level later in this chapter), or (3) uplift—either coseismic or gradual—of the land (Figure 6.6). Tectonic uplift of a clastic coast will tend to shift the beach profile upward and out of equilibrium with the forces that formed it. The upper portion of the profile, particularly the high winter storm berm, is removed from wave erosion and may be preserved as a **beach ridge** (sometime called a "strandline"), landward of and higher than the active berm that reforms. Inactive berms also can be formed by oceanward progradation of the shore (see Figure 6.5), but these ridges will be at approximately the same elevation as the ac-

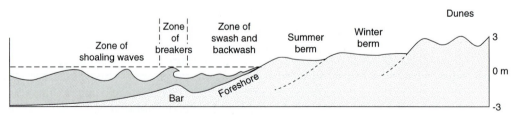

Figure 6.4. General morphology of a clastic shoreline. (After Easterbrook, 1999 [53].)

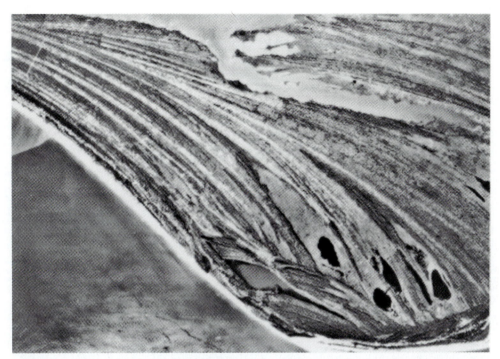

Figure 6.5.   Multiple prograded beach ridges developed at St. Vincent Island, Florida. Prograded beach ridges are formed by a massive influx of sediment, and not necessarily by either uplift or sea-level change.

tive berm. Because they are formed of unconsolidated sand at the surface, few pre-Holocene beach ridges are preserved.

CARBONATE COASTS

Coral reefs are constructed by biological activity in warm (>18°C), clear, oxygenated water (Figure 6.7) [4]. Modern theories of coral reef development trace back to Charles Darwin, who made many observations during the voyage of the *Beagle* from 1832 to 1835. He characterized different reefs around the Pacific, including circular "atolls", and suggested a hypothesis to explain their origin by subsidence of extinct island volcanoes and upward growth of the coral.

The growth of coral is closely tied to sea level (Figure 6.8). Active colonies can grow only as high as the lowest low tide. Reef corals live no deeper than permitted by the penetration of light, and most species live at depths of less than 25 m, with maximum growth shallower than 10 m [5]. Coral reefs are a complex ecosystem, and particular species—for example, *Acropora palmata*—predominate at shallow depths. In fossil reefs, these species are the best indicators of the position of sea level when the reef formed.

Some of the best documented examples of tectonic coastlines have uplifted coral reefs [6, 7]. Reefs have some characteristics that make them ideal for studying active

Figure 6.6. Emerged Holocene beach ridges on Norsholmen, Sweden. Ongoing uplift is driven by isostatic compensation for the melting of the ice cap that covered Scandinavia during the Pleistocene. Although the ice has been gone for several thousand years, isostatic compensation continues and causes very rapid uplift, at a rate of about 2 m/ka at this location. (Photo courtesy of Arne Philip, Visby, Sweden.)

tectonics. First, uplifted coral-reef surfaces are good timelines because the first vertical movement carrying the reef crest out of water kills the organisms and preserves the surface as a record of sea level at that moment. Second, the carbonate material of which coral reefs are constructed usually can be dated. Radiocarbon and uranium-series analyses are used to date uplifted reefs as old as 350 ka (for more details see the section titled "Dating Coastal Landforms" later in this chapter).

COSEISMIC DEFORMATION

Many of the large and great earthquakes of historical time have been accompanied by measurable deformation of the surface. In 1964, approximately 250,000 km^2 of the Alaskan coast experienced vertical deformation, with uplift of up to 10 m and subsidence of up to 2.4 m (see Figure 1.26) [8]. Such episodic pulses of uplift, summed over geologic time spans, are one mechanism by which the great topography of the Earth is

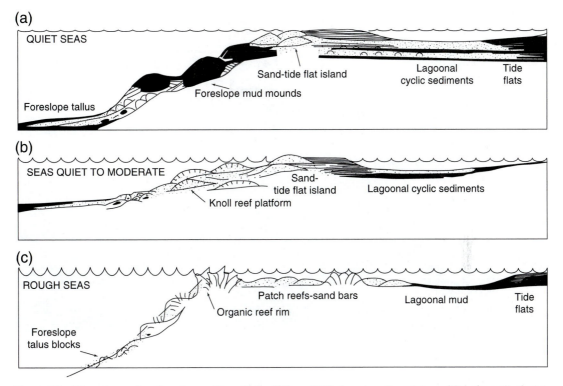

Figure 6.7.   Three types of carbonate coastlines. (After Wilson, 1974. *American Association of Petroleum Geologists Bulletin*, 58:811.)

created (more on this in Chapter 9). Where earthquakes strike near a coastline, it is often possible to measure the magnitude and pattern of the coseismic deformation by virtue of the nature of coastal landforms and their link to the global sea-level datum. The different types of coastlines provide a range of tools for studying coseismic vertical deformation:

- Tide-gauge data
- Mortality of intertidal organisms ("kill zones")
- Uplifted Holocene coastal landforms
- Tsunami deposits
- Submerged stratigraphy

TIDE GAUGES

It may be a bit counterintuitive, but the surface of an ocean is a very precise horizontal level. Although at any one time the water is tossed by waves, drawn in and out with the tides, and pushed about by currents and weather, when all these variations are averaged over a long period of time, the result is a fixed elevation precise to within millimeters. Where monitoring stations are set up to systematically measure the average level of the

Figure 6.8. A coral reef at low tide. Note that the upper limit of growth is approximately equal to the low-tide water level.

ocean surface at a point on the coast, at ports or at scientific stations, then it becomes possible to precisely determine the elevation of that point through time relative to sea level (Figure 6.9). These monitoring stations are known as **tide gauges**, and several hundred stations worldwide have a long enough duration of record and have proven sufficiently stable to provide geodetic-quality measurements of mean sea level [9].

Where an earthquake causes coseismic deformation near one of these tide gauges, the amount of vertical movement is recorded as an abrupt change in the elevation of the land relative to mean sea level. In fact, coseismic movement is one of the most straightforward applications of tide-gauge data. Long-duration tide-gauge records allow measurement of changes in global sea level as well as uplift of the land due to subduction, glacial unloading, and volcanism [9]. Like permanent GPS stations (see Chapter 3), tide gauges are especially valuable in documenting deformation because they are a continuous record of local sea level through time, capable of distinguishing between preseismic, coseismic, and postseismic movements.

There are, however, drawbacks to tide-gauge data. Coseismic deformation is local and variable in magnitude. Because tide gauges measure movement only at a few scattered points, such measurements alone (1) are only minimum estimates of the maximum uplift magnitude, and (2) provide no information regarding the pattern or distribution of uplift.

MORTALITY OF INTERTIDAL ORGANISMS

The uplift and death of organisms, such as after the 1992 event at Cape Mendocino described earlier, have also been observed after other subduction-zone earthquakes.

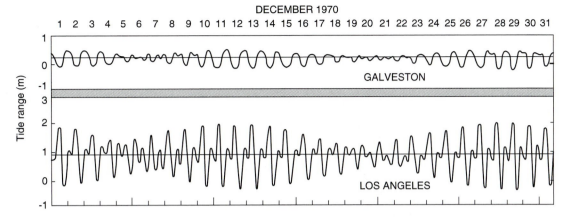

Figure 6.9.   Tidal curves for two stations along the coasts of the United States. Note that although the water level is quite variable, the average sea level can be determined with great accuracy. (After Emery and Aubrey, 1991 [9].)

Such kill zones have been used to measure the coseismic uplift associated with the 1964 Alaska earthquake [10], the 1985 Mexico City earthquake [11], the 1960 Chile earthquake [12], a magnitude 7.2 earthquake in New Guinea in 1992 [13], and others. The principal advantage of studying the mortality of intertidal organisms is that data are available over a wide area—wherever uplift coincides with the coastline—and the approach does not require a set of measurements that predate the deformation as geodetic techniques do.

Looking at the kill zone along the Mexican coast after the 1985 earthquake, Bodin and Klinger [11] defined the **vertical extent of mortality** (VEM) as the average height of mortality of attached intertidal organisms, measured by surveying equipment. Just as tide gauges average the short-term fluctuations of the ocean surface, organisms living in the intertidal zone are adapted to a particular position relative to sea level—exposed to air no more nor less than a certain portion of the time, and just out of reach of their fiercest predators. For example, mussels form a dense band along the intertidal zone of some rocky shores, a band with abrupt upper and lower boundaries. Any mussel that takes a position too high is stranded out of water too long and dies; any mussel that is too low quickly falls victim to hungry starfish inhabiting the subtidal zone below. Organisms that prove particularly useful in measuring VEM include species of corraline algae, barnacles, mussels, colonial anemone, surf grass, and sea lettuce [1]. After the Mendocino earthquake, mortality was evident within one to two weeks. Measurements taken after one month and again two months after the earthquake showed a slight increase in the VEM, but little more than the range of measurement uncertainty [1].

UPLIFTED HOLOCENE COASTAL LANDFORMS

Sequences of late Holocene coastal landforms—beach ridges, wave-cut notches, small terraces, or near-shore sedimentary packages—that rise upward and/or step landward from the active shoreline are characteristic of many tectonically active coasts [e.g., 14–19] (see Figure 8.8a). An important question about any such sequence is why it has the particular number and spacing of ridges, notches, and so forth that it has. Three mechanisms are possible: (1) small-scale fluctuations in Holocene sea level; (2) gradual uplift of the land, punctuated by severe storms; and (3) episodic coseismic uplift [15]. Given the variable number and ages of such uplifted Holocene coastal features and the lack of independent evidence of Holocene sea-level fluctuations, most researchers reject the first explanation. Although storms are a reasonable explanation for sandy beach ridges, the length of time necessary to form distinct reef or terrace features generally precludes that explanation for erosional or carbonate coasts. Where uplifted Holocene coastal features formed coseismically, they provide good information regarding the number of events and earthquake recurrence over time periods of hundreds to thousands of years. On the Boso Peninsula of Japan, an earthquake in 1703 lifted the active shoreline up and out of contact with the ocean. Three other paleo-shorelines step upward from the pre-1703 shore, the highest dated at about 6500 years before present (BP) [14]. These landforms reveal that at least four coseismic uplift events of the magnitude of the 1703 earthquake have occurred on the Boso Peninsula in the past 6500 years. Similarly, on the Huon Peninsula of New Guinea, careful dating of Holocene reef strata in a 52-m borehole documents six uplift events during the past 9000 years and confirms that long-term uplift in this famous terrace locility seems to be occurring during distinct coseismic pulses [20].

TSUNAMI DEPOSITS

As discussed in Chapter 1, the association between earthquakes and tsunami is well known. Several historical earthquakes have generated tsunami, with waves up to several meters in height and run-up onto the shore of up to a few tens of meters. Recently, it has increasingly become clear that tsunami-related deposits can be recognized in Holocene (and perhaps older) coastal deposits and used to document paleoseismic events [17, 21–27]. The on-rushing waters of a tsunami can carry sediment and debris several kilometers inland, depending on the topography of the coastal zone. These deposits are commonly preserved as sheets of sand that become thinner and finer inland, fine upward, and incorporate marine macro- and microfossils as well as rafted organic debris. Although tsunami sand sheets may be fairly pervasive in affected areas, the deposits are most commonly recognized as anomalous sandy units within finer-grained environments such as coastal marshes [e.g., 21]. On North Island, New Zealand, three fining-upward sedimentary packages widespread in near-coastal deposits are recognized as the result of three tsunami waves generated by a large earthquake, probably one that struck New Zealand in 1855 [27]. As is shown in the next section, tsunami deposits are playing a growing role in determining the occurrence, location, magnitude, and even precise timing of coastal-zone earthquakes in different locations.

COSEISMIC SUBSIDENCE

In most cases, downward movement of the land relative to the ocean is much more difficult to discern than upward movement. Subsidence covers the coastal landscape and deposits with water and sediment, hidden from the scrutiny of air-breathing geologists.

In contrast to the strike-slip motion on the San Andreas fault to the south and along the Pacific margin of Canada to the north, the western margin of the North Amer-

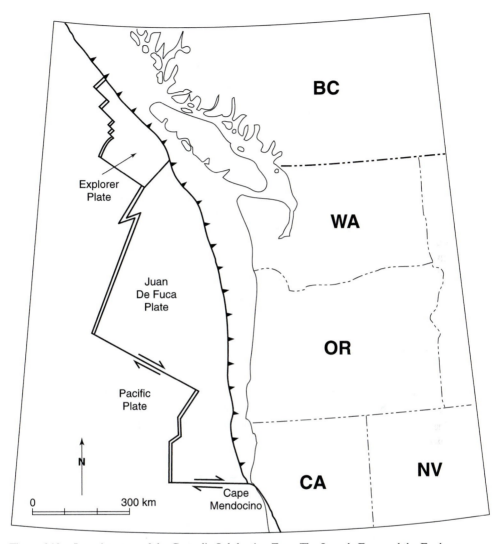

Figure 6.10.   Location map of the Cascadia Subduction Zone. The Juan de Fuca and the Explorer Plates are being carried beneath North America. In contrast, where the Pacific Plate is in direct contact with the North American Plate, south of Cape Mendocino and north of the Explorer Plate, the plate boundary is a transform fault. The April 25, 1992 earthquake discussed at the beginning of the chapter occurred near Cape Mendocino, in northern California.

ican Plate in northern California, Oregon, Washington, and southern British Columbia is characterized by convergent tectonics. In this area, called the **Cascadia Subduction Zone**, the small Juan de Fuca Plate (Figure 6.10) is converging with North America. Convergent plate boundaries are responsible for most of the largest earthquakes worldwide. Although it was demonstrated early on that subduction is indeed occurring in Cascadia [28], there have been no damaging earthquakes there in historical times. This area and this problem has developed into one of the most interesting paleoseismic puzzles in memory.

Is convergence along the Cascadia Subduction Zone being stored as accumulating strain on a locked plate-boundary megafault [29], or is strain being released through gradual aseismic slip [30]? The answer to this question has come in the form of the following evidence:

- Coastal marsh stratigraphy
- Tsunami deposits
- "Ghost forests"
- Lique faction features

Along the coast of the U.S. Pacific Northwest and the adjacent Canadian coast, generally fine-grained deposits in coastal marshes and similar environments record gradual emergence of the land from the ocean during late Holocene time [31, 32], forming stable coastal landforms capped by soil and vegetation. At widely separated localities, these stable surfaces were resubmerged beneath sea level, apparently catastrophically (Figure 6.11). During this resubmergence, marsh grasses were buried either by intertidal mud or, in some locations, by a layer of coarse sand interpreted as a tsunami deposit [21, 33]. Coastal forests were inundated, killing the trees and forming so-called "ghost forests" (see Figure 8.20). The stumps of those trees, still rooted in the soil from which they grew, were buried by marine sediments. Examination of the outermost growth rings of the trees shows that growth was normal during the last few years before the death of the trees, suggesting that submergence was sudden [33]. At all sites, submergence was followed by renewed emergence from the ocean, soil formation, and vegetation growth. Tree-ring analyses of the catastrophically submerged trees revealed that they died some time during the last years of the seventeenth century or the first years of the eighteenth century [34–36]. In addition, coeval liquefaction deposits (see Chapters 1 and 8) have been identified at a number of sites along the coast, further confirming the coseismic nature of the subsidence [37, 38]. All of this accumulating evidence is consistent with the occurrence of a very large earthquake on the Cascadia Subduction Zone that caused sud-

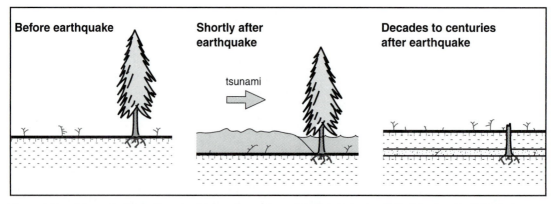

Figure 6.11.   Schematic illustration of submerged coastal stratigraphy along the Pacific Northwest of the United States and the southwestern coast of Canada. Such sequences are interpreted as coseismic submergence, resulting in the death of trees and other freshwater vegetation, accompanied by tsunami waves that deposit a sheet of sand. These deposits are then overlain by marine deposits that gradually shift back to terrestrial as the coast begins to reemerge from the ocean. (After Atwater et al., 1995 [31].)

den and dramatic inundation of the coastline around 1700 A.D., just prior to the arrival of the first Europeans to the area.

Evidence of widespread coseismic subsidence in Cascadia was somewhat surprising because historical earthquakes on other subduction zones around the world have resulted in more dramatic uplift than sinking [39]. Furthermore, the Cascadia margin is undoubtedly experiencing long-term uplift as recorded by Pleistocene coastal terraces and other features formed over longer periods of geologic time. The model that is now widely being embraced and is supported by mounting geodetic measurements [40, 41] is illustrated in Figure 6.12. Locking on the Cascadia plate boundary leads to flexure of the overriding plate, causing interseismic up-arching of the coast and sudden release when the next earthquake occurs. The subsidence is less than the cumulative uplift, so that the coast is gradually rising over multiple seismic cycles (see discussion of the **seismic deformation cycle** in Chapter 3). Coastlines near other subduction zones around the

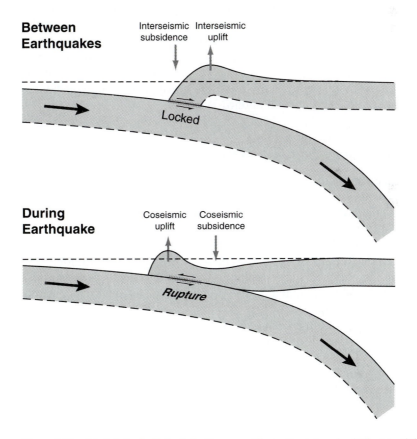

Figure 6.12.   Model of a locked subduction zone, illustrating interseismic flexure of the upper plate and coseismic deformation in the opposite direction. Most of the coastal zone of Oregon, Washington, and British Columbia lie in the area characterized by interseismic uplift and coseismic subsidence. (After Hyndman and Wang, 1993. *Journal of Geophysical Research*, 98:2039–2060.)

world include broader areas on the down-flexed margin of the plate—where the opposite pattern of interseismic subsidence and coseismic uplift predominates—but in Cascadia, such areas lie offshore so that this deformation pattern is not observed.

The last piece in the Cascadia earthquake puzzle is perhaps the most startling. One additional line of evidence supports not only the occurence of a great earthquake near the beginning of the eighteenth century, but specifies its time to the year, day, and hour [42]. Although there was no written history in the Pacific Northwest at the time to document such an event, meticulous public records were kept in Japan, including records documenting the occurrence of earthquakes and earthquake-related phenomena. Located at the western margin of the Pacific's "Ring of Fire," the Japanese were familiar with tsunami. Most tsunami that struck Japan were associated with earthquakes documented there in Japan, Korea, the Philippines, or other nearby regions ... with the exception of one tsunami that hit late in January of 1700. Other areas around the Pacific Rim, including the west coasts of Central and South America and the Kamchatka peninsula, were also keeping earthquake records at that time, and none of them document a large earthquake. What researchers believe they see is the tsunami generated by the last great Cascadia earthquake, a tsunami recorded by deposits at many sites in Oregon and Washington that also traversed the Pacific Ocean. Furthermore, the travel times of tsunami waves across the Pacific are well known, and given the time of arrival in Japan, it has been calculated that the Cascadia earthquake struck on January 26, 1700 at approximately 9 P.M. Pacific Standard Time [42].

COASTAL GEOMORPHOLOGY AND SEA LEVEL

No discussion of coastal geomorphology or movement of the land relative to the ocean can be separated from the dramatic fluctuations in sea level that have occurred during at least the last 2 million or so years. Changes in sea level (**eustatic** change) during this interval primarily reflect the growth and decay of the great ice sheets that repeatedly covered most of North America north of 40° latitude and northern Europe. These ice sheets reached thicknesses of up to 3000 m [4], storing up to 100 billion cubic meters of water [43] on the continents and depleting the world's oceans proportionally. At the time of the last glacial maximum about 18 ka, global sea level was approximately 125 m lower than it is today [6].

Coastal landforms can only record changes in the position of the land relative to the position of sea level. The number of coastal features, their spacing, and their character are a function of both tectonic movement and sea-level change. In order to say anything about the history of tectonics, we must first know something about the history of sea level. A full discussion of how paleo–sea levels are derived and what their implications are is far beyond the scope of this book; instead, we will focus on a few aspects of sea-level research directly relevant to coastal geomorphology and active tectonics.

The best records of sea level come, not from stable coastlines, but from coasts experiencing steady long-term uplift. Sea-level change on stable coastlines repeated itself during the glacial advances and interglacial retreats of the Pleistocene ice sheets, destroying evidence of older highs and lows. On an uplifting coasts, though, shorelines were progressively lifted out of the range of eustatic change. Sea-level data have been

Figure 6.13.   Uplifted coral reef terraces at the Huon Peninsula, New Guinea. Relatively rapid fluctuations of global sea level during the Pleistocene have been superimposed on long-term tectonic uplift of the New Guinea coast. Each terrace that you can see in the photo indicates a former position of the shore. Because coral can be dated, each terrace also gives the age at which the shore was at that position. (Photo courtesy of Arthur L. Bloom.)

collected around the world, but among the most complete records are in Barbados [7] and New Guinea [6] (Figure 6.13). At both of these sites, coral reefs formed at approximately the times of sea-level high stands and were uplifted by long-term tectonic activity. The agreement between these two sites, located at completely different areas of the globe, has led to widespread acceptance of the nature and timing of Pleistocene sea-level changes.

In addition to the timing of past sea-level events, it is important to know the elevations of the different high stands and low stands. In Barbados and New Guinea, this information was obtained by calculating the long-term uplift rate. For a reef or other feature formed at sea level, its present elevation is equal to

$$\text{elevation} = (\text{uplift rate}) \cdot \text{time} + (\text{sea level}), \qquad \textbf{(6.1)}$$

where **time** refers to the age of the feature and **sea level** to the level of the ocean surface at that time, measured in distance above (+) or below (−) modern sea level. By assuming that the sea level of the last interglacial reef (125 ka old) was about 6 m higher than modern sea level (an assumption supported by data from stable, nontectonic coasts),

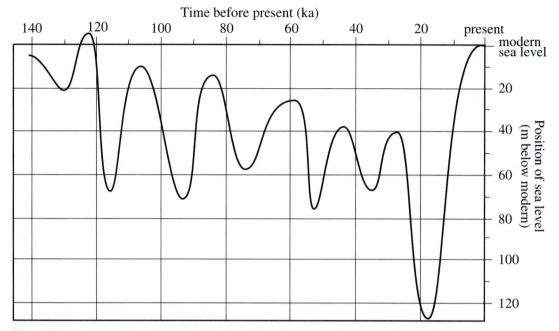

Figure 6.14. A graphical representation of global sea level during the past 140 ka. This sea-level model utilizes the data from New Guinea coral reefs (Bloom et al., 1974. *Quaternary Research* 4:185–205) and an analysis technique modified after Pinter and Gardner (*1989, Geology*, 17:295–298). The formula for calculating this model numerically is outlined in Table 6.1.

it became possible to calculate the uplift rates for the sites at Barbados and New Guinea and then to determine the position of sea level at the time of the formation of each reef. A summary of the New Guinea data is shown in Figure 6.14, as well as a mathematical expression to estimate sea level back to about 125 ka (Table 6.1).

Uplifted coral reefs are not the only features used to infer ancient sea levels. The slow, steady buildup of sediment on the deep ocean floor and of ice atop the Greenland and Antarctic ice caps records variations in naturally occurring isotopes of oxygen and carbon that are approximately synchronous with the coral-reef eustatic signal. Meteoric water (rainfall and snowfall) is depleted in the heavy isotope of oxygen (^{18}O) relative to ocean water; therefore, the great accumulations of continental ice sheets that lowered sea level during the Pleistocene concentrated ^{18}O in the world's oceans [44]. A different sea-level driving mechanism operates for carbon. The light isotope of carbon (^{12}C) is concentrated in plants and other organic matter compared with its heavy stable isotope (^{13}C). Low sea levels expose the shallow continental shelves to erosion, flushing large amounts of light carbon into the oceans [44]. In fact, these oxygen and carbon isotope records provide a much more detailed history than do coral-reef sequences, suggesting a continuous record of sea-level changes through time. But the isotope histories are necessarily *indirect* measures of sea level. For example, the oxygen isotope ratio also is affected by the average temperature of the oceans as well as the latitude in which most precipitation falls; it is very difficult to translate a definite isotopic value into a

Table 6.1

SEA LEVEL (SL) IS CALCULATED AS FOLLOWS: SL = AT^4 + BT^3 + CT^2 + DT + E.

Age Range (in ka)	CF	A	B	C	D	E
0–18	0	-0.005026	0.137217	-0.447917	0	0
18–28	0	0	0.1895	-12.993	283.554	-1871.9
28–35	0	-0.0336253	4.11303	-187.787	1796.45	-28655.8
35–45	30	0	0.0745	-2.1525	15.938	32.81
45–53	30	-0.0227332	1.60311	-41.513	471.838	-1957.77
53–58	50	0	0.906	-14.784	64.242	-6.63
58–74	50	-0.00278963	0.168136	-3.48014	30.763	-73.04
74–85	70	0	0.0812547	-2.24076	14.026	34.05
85–93	70	-0.0231131	1.55582	-37.982	402.962	-1564.75
93–107.5	90	0	0.0478658	-1.41498	7.197	62.35
107.5–116.5	90	-0.0159186	1.26071	-36.4263	459.548	-2138.6
116.5–122.5	110	0	0.731019	-20.6965	176.398	-402.92
122.5–130	110	-0.027282	1.67703	-37.8731	375.51	-1389.59
130–142	110	0	0.0282407	-2.13406	51.472	-379.26

In each interval, the age t (in ka) is modified by a correction factor (CF): t - CF = T.

particular sea-level elevation [45]. Uplifted coastal features are much more direct indicators of paleo–sea level.

LONG-TERM UPLIFT

Long-term vertical movement of the land is the sum of all coseismic, interseismic, and aseismic movement averaged over a long period of time (these terms are formally defined in the discussion of the **seismic deformation cycle** in Chapter 3). Long-term rates of uplift can, under favorable conditions, be determined on clastic, carbonate, or erosional coasts. Favorable conditions consist of coastal landforms that are (1) durable and long-lived and (2) capable of being dated.

As a general rule, clastic coasts are the least likely to meet these two criteria. Beach ridges are loose and unconsolidated when they form, and any datable material originally incorporated into the deposits decays rapidly in this porous and unprotected setting. A good example of the limitations of and possibilities for using clastic coastal features for studying long-term uplift comes from the Atlantic Coastal Plain of the United States. The Coastal Plain has been slowly emerging from the Atlantic Ocean through the Pliocene and Pleistocene, leaving a path of old beach and dune ridges, escarpments, and marine-terrace remnants behind the retreating coast [46]. The shoreline features are traceable over long distances and, in places, can be correlated with the underlying marine stratigraphy to provide a rough chronology of their emergence. The slow but steady emergence of the Atlantic Coastal Plain is understood to be a result of erosion of the North American continent, deposition on the Atlantic continental shelf, and the isostatic response [47].

In general, though, carbonate and erosional coasts tend to preserve a clearer picture of long-term uplift. The coral-reef sequences of Barbados and New Guinea are classic examples of emergent coastlines, reflecting the combined effects of long-term uplift and Pleistocene sea-level fluctuations. Flights of erosional coastal terraces, such those along the Pacific coast of North America, reflect the same processes at work.

Where such sequences of uplifted shorelines are used to infer former sea levels, a rate of uplift must be assumed; where the shorelines are used to infer the uplift rate, global sea-level history must be assumed (for example, Figure 6.14). Uplift rate can be either calculated using Equation 6.1 or determined graphically, as shown in Figure 6.15. The graphical solution combines sea-level history plotted against time, the ages of preserved shoreline features, and the elevations at which those features are currently found. For example, a terrace formed 85 ka ago when global sea level was 15 m below its current level (point A in Figure 6.15) is now found 88 m above modern sea level (point C at elevation B in Figure 6.15). The line connecting points A and B is the **uplift path** for that terrace. The uplift paths for all the terraces preserved in the same sequence should be parallel, assuming that the sequence was uniformly uplifted. The average slope of the uplift paths is the uplift rate [48].

Where many terraces can be identified in a single sequence, and where their elevations can be precisely determined, the uplift rate can be determined without dating every terrace in the sequence. Using the principles of the graphical correlation method illustrated in Figure 6.15, the **relative spacing** of the terraces is used to assign ages to each terrace in the sequence and an uplift rate to the site [48]. Extrapolating age estimates beyond the limits of available age control using the relative-spacing technique requires that one assume that the uplift rate in a given area has remained constant over time. The relative-spacing technique provides some internal confirmation when a reasonable solution has been reached, but the assumption of constant uplift rate is difficult

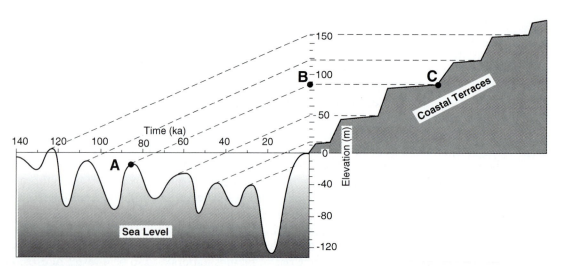

Figure 6.15.   Graphical determination of the rate of uplift for a sequence of coastal terraces (the terraces are shown on the upper right of the figure). The terraces (elevations at time = 0) are correlated to sea-level high stands during the Late Pleistocene. The lines connecting the modern elevation of each terrace to the time and position at which it formed is its "uplift path," the slope of which is the rate of uplift.

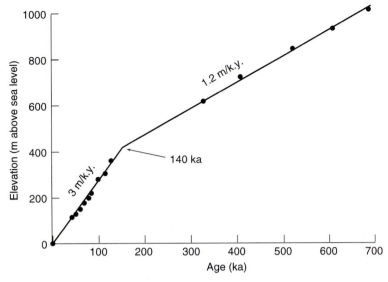

Figure 6.16. Plot of uplift versus time for a sequence of coastal terraces at Fiordland, New Zealand. Terraces formed prior to 135 to 140 ka BP indicate uniform uplift at a rate of 1.2 m/ka, and terraces formed later indicate a rate two to three times more rapid. (After Bishop, 1991. *Special Publications of the International Association of Sedimentology*, 12:69–78.)

to test independently because coastal features older than the late Pleistocene often prove very difficult to date. It is certainly fair to say that the character and order of magnitude of earlier deformation is in general agreement with late Pleistocene uplift rates at many locations [e.g., 49, 50], but quantitative continuity is difficult to prove. Conversely, three coastal-terrace sequences at widely separated sites along the Alpine fault in New Zealand were used to document that uplift rates can and do change over time (Figure 6.16) [49]. Prior to about 140 ka ago, uplift was uniform (1.2 m/ka). However, between 140 and 135 ka, the uplift rate doubled to its present value of 5 to 8 m/ka. This change is interpreted to reflect accelerated convergence between the Australian and Pacific Plates [51].

Long-term rates of uplift worldwide, measured over intervals of tens of thousands to hundreds of thousands of years, are greatest at convergent plate boundaries, where sustained uplift rates of up to 6 to 10 m/ka have been calculated [52]. Transform boundaries show moderate to low uplift, generally less than 0.3 m/ka, and intraplate coasts show no movement to very low rates of uplift associated with erosion and isostatic rebound (see Chapter 9). The fastest uplift rates are ephemeral events related to temporary disequilibria in the Earth's crust—for example, **postglacial rebound** (uplift due to melting ice caps and the resulting isostatic compensation) and **volcanic tumescense** (uplift caused by magma-chamber inflation beneath a volcano). Whereas volcanic tumescense is a short-lived phenomenon of weeks to months, glacial rebound has occurred during the past 18 ka, uplifting Hudson Bay by over 250 m and still lifting Scandinavia at rates of up to 1 m per century (Figure 6.17) [53].

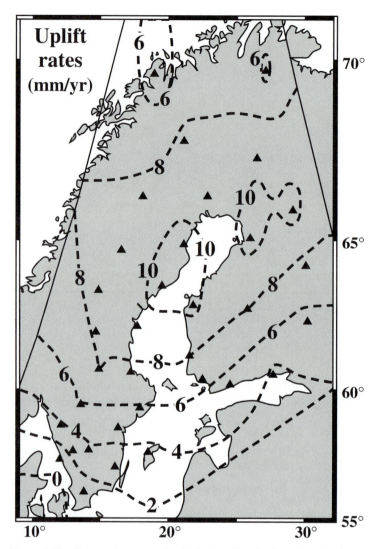

Figure 6.17.   Present-day rates of uplift in Scandinavia, determined by geodetic GPS positioning. The highest uplift rates coincide with the center of Pleistocene Fennoscandinavian ice sheet, suggesting that modern uplift reflects continuing isostatic compensation to postglacial unloading. (After Milne, et al., 2001. Science, 291: 2381–2385)

DEFORMATION OF COASTAL TERRACES

In addition to the information that coastal terraces and other coastal landforms provide regarding uplift, either coseismic or long-term, these features can also reveal patterns and sometimes rates of tectonic deformation of the surface. Most shoreline features make good timelines, having emerged from the geomorphic maelstrom of the coastal zone in an instant of geologic time. Furthermore, the shoreline angle of a coastal terrace and the crest of a beach ridge form in close association with sea level. Together these two

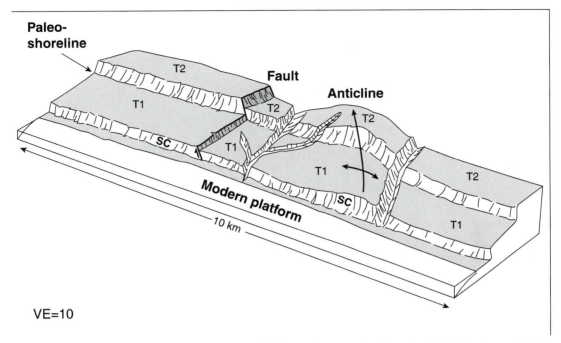

Figure 6.18.   Faulting, warping, and erosion of an uplifted coastal terrace after formation. (After Trecker et al., 1998 [65].)

facts make coastal landforms very sensitive indicators of deformation, including both faulting and warping (Figure 6.18).

Faults are as likely to cut the surface of a coastal terrace as they are to cut any other landform. Unlike on most of the Earth's surface, however, it is sometimes possible to assign an age to coastal terraces. If a fault cuts the surface of a given terrace, then the most recent movement on the fault is known to *postdate* the age of the terrace [54]. The amount of vertical motion on the fault is best estimated by measuring the displacement on the wave-cut platform, which is subject to neither the erosion nor the mantling that may affect the surface of the terrace. On the other hand, when the trace of a fault crosses a dated terrace but offsets neither the surface nor the platform, then regardless of any quantity of displacement on the underlying bedrock, the latest movement on the fault must *predate* the age of the terrace. For example, examination of coastal terraces along the Ionian coast of southern Italy confirms that one fault (the Castrovillari fault) has moved during late Pleistocene time, that another fault (the Pollino fault) has not moved, and that a previously unmapped structure exists and is active [55].

Because uplifted coastal landforms often can be traced or correlated over long distances, variations in elevation can be used to quantify variations in the rate of uplift. Along the Atlantic coast of the United States, the same traceable Pleistocene strandline ranges in altitude from less than 50 m to about 100 m (Figure 6.19) in a systematic pattern reflecting a wave of postglacial rebound that has migrated northward during the past 18 ka [46]. On the Osa Peninsula of Costa Rica, Late Pleistocene beach ridges and near-shore marine deposits are uplifted by amounts that decrease with distance away from

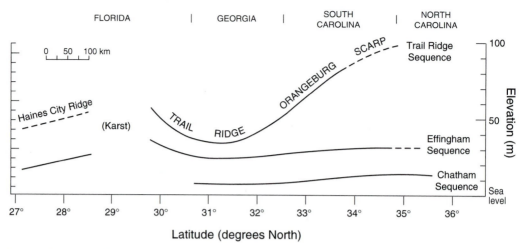

Figure 6.19. Elevations of strandlines along the Atlantic coastal plain south of the Cape Fear River, North Carolina. Strandlines reflect ongoing uplift in the Cape Fear area. The magnitude of recorded uplift increases with increasing age of the landforms. (From Winker and Howard, 1977. *Geology*, 5:123–127.)

the Middle America Trench and merge with active subsidence in the gulf behind the peninsula [56]. Such landward tilting, across zones ranging in width from 30 to 50 km, seems to be a characteristic process in the outer arcs of many subduction zones [14].

LAKE SHORELINES

It should also be noted that most tectonic analyses of coastal landforms can be used for the shores of lakes as well as of oceans. The Basin and Range province of the United States, for example, is characterized both by repeated formation of Pleistocene **pluvial lakes** and by ongoing tectonic activity. Although lake formation and lake levels are primarily functions of climate, lake shorelines have been used as timelines to date fault movement [e.g., 57] and as horizontal benchmarks to determine vertical deformation [e.g., 58]. Some of the groundbreaking work in this area was done by G. K. Gilbert, a founding father of modern geology. Gilbert recognized that the shorelines of Pleistocene Lake Bonneville, which formed as horizontal lines, are now substantially deformed [59]. In the eastern half of the Basin and Range, Pleistocene Lake Lahontan covered an area of 22,500 km^2, encompassing about 2100 km^3 of water and covering what is now desert floor to depths up to 270 m [60]. Mapping and measurement of the highest shorelines of Lake Lahontan, which formed about 13 ka, revealed that this paleo-horizontal datum now varies by as much as 22 m (Figure 6.20). Most of this deformation resulted from isostatic compensation of the crust to the removal of the water load, but an additional component of regional down-to-the-north tectonic tilt is evident [60]. In addition, the 13 ka shorelines are shown to be cut by some faults that cross the area and are uncut by others, documenting the degree of activity on these structures over this span of time.

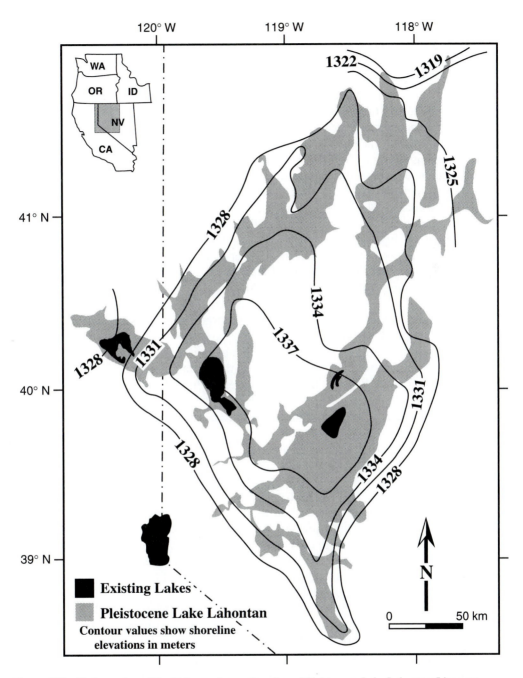

Figure 6.20.   Deformation of the 13 ka maximum shoreline of Pleistocene Lake Lahontan. Lines are contours (values indicate meters elevation above sea level) of the shoreline, which, of course, was a horizontal surface at the time of formation. The present-day relief on the remains of this shoreline is primarily the result of isostatic response to the removal of the water that filled the lake. A smaller component of additional, possibly tectonic deformation is also suggested. (After Adams et al., 1999 [60].)

DATING COASTAL LANDFORMS

Like all studies in active tectonics, work on coastal landforms requires absolute dates in order to calculate the rates of uplift or deformation. Fortunately, coastal landforms sometimes incorporate material, such as shells and coral, that can be dated by one or more of several possible methods. Quaternary dating methods are outlined in detail in Appendix A of this book, but the application of some of these techniques specifically to marine and coastal deposits will be reviewed briefly here. Depending on the nature and age of the material incorporated into a coastal landform or deposit, one or more of the following dating techniques may be applicable: radiocarbon, uranium-series analyses, amino acid racemization, and thermoluminescence dating.

The most straightforward coastal landform for dating is an uplifted coral reef. Unaltered coral as old as 350 ka can be dated using the uranium-series method [e.g., 61]. In fact, the technique is considered sufficiently precise that it has been used as an independent calibration of the radiocarbon method [62]. Dating other coastal features, such as wave-cut terraces and beach ridges, depends largely on the likelihood that carbonate material has been incorporated into the feature or into deposits contemporaneous with it. For example, the marine veneer that usually overlies coastal terraces may contain solitary corals that can also be dated using uranium-series [63]. The uranium-series technique also should be able to date late Quaternary mollusc shells, but experience has shown that shell carbonate almost invariably acts as an open system with respect to uranium, incorporating new uranium from the environment and invalidating the resulting age determinations in most cases [64]. An alternative approach for dating young shell material (or corals) is radiocarbon analysis, which utilizes the carbon in calcium carbonate to determine ages back, at least theoretically, to the maximum limits of radiocarbon at 40,000+ years. When dating material of marine origin, a correction of a few hundred years typically is necessary in order to account for the reservoir effect of the Earth's oceans, in which modern carbon appears isotopically old because of century-scale ocean mixing times, but this correction is straightforward and any supplementary analysis (measurement of the ratio of ^{13}C to ^{12}C) is inexpensive.

A note of caution must be added, however, to the dating of shell material near or beyond the limits of radiocarbon dating. Growing anecdotal evidence suggests that minute contamination of old shells with small amounts of younger carbon results in false age determinations in many cases. The authors of this book attempted, as a control, to use accelerator mass spectrometry (AMS) radiocarbon analysis on shells from California coastal terraces at least 400 ka in age. Instead of radiocarbon "dead" results, however, these samples yielded results of just over 40 ka. Subsequent scanning electron microscopy determined that this shell material was indeed recrystallized and therefore open to contamination, but additional AMS ^{14}C dating of *unaltered* shells from a terrace dated by uranium-series at 125 ka also yielded ages in the 35 to 40 ka range. Theoretical calculations show that a very old carbon sample, even millions of years old, will yield a "finite" age of 38 ka as a result of just 1.0% contamination by modern carbon. Apparently, shells may be susceptible to this level of contamination. Additional calculations show, however, that the same 1.0% contamination will perturb a 20-ka sample only by about 800 years and a 10-ka sample by less than 200 years, meaning that ^{14}C dating of young shells remains a viable approach.

Additional approaches to dating Pleistocene coastal terraces and other coastal features include paleontology, stable-oxygen isotopic analyses, and morphological dating. Among some fossil shell assemblages, the variety of organisms preserved may contain either (1) index fossils that help constrain the age of the deposit, or (2) species or varieties indicating conditions either warmer or colder than the present, and thereby may suggest a landform age [15]. True index fossils are rare in Quaternary coastal deposits, but assemblages may contain species that allow loose age discrimination, such as Early versus Middle versus Late Pleistocene. A new and promising approach to dating uplifted coastal deposits involves using the ratio of stable isotopes of oxygen (^{18}O and ^{16}O) to help differentiate features of different age. This isotopic ratio is the same one documented in microfossils from deep marine cores, and similar variations over time show up in shallow-water shells in terrace deposits. By determining the oxygen-isotope ratios of shells from both dated and undated coastal terraces, it was found that the undated deposits could be calibrated to one or the other of the dated features [65]. Finally, systematic processes of erosion may permit morphological dating of coastal landforms by comparing the degree of degradation (see discussion of morphological dating in Chapter 8). At the least, this last method is a useful tool for correlating from dated coastal landforms to undated equivalents in the same region [51, 66].

In the case of lakes and lake-related deposits, many of the techniques used to date marine features, such as uranium-series and radiocarbon analysis, can be used as well. Isolated lakes, however, may have different baseline geochemical and/or isotopic conditions than in the much larger reservoir of the world's oceans. In addition, techniques are available for dating lake levels that do not exist for the oceans. For example, pack-rat middens (the nest and waste at the bottom of pack-rat dwellings) provide minimum ages for the extent of lakes in the Basin and Range—because rats don't live under water, the oldest material accumulated in their nests must postdate the time that a lake retreated and exposed the site to air [67]. Other methods used to estimate the ages of different lake levels include measuring the development of rock varnish and soils.

COASTAL TECTONICS AND TIME SCALE

This chapter has discussed a number of geomorphic features of coasts that are formed by or are useful in studying active tectonics. The particular types of landforms preserved and the types of information they can yield are functions of the time scale over which the formative processes have operated. Figure 6.21 summarizes this information.

Processes that operate over geologically short scales of time—for example, coseismic pulses of uplift or catastrophic storms—leave their mark on the coastal area as beach ridges, wave-cut notches, or zones of uplifted intertidal organisms. Larger and more regional features, such as coastal- and coral-terrace sequences, are the results of many smaller incremental events repeated over time or the results of processes that operate over time spans of several thousand to tens of thousands of years, such as glacial-interglacial sea-level changes. Finally, the end product of sustained tectonic and geomorphic activity, the sum of both short-term and medium-term processes repeated through geologic time, is construction of whole physiographic regions, from mountain chains to rift valleys. In time, the individual landforms that were the record of short

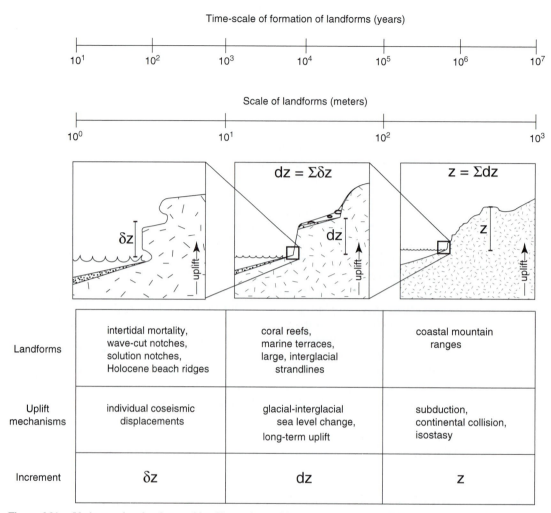

Figure 6.21. Variety and scale of coastal landforms formed by tectonic activity at different time scales. Processes that operate at longer scales of time are generally the sum of processes that operate at short time scales.

time-scale events are obliterated by erosion and are preserved only as anonymous increments in the construction of the topography of the Earth.

SUMMARY

At the boundary between land and water, coastlines form a distinct group of landforms. The particular landforms that occur depend on the sediment supply and the history of sea-level rise or fall relative to the land. Through time, coastal landforms may be deformed by tectonic activity. They can be uplifted, downdropped, faulted, or warped. Coastal landforms are particularly useful tools for studying tectonics for two reasons: (1) sea level acts as an absolute frame of reference, and (2) it is often possible to determine the age of coastal features. In addition, coastal landforms illustrate very clearly the role of time scale in landscape evolution—processes that act over short periods of time blur together over longer and longer time intervals and create more regional elements of the landscape.

REFERENCES CITED

1. Carver, G. A., A. S. Jayko, D. W. Valentine, and W. H. Li, 1994. Coastal uplift associated with the 1992 Cape Mendocino earthquake, northern California. *Geology*, 22:195–198.

2. Fischer, R., 1980. Recent tectonic movements of the Costa Rican Pacific coast. *Tectonophysics*, 70:T25–T33.

3. Norris, R. M., 1990. Sea cliff erosion: a major dilemma. *California Geology*, 43:171–177.

4. Bloom, A. L., 1998. *Geomorphology: a Systematic Analysis of Late Cenozoic Landforms*, 3rd ed. Englewood Cliffs, NJ: Prentice Hall.

5. Stoddart, D. R., 1969. Ecology and morphology of recent coral reefs. *Cambridge Philospohical Society Biological Review*, 44:433–498.

6. Bloom, A. L., W. S. Broecker, J. M. A. Chappell, R. K. Matthews, and K. J. Mesolella, 1974. Quaternary sea level fluctuations on a tectonic coast: New ^{230}Th/^{234}U dates from the Huon Peninsula, New Guinea. *Quaternary Research*, 4:185–205.

7. Bender, M. L., R. G. Fairbanks, F. W. Taylor, R. K. Matthews, J. G. Goddard, W. S. Broecker, 1979. Uranium-series dating of the Pleistocene reef tracts of Barbados, West Indies. *Geological Society of America Bulletin*, 90:577–594.

8. Eckel, E. B., 1970. The Alaska earthquake, March 27, 1964—lessons and conclusions. *U.S. Geological Survey Professional Paper* 546.

9. Emery, K. O., and D. G. Aubrey, 1991. *Sea Levels, Land Levels, and Tide Gauges*. New York: Springer-Verlag.

10. Plafker, G., 1965. Tectonic deformation associated with the 1964 Alaska earthquake. *Science*, 148:1675–1687.

11. Bodin, P., and T. Klinger, 1986. Coastal uplift and mortality of intertidal organisms caused by the September 1985 Mexico earthquakes. *Science*, 233:1071–1073.

12. Castilla, J. C., 1988. Earthquake-caused coastal uplift and its effects on rocky intertidal kelp communities. *Science*, 242:440–443.

13. Pandolfi, J. M., M. R. B. Best, and S. P. Murray, 1994. Coseismic event of May 15, 1992, Huon Peninsula, Papua new Guinea: Comparison with Quaternary tectonic history. *Geology*, 22:239–242.

14. Ota, Y. 1986. Marine terraces as reference surfaces in late Quaternary tectonics studies; examples from the Pacific rim. In W. I. Reilly and B. E. Harford (eds.), Recent crustal movements of the Pacific region, *Bulletin—Royal Society of New Zealand*, 24:357–375.

15. Lajoie, K. R., D. J. Ponti, C. L. Powell II, S. A. Mathieson, and A. M. Sarna-Wojcicki, 1991. Emergent marine strandlines and associated sediments, coastal California; a record of Quaternary sea-level fluctuations, vertical tectonic movements, climatic changes, and coastal processes. In R. B. Morrison (eds.), *Quaternary Nonglacial Geology: Conterminous U.S., DNAG*, vol. K-2. Boulder, CO: Geological Society of America, 190–214.

16. Pirazzoli, P. A., J. Ausseil-Badie, P. Giresse, E. Hadjidaki, and M. Arnold, 1992. Historical environmental changes at Phalasarna harbor, West Crete. *Geoarchaeology*, 7:371–392.

17. Berryman, K. R., 1993. Age, height, and deformation of Holocene marine terraces at Mahai Peninsula, Hikurangi Subduction margin, New Zealand. *Tectonics*, 12:1347–1364.

18. Merritts, D., 1996. The Mendocino Triple Junction: active faults, episodic coastal emergence, and rapid uplift. *Journal of Geophysical Research*, 101:6051–6070.

19. Chappell, J., Y. Ota, and K. R. Berryman, 1996. Holocene and late Pleistocene coseismic uplift of Huon Peninsula, Papua New Guinea. *Quaternary Science Reviews*, 15:7–22.

20. Chappell, J., Y. Ota, and C. Campbell, 1998. Decoupling post-glacial tectonism and eustasy at Huon Peninsula, Papua New Guinea. In I. S. Stewart and C. Vita-Finzi (eds.), *Coastal Tectonics*. Geological Society of London, Special Publications, 146:31–40.

21. Clague, J. J., and P. T. Bobrowsky, 1994. Evidence for a large earthquake and tsunami 100–400 years ago on western Vancouver Island, British Columbia. *Quaternary Research*, 41:176–184.

22. Ota, Y., P. A. Pirazzoli, T. Kawana, and H. Moriwaki, 1985. Late Holocene coastal morphology and sea-level records on three small islands, the South Ryukyus, Japan. *Geographical Review of Japan*, 58B:185–194.

23. Moore, J. G., W. B. Bryan, and K. R. Ludwig, 1994. Chaotic deposition by a giant wave, Molokai, Hawaii. *Geological Society of America Bulletin*, 106:962–967.

24. Minoura, K., S. Nakaya, and M. Uchida, 1994. Tsunami deposits in a lacustrine sequence of the Sanriku coast, Northeast Japan. *Sedimentary Geology*, 89:25–31.

25. Bryant, E. A., R. W. Young, and D. M. Price, 1996. Tsunamis as a major control on coastal evolution, southeastern Australia. *Journal of Coastal Research*, 12:831–840.

26. Young, R. W., E. A. Bryant, and D. M. Price, 1996. Catastrophic wave (tsunami?) transport of boulders in southern New South Wales, Australia. *Zeitschrift für Geomorphologie*, 40:191–207.

27. Goff, J. R., M. Crozier, V. Sutherland, U. Cochran, and P. Shane, 1998. Possible tsunami deposits from the 1855 earthquake, North Island, New Zealand. In I. S. Stewart and C. Vita-Finzi (eds.), *Coastal Tectonics*. Geological Society of London, Special Publications, 146:353–374.

28. Riddihough, R. P., and R. D. Hyndman, 1976. Canada's active western margin—the case for subduction. *Geoscience Canada*, 3:269–278.

29. Savage, J. C., M. Lisowski, and W. H. Prescott, 1981. Geodetic strain measurements in Washington. *Journal of Geophysical Research*, 86:4929–4940.

30. Ando, M., and E. I. Balazs, 1979. Geodetic evidence for aseismic subduction of the Juan de Fuca plate. *Journal of Geophysical Research*, 84:3023–3028.

31. Atwater, B. F., et al., 1995. Summary of coastal geologic evidence for past great earthquakes at the Cascadia subduction zone. *Earthquake Spectra*, 11:1–18.

32. Clague, J. J., and P. T. Bobrowski, 1999. The geological signature of great earthquakes off Canada's west coast. *Geoscience Canada*, 26:1–15.

33. Atwater, B. F., and D. K. Yamaguchi, 1991. Sudden, probably coseismic submergence of Holocene trees and grass in coastal Washington State. *Geology*, 19:706–709.

34. Atwater, B. F., M. Stuiver, and D. K. Yamaguchi, 1991. Radiocarbon test of earthquake magnitude at the Cascadia Subduction Zone. *Nature*, 353:156–158.

35. Nelson, A. R., et al., 1995. Radiocarbon evidence for extensive plate-boundary rupture about 300 years ago at the Cascadia Subduction Zone. *Nature*, 378:371–374.

36. Yamaguchi, D. K., G. C. Jacoby, B. F. Atwater, D. E. Bunker, B. E. Benson, M. S. Reid, and C. Woodhouse, 1997. Tree-ring dating of an earthquake at the Cascadia subduction zone to within several months of January 1700. In *Proceedings of the First Joint Meeting of the U.S.-Japan Conference on Natural Resources (UJNR) Panel on Earthquake Research*. U.S. Geological Survey Open-File Report, 97-0467:143–150.

37. Atwater, B. F., 1994. Geology of liquefaction features about 300 years old along the lower Columbia River at Marsh, Brush, Price, Hunting, and Wallace Islands, Oregon and Washington. *U.S. Geological Survey Open-File Report*, 94–209.

38. Obermeier, S. F., 1995. Preliminary estimates of the strength of prehistoric shaking in the Columbia River valley and the southern half of coastal Washington, with emphasis for a Cascadia subduction zone earthquake of about 300 years ago. *U.S. Geological Survey Open-File Report*, 94–589.

39. West, D. O., and D. R. McCrumb, 1988. Coastline uplift in Oregon and Washington and the nature of Cascadia subduction-zone tectonics. *Geology*, 16:169–172.

40. Dragert, H., and R. D. Hyndman, 1995. Continuous GPS monitoring of elastic strain in the northern Cascadia subduction zone. *Geophysical Research Letters*, 22:755–758.

41. Khazaradze, G., A. Qamar, and H. Dragert, 1999. Tectonic deformation in western Washington from continuous GPS measurements. *Geophysical Research Letters*, 26:3153–3156.

42. Satake, K., K. Shimazaki, Y. Tsuji, and K. Ueda, 1996. Time and size of a giant earthquake in Cascadia inferred from Japanese tsunami records of January 1700. *Nature*, 378:246–249.

43. Selby, M. J., 1985. *Earth's Changing Surface: An Introduction to Geomorphology*. Oxford, U. K.: Clarendon Press.

44. Broecker, W. S., 1982. Glacial to interglacial changes in ocean chemistry. *Progress in Oceanography*, 11:151–197.

45. Pinter, N., and T. W. Gardner, 1989. Construction of a polynomial model of sea level: estimating paleo-sea levels continuously through time. *Geology*, 17:295–298.

46. Winker, C. D., and J. D. Howard, 1977. Correlation of tectonically deformed shorelines on the southern Atlantic coastal plain. *Geology*, 5:123–127.

47. Colquhoun, D. J., G. H. Johnson, P. C. Peebles, P. F. Huddlestun, and T. Scott, 1991. Quaternary geology of the Atlantic Coastal Plain. In R. B. Morrison (ed.), *Quaternary Nonglacial Geology: Conterminous U.S., DNAG*, vol. K-2. Boulder, CO: Geological Society of America, 629–650.

48. Bull, W. B., 1985. Correlation of flights of global marine terraces. In M. Morisawa and J. T. Hack (eds.), *Tectonic Geomorphology*. The Binghamton Symposia in Geomorphology: Internat. Series, No. 15. Boston: Allen & Unruh, 129–152.

49. Bull, W. B., and A. F. Cooper, 1986. Uplifted marine terraces along the Alpine Fault, New Zealand. *Science*, 234:1225–1228.

50. Orme, A. R., 1998. Late Quaternary tectonism along the Pacific coast of the Californias: A contrast in style. In I. S. Stewart and C. Vita-Finzi (eds.), *Coastal Tectonics*. Geological Society of London, Special Publication, 146:179–197.

51. Bishop, D. G., 1991. High-level marine terraces in western and southern New Zealand: indicators of the tectonic tempo of an active continental margin. *Special Publications of the International Association of Sedimentology*, 12:69–78.

52. Lajoie, K. R., 1986. Coastal tectonics. In R. E. Wallace (ed.), *Active Tectonics*. Washington, DC: National Academy Press, pp. 95–124.

53. Easterbrook, D. J., 1999. *Surface Processes and Landforms*, 2nd ed. New York: Macmillan.

54. Keller, E. A., 1988. Estimating timing of fault activity on uplifted wave-cut platforms. *Bulletin of the Association of Engineering Geologists*, 25:505–507.

55. Cucci, L., and F. R. Cinti, 1998. Regional uplift and local tectonic deformation recorded by the Quaternary marine terraces on the Ionian coast of northern Calabria (southern Italy). *Tectonophysics*, 292:67–83.

56. Gardner, T. W., D. Verdonck, N. Pinter, R. L. Slingerland, K. P. Furlong, T. F. Bullard, and S. G. Wells, 1992. Quaternary uplift astride the aseismic Cocos Ridge, Pacific coast, Costa Rica. *Geological Society of America Bulletin*, 104:219–232.

57. Hanks, T. C., and R. E. Wallace, 1985. Morphological analysis of the Lake Lahontan shoreline and beachfront fault scarps, Pershing County, Nevada. *Bulletin of the Seismological Society of America*, 75:835–846.

58. Bills, B. G., and G. M. May, 1987. Lake Bonneville; constraints on lithospheric thickness and upper mantle viscosity from isostatic warping of Bonneville, Provo, and Gilbert Stage shorelines. *Journal of Geophysical Research*, 92:11,493–11,508.

59. Gilbert, G. K., 1890. Lake Bonneville. *U.S. Geological Survey Monograph*, 1.

60. Adams, K. D., S. G. Wesnousky, and B. G. Bills, 1999. Isostatic rebound, active faulting, and potential geomorphic effects in the Lake Lahontan basin, Nevada and California. *Geological Society of America Bulletin*, 111:1639–1756.

61. Edwards, R. L., J. H. Chen, T. L. Ku, and G. J. Wasserburg, 1987. Precise timing of the last interglacial period from mass spectrometric determination of thorium-230 in corals. *Science*, 236:1547–1553.

62. Bard, E., B. Hamelin, R. G. Fairbanks, and A. Zindler, 1990. Calibration of the ^{14}C timescale over the past 30,000 years using mass spectrometric U-Th ages from Barbados corals. *Nature*, 345:405–409.

63. Stein, M., G. J. Wasserburg, K. R. Lajoie, and J. H. Chen, 1991. U-series ages of solitary corals from the California coast by mass spectrometry. *Geochimica et Cosmochimica Acta*, 55:3709–3722.

64. Kaufman, A., W. S. Broecker, T. L. Ku, and D. L. Thurber, 1971. The status of U-series methods of mollusk dating. *Geochimica et Cosmochimica Acta*, 35:1155–1183.

65. Trecker, M. A., L. D. Gurrola, and E. A. Keller, 1998. Oxygen-isotope correlation of marine terraces and uplift of the Mesa hills, Santa Barbara, California, USA. In I. S. Stewart and C. Vita-Finzi (eds.), *Coastal Tectonics*. Geological Society of London, Special Publication, 146:57–69.

66. Anderson, R. S., A. L. Densmore, and M. A. Ellis, 1999. The generation and degradation of marine terraces. *Basin Research*, 11:7–19.

67. Benson, L. V., D. R. Currey, R. I. Dorn, K. R. Lajoie, C. G. Oviatt, S. W. Robinson, G. I. Smith, and S. Stine, 1990. Chronology of expansion and contraction of four Great Basin lake systems during the past 35 ka. *Palaeogeography, Palaeoclimatology, Palaeoecology*, 78:241–286.

7

Active Folding and Earthquakes

INTRODUCTION

Folds in layered rock form some of the most beautiful geologic structures on Earth (Figure 7.1). Folds are found in a variety of shapes and sizes, from microscopic to regional warps hundreds of kilometers in length. Folds are described by a few simple terms (Figure 7.2). Each fold consists of two **limbs**, and the limbs join at the **hinge** of the fold. A fold hinge is defined as the line that connects points of maximum curvature along a fold. In a fold consisting of several rock layers, the surface that connects the different hinge lines is known as the **axial surface**.

Folds are classified into several different categories, depending on the orientation of their limbs. In the simplest case, where young rock layers overlie older layers, an **anticline** is a fold in which the limbs slope down away from the hinge, and a **syncline** is a fold in which limbs slope toward the hinge (Figure 7.2). Anticlines and synclines are the two most basic and most common types of folds. Three additional complications also need to be considered; Figure 7.3 illustrates folds that are overturned, recumbent, and asymmetric. In an **overturned fold**, the axial surface slopes such that both limbs dip in the same direction. In a **recumbent fold**, the axial surface is horizontal or nearly horizontal. Finally, in an **asymmeteric fold**, one of the limbs dips significantly more steeply than the other.

In Figures 7.2 and 7.3, all of the hinge lines for the folds are horizontal; this is seldom the case in nature. When the hinge line is inclined, we say that the fold **plunges**. A

Figure 7.1 Folded rocks associated with the San Andreas fault near Palmdale, California. (Photograph courtesy of A. G. Sylvester.)

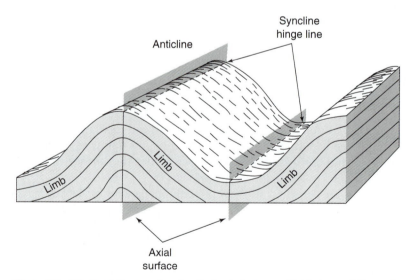

Figure 7.2 Idealized diagram showing the basic features of folds (an anticline and syncline). (After Skinner and Porter, 1989. *The Dynamic Earth*. New York: John Wiley & Sons.)

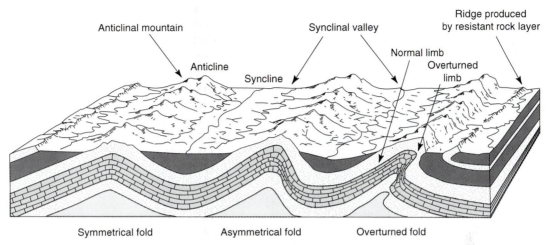

Figure 7.3   Block diagram illustrating several types of common folds, with possible surface expressions such as anticlinal mountains and synclinal valleys. (After Lutgens and Tarbuck, 1992. *Essentials of Geology*. New York: Macmillan.)

block diagram illustrating the concept of a plunging fold is shown in Figure 7.4a. Folds on geologic maps are represented by the hinge lines, as well as the **strike** (the compass direction of the intersection of the rock layer with a horizontal plane) and the **dip** (the maximum angle that the rock layer makes with the horizontal) of rock layers that crop out at the surface. A simple example of a geologic map is shown on Figure 7.4c. The hinge line and strike and dip symbols on the map show an anticline plunging to the northwest. The fold is slightly asymmetrical, with the northeast limb dipping more steeply than the northwest limb.

Geologic structure can be evaluated by constructing geologic cross sections (Figure 7.4d), which aid in the interpretation of structures such as folds and faults. Subsurface control may come from well logs and geophysical data. Interpretation of cross sections of folds and faults is facilitated by using specific models to explain folding and to construct quantitative **balanced solutions** to cross sections. Such cross sections balance (maintain a constant total length and thickness of rock units) before and after deformation [1, 2]. Other models have been developed to allow for deformation with changes in thickness of rock units [3]. Solutions to cross sections can be tested by **retrodeformation**, which removes the deformation, restoring the section to geologic conditions prior to folding and faulting. Another purpose of retrodeformation is to ensure that interpretation makes sense—to make sure that there are no voids or overlaps in the undeformed section. For example, if one interprets the cross section in a particular geologic setting to be a tapered wedge of sediments that have been folded and faulted, then retrodeformation may confirm that the balanced section is viable. Comparison of deformed to retrodeformed conditions allows the amount of shortening due to folding and faulting to be estimated. If the ages of the deformed rocks are also known, then rates of tectonic processes can be estimated [4].

Construction of balanced cross sections is becoming a powerful tool for evaluating deformation and rates of folding and faulting. Having said that, it is important to

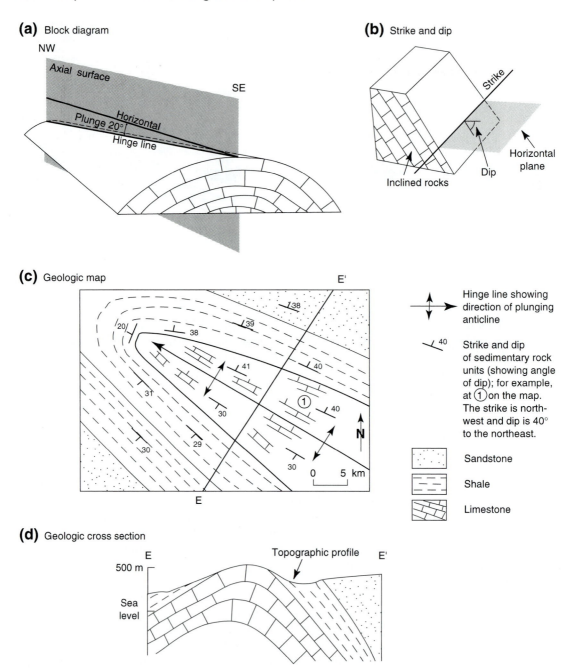

Figure 7.4   Idealized block diagrams of (a) a plunging fold, (b) strike and dip, (c) geologic map of plunging fold, and (d) geologic cross section.

understand that the use of balanced cross sections to produce models of subsurface geological structures (i.e., folds and faults) is controversial. There are serious issues related to this form of modeling, including the following:

- There may not be sufficient availability of field, well log, and geophysical data to ensure accuracy of cross-section models.
- Strike-slip motion may make it difficult to produce viable cross sections.
- The models may infer the existence of faults that, lacking surface deformation, may be difficult to evaluate.
- The models do not produce unique solutions (several solutions are possible that balance the cross section).
- Methods of deriving rates of tectonic processes from the models depend upon assumptions inherent in the models.

Nevertheless, models produced by construction of balanced cross sections are an exciting area of research that provides an important new dimension in understanding active tectonic processes.

FOLD-AND-THRUST BELTS

Folds are seldom found in isolation. Rather, folds are often found as belts of anticlines and synclines. Folds in an individual belt may plunge in one direction or another, and the hinge lines tend to be long relative to the width of individual folds. In addition, the fold belts themselves tend to be long and relatively narrow features. Folding of sedimentary layers is directly related to compressional stress and crustal shortening. Three principal causes of crustal shortening are as follows:

- Plate motion at convergent plate boundaries
- Rotation of crustal blocks that cause one block to converge with another
- Strike-slip faulting where restraining bends and steps occur (see Chapter 2)

Shortening related to plate motion at a convergent plate boundary is illustrated in a cross section of the Cascadia Subduction Zone and associated fold-and-thrust belt of the Olympic Mountains (Figure 7.5). Farther south, in Northern California, the fold-and-thrust belt bends and comes onshore near Eureka, where anticlines form hills and small mountains parallel to the faulting, while synclines are topographically low areas such as Humboldt Bay and Big Lagoon (review Figure 2.22 and accompanying discussion in Chapter 2).

Of particular importance to fold belts is the fact that the folds are commonly accompanied by reverse faulting. Many of these reverse faults are low-angle (less than 45°) and are called **thrust faults**, so that the resulting structural domain is called a **fold-and-thrust belt**. Some of the thrust faults and high-angle reverse faults may break the surface, but many others remain hidden within the cores of anticlines and are termed **buried reverse faults**.

Buried active faults present a significant earthquake hazard—that large damaging earthquakes can occur on faults located entirely beneath or within folded rocks. The tips of such faults may be buried at depths of several kilometers, and when they rupture during earthquakes, uplift and folding occur at the surface [5]. The recognition of buried

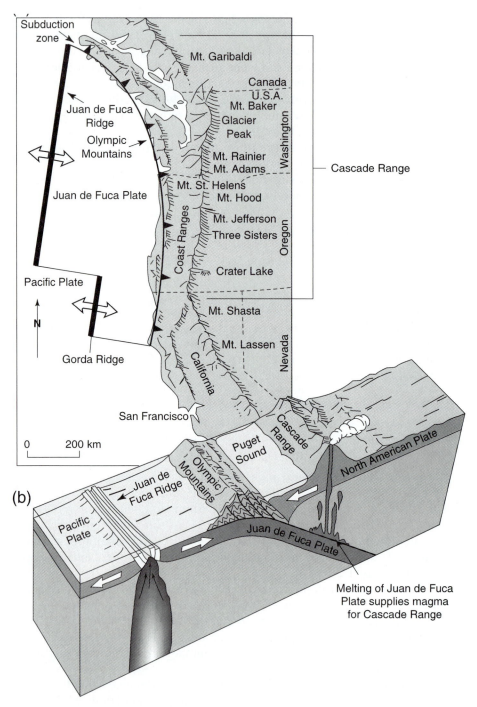

Figure 7.5 Cascadia Subduction Zone. The fold-and-thrust belt forms the Olympic Peninsula, just west of Puget Sound. (From Coch and Ludman, 1991. *Physical Geology*. New York: Macmillan.)

reverse faults and their accompanying earthquake hazard has resulted from the observation of several earthquakes in California and other areas that occurred on buried faults in anticlinal folds, without accompanying surface rupture. These events include the 1983 M_W 6.5 Coalinga earthquake, which caused $31 million in property damage; the 1987 M_W 5.9 Whittier earthquake, which claimed 8 lives and caused $358 million in damage; and the 1994 M_W 6.7 Northridge (Los Angeles) earthquake, which killed 61 people and caused more than $40 billion in damage.

Other large earthquakes in fold-and-thrust belts produce surface rupture as well as uplift and folding. For example, the 1980 M_W 7.3 El Asnam, Algeria, earthquake that killed 3500 people was produced by 3 to 6 m of slip on a reverse fault several kilometers below the surface [5]. Ground rupture (as much as 2 m) reached the surface over a significant part of the fault, but the greatest deformation at the surface was anticlinal uplift of approximately 5 m. Figure 2.24 shows a schematic drawing of the fold as well as a topographic profile showing the uplift produced by the earthquake. The good correlation between the shape of the topographic uplift and the shape of the fold suggests that faulting during this earthquake and other events like it does produce folds [5].

The 1980 El Asnam earthquake produced a variety of near-surface deformations (Figure 7.6) [6]. The dominant deformation was hanging-wall folding (folding above the fault plane; Figure 7.6a, b, d) but also included a variety of extensional features produced by a component of left-lateral shear, such as grabens (Figure 7.6d), tension fractures (Figure 7.6c), and elongated en echelon depressions (Figure 7.6a). Deformation also included foot-wall folding (folding beneath the fault plane) and flexural-slip faulting (Figure 7.6b), which is discussed later in this chapter. During the earthquake, the uplift and faulting blocked the Ech Cheliff River, producing a lake upstream of the epicentral area. The river flowing into the lake deposited silt particles, producing a layer approximately 0.4 m thick. Small tectonic dams on rivers are not permanent features, and when the dam is breached, the silt deposits remain as evidence that a lake existed upstream of the fold. Excavations in the El Asnam area revealed that there have been six such lakes in the past 6 ky. Of course, additional earthquakes may have occurred but did not produce lakes. Smaller earthquakes might not cause sufficient uplift, and if a flood occurred very soon following the earthquake, then the dam might have been too short-lived for appreciable silt to have been deposited [5, 6].

As a result of earthquakes in the past 20 years on buried faults, intensive study was begun on active folds to learn more about the relationship between buried faults and folds as well as the earthquake hazard that these structures present. For example, one of these belts occurs along the southern flank of the San Gabriel Mountains extending into the Los Angeles Basin. Buried faults present a significant earthquake hazard to the millions of people who live in the Los Angeles Basin [7, 8] and have produced several damaging earthquakes in the last 25 yr, including the 1994 M_W 6.7 Northridge earthquake, which caused damage to or destruction of thousands of buildings (Figure 7.7) and collapse of several freeway overpasses. The geometry of the Northridge earthquake is shown in Figure 7.8. The fault that caused the earthquake is inferred to be a reverse fault dipping about 40° to the south. Rupture was initiated at a depth of about 18 km and quickly propagated upward (northward) up to a depth of about 4 km and laterally (mostly westward) about 20 km at a rate of about 3 km per second. The earthquake deformed the Earth's crust over an area of about 4000 km^2, producing a broad doming above the

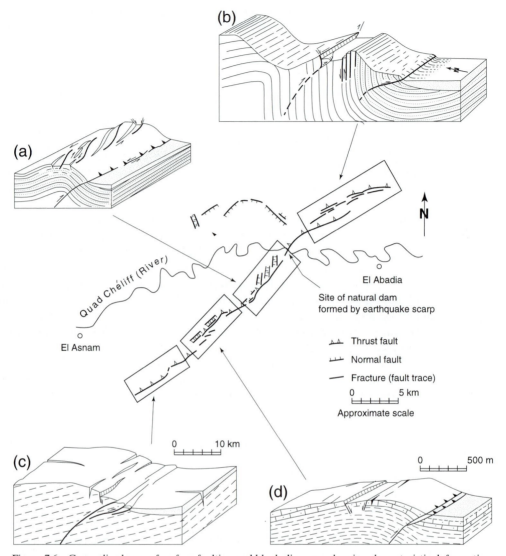

Figure 7.6 Generalized map of surface faulting and block diagrams showing characteristic deformation produced by the October 10, 1980, El Asnam, Algeria, earthquake in North Africa. (After Philip and Meghraoui, 1983 [6].)

buried fault. The faulting and folding uplifted part of the Santa Susana Mountains (a few kilometers north of Northridge) about 40 cm and moved them 22 cm to the northwest. Maximum fault displacement (slip) on the fault plane was about 3 m [9]. Two other Southern California fold-and-thrust belts (Figure 7.9) are located south of Bakersfield, on the north flank of the western Transverse Ranges, and near Ventura, on the southern flank of the ranges; these two belts are discussed in detail at the end of this chapter.

In order to understand relationships between faulting, folding, and fold-and-thrust belts, geologists have developed models for predicting the behavior of folding as the re-

Figure 7.7   Damage to a parking structure at California State University, Northridge, produced by the 1994 Northridge earthquake. (Photograph courtesy of F. Hopson.)

sult of buried faults. Several fold models have been suggested [10], but it is important to recognize that models are only approximations of how the Earth works.

FAULT-PROPAGATION FOLDS

Fault-propagation folds are folds that form on the upper plate (**hanging wall**) of a thrust or reverse fault. Remember that a thrust fault is a reverse fault with a dip of less than 45°; the terms are loosely used and often are interchangeable (see Chapter 1). Fault-propagation folds result from deformation in front of a propagating fault [4]—that is, the tip of the fault is advancing the same way that a newly formed crack advances across a car's windshield. Fault propagation folding at the tip of a thrust fault is idealized in Figure 7.10a. The propagating fault tip (time 1 to time 3) diverges from a **décollement**, or **detachment fault**, that runs along a weak stratigraphic horizon [4] and steps up to form a fault ramp. Dashed lines are axial surfaces of the developing fold. Rocks below a décollement are often relatively undeformed compared to rocks above the décollement. As the fold develops, it becomes more asymmetric. Figure 7.10b shows a fault-propagation anticline (the Meilin anticline) that is part of a fold-and-thrust belt in western Taiwan, deforming Pleistocene strata. An important observation associated with fault-propagation folding is that thrust faults and folding are intimately related. Thrust

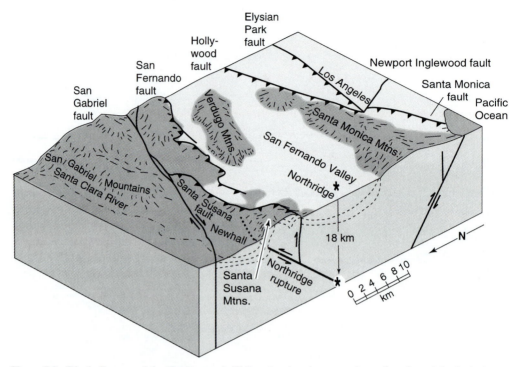

Figure 7.8 Block diagram of the San Fernando Valley showing the approximate location of the fault that generated the 1994 (M_W 6.7) Northridge earthquake. Solid lines are strike-slip faults, and barbed lines are reverse or thrust faults, some of which, such as the Elysian Park and Hollywood faults, are buried. (Courtesy of P. Williams and P. Holland, Lawrence Berkeley Laboratory.)

faults commonly die out in folds that terminate a short distance from the end of the fault [4]. Furthermore, the rate of lateral propagation of the fault and fold may be several times greater than the vertical slip rate of the fault [11, 12].

FAULT-BEND FOLDS

Another inferred geometry from which slip on buried reverse faults can create folds is **fault-bend folding**. Like the fault-propagation model, the reverse fault in a fault-bend fold is inferred to consist of several planar panels separated by sharp *kinks* (Figure 7.11a). As material in the hanging wall crosses the first kink, it begins to be uplifted as a result of the dip of the fault ramp beneath. Continued slip on the fault increases the distance between neighboring kink bands (Figure 7.11b) until slip equals the width of the ramp, and the growing fold reaches its equilibrium shape (Figure 7.11c). Such flat-topped anticlines were described in 1934 by J. L. Rich [13], while interpreting low-angle thrust faulting of the Cumberland fault block in Virginia, Kentucky, and Tennessee. The geometry above the thrust fault for the Cumberland fault block is an anticline with a long, flat top, bordered by much shorter limbs. The fault-bend-fold geometry is extremely useful because total fault slip can be inferred from the distance between kink bands,

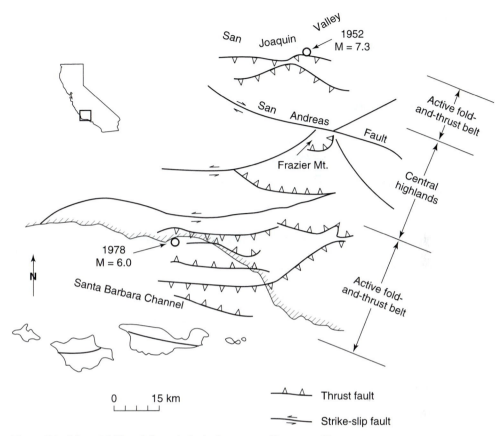

Figure 7.9   Map of fold-and-thrust belts in the western Transverse Ranges.

but alternative geometries have been proposed that do not require the underlying re-verse fault to be everywhere perfectly planar [14].

ROLLOVER FOLDS

Another type of folding is also associated with changes in dip of a buried normal fault. Displacement on normal faults that bend and become flatter with depth (**listric faults**) are also associated with folding on the downthrown (hanging wall) block of the fault [4, 15]. The term for such folding is **rollover** and the phenomenon is common in the Gulf Coast region and the Colorado Plateau [15], as well as in other areas associated with ex-tensional tectonics. The progressive development of a rollover fold along the Hurricane fault is shown in Figure 7.12. Folding had already started when extrusion of Quaternary age basalt (Q_{b1}, Figure 7.12a) covered the fault. Subsequently, the basalt was displaced (faulted) by about 400 m and folded (Figure 7.12b), and a younger basalt (Q_{b2}) was ex-truded and is also faulted and folded (Figure 7.12c and d).

Development of the rollover anticline is clearly related to collapse of the down-thrown (hanging wall) block with time. Older strata are more strongly folded than the

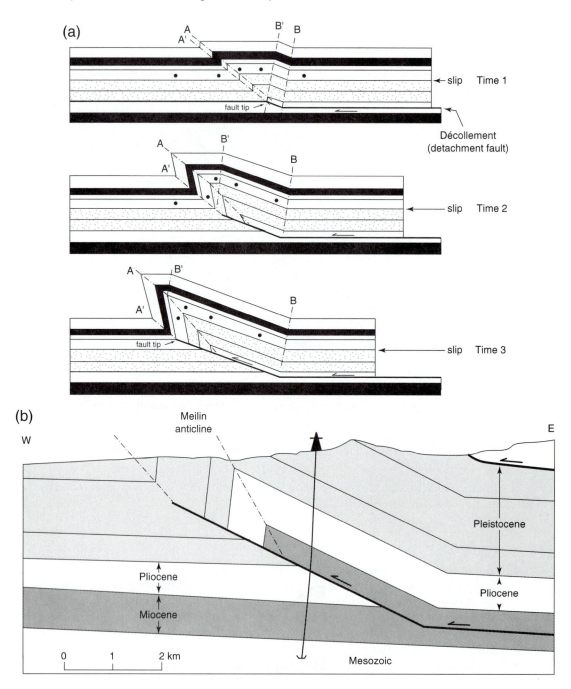

Figure 7.10 (a) Development of a fault-propagation fold as a result of slip along a décollement and a steeper thrust fault. (b) Fault-propagation fold and the Meilin anticline in western Taiwan. (From Suppe, 1985 [4].)

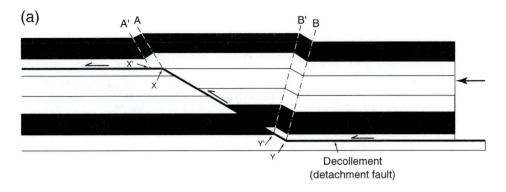

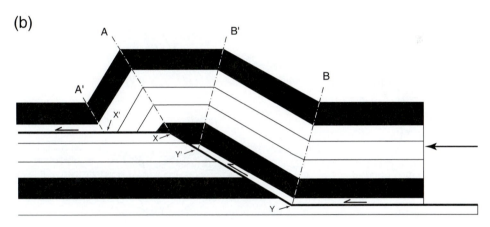

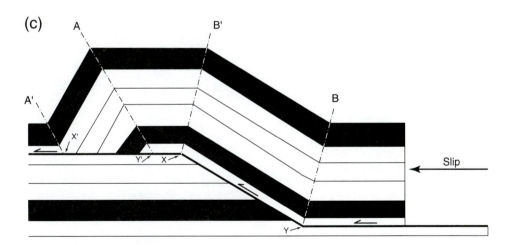

Figure 7.11   Progressive development (time 1 to time 3) of a fault-bend anticline. (After Suppe, 1983 [2].)

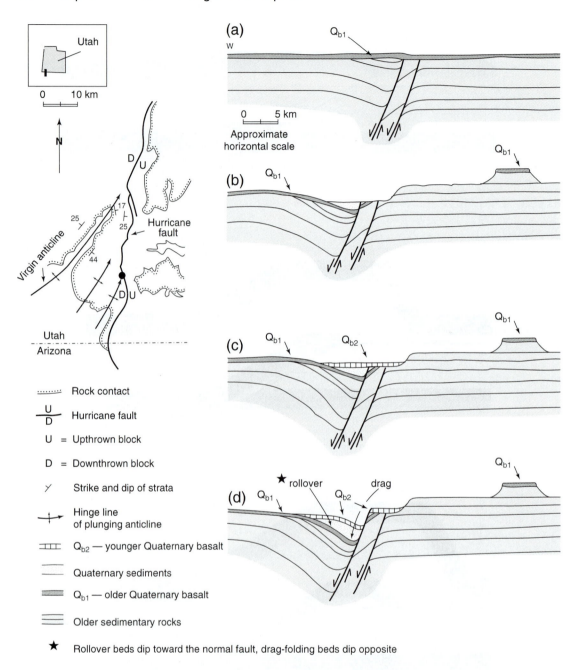

Figure 7.12   Progressive development of a rollover fold by collapse of the downthrown block (hanging wall) on the Hurricane normal fault, Utah and Arizona. (After Hamblin, 1965 [15].)

Quaternary basalt, indicating that faulting and folding occurred at the same time. Notice in Figure 7.12 that while the rollover anticline has beds dipping toward the fault, reversal of dip occurs at the fault due to drag. There is hot debate as to whether or not listric normal faults are capable of generating large earthquakes. In the Basin and Range province, listric faults are interpreted from seismic-reflection profiles, but to date all large normal earthquakes have been on steeply dipping fault planes. If large, low-angle normal faults are capable of generating large earthquakes, then the earthquake hazard of many regions has been grossly underestimated [16]! Of course, this is exactly what was done with respect to buried reverse faults in the Los Angeles area and other regions before the significance of active folding and the faults they conceal was recognized. Our historical record of large earthquakes on normal faults is too short to answer this important question.

Folds may also form over a buried normal fault, as idealized in Figure 7.13. Such folds are called **drape folds**. A good example of a young and active drape fold occurs on the east flank of the Sierra Nevada of southern Spain, near Granada. Normal faults are common along the range front, and several displace Quaternary deposits [17]. A drape fold is present near the toe of a prominent alluvial fan. Fan gravels generally dip 3° to 4°, but steepen to 10° to 12°, forming a scarp several tens of meters in height. This simple structure is a **monocline**, defined by steeply inclined strata bounded by more gently inclined strata. The pattern of deformation in the fan gravels suggests the presence of a

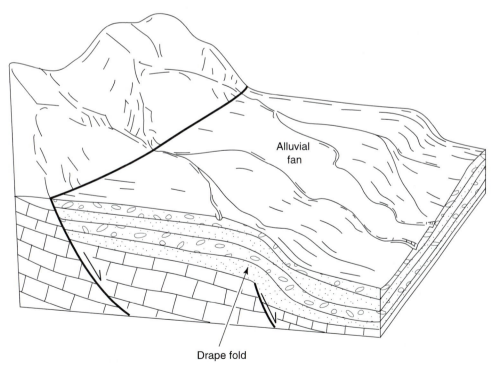

Drape fold

Figure 7.13   Idealized block diagram showing the development of a drape fold on alluvial-fan gravels over a buried normal fault.

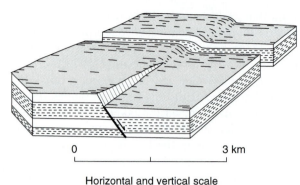

Figure 7.14 Block diagram showing how a normal fault can change laterally into a monocline, which is a simple fold characterized by a local steepening of an otherwise generally gentle dip. (After Skinner and Porter, 1989. *The Dynamic Earth*. New York: John Wiley & Sons.)

0   3 km

Horizontal and vertical scale

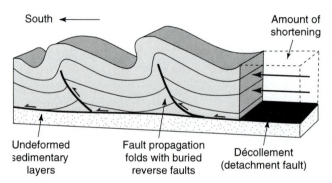

South ◄————

Amount of shortening

Undeformed sedimentary layers

Fault propagation folds with buried reverse faults

Décollement (detachment fault)

Figure 7.15 Asymmetric fault-propagation folds developed over a décollement.

buried normal fault below the alluvial fan. Figure 7.14 is a block diagram showing how displacement along a normal fault might change laterally to a monocline.

FOLD-AND-THRUST BELTS: SELECTED PROCESSES

Faulting with displacement along a décollement (detachment fault) is an important process in fold-and-thrust belts. For example, the décollement shown in Figure 7.15 displaces earth material in the opposite direction as the dip of the faults. Two asymmetric anticlines, formed by buried reverse faults, accommodate the shortening. The two fault-propagation folds in Figure 7.15 **verge** to the south; vergence refers to the dominant direction of transport of material along both the décollement and the reverse faults that merge with the décollement. However, the situation is often complex, and there may be other reverse faults, or back thrusts, off the detachment surface that verge in the opposite direction. For example, the north-verging fault that generated the 1994 Northridge earthquake is probably a **back thrust** of the south-verging Santa Monica Mountains thrust fault. With this interpretation we assume that the fault that produced the Northridge rupture intersects the north-dipping reverse fault at greater depth than that shown in Figure 7.8. At convergent plate boundaries, the dominant direction of transport is toward adjacent basins. For example, in the continental collision forming the Himalayan Mountains, the dominant direction of transport is away from the mountain mass and toward the adjacent basins. Furthermore, the locus of tectonic activity usually migrates toward adjacent sedimentary basins. That is, as a mountain range forms, the locus of tectonic activity migrates away from the highlands of the range toward the adjacent flanks of the ranges. This migration widens the fold-and-thrust belt with time, and interior faults of the system may become inactive as the active tectonic processes are transferred to frontal fault systems [18, 19]. The pattern of deformation just described has been observed in fold-and-thrust belts in Taiwan (subduction zone) [20], India and Pakistan (continental collision forming the Himalaya) [18], Japan (subduction zone) [18], and California Transverse Ranges (shortening associated with San Andreas fault boundary) [8, 18, 19, 21, 22]. The observed pattern of thrust-fault migration is consistent with a mechanical model (Figure 7.16) proposed to explain the observed pattern of folding and faulting in Taiwan [20]. This model is based on earlier observations of onshore and offshore fold-and-thrust belts [23] that emphasize the following:

- Existence of a basal décollement (detachment fault) sloping toward the interior of the mountain belt below which there is little deformation
- The shape of the fold-and-thrust belt in cross section, which resembles a tapering wedge

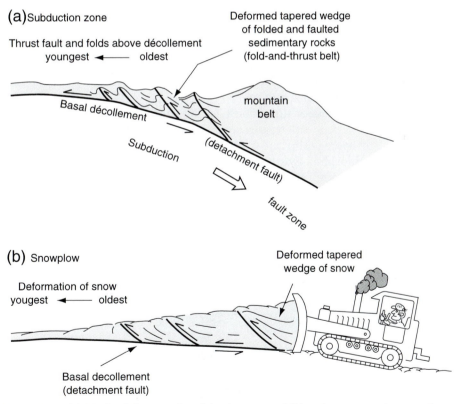

Figure 7.16   (a) Idealized diagram of a subduction zone and (b) analogy to a moving snowplow. With time, new thrust faults form above the detachment near the narrower part of the tapered wedge. (After Davis et al., 1983 [20].)

- Existence of shortening in the tapering wedge

The mechanics operating in the model are analogous to those observed in the wedge of snow in front of a moving snowplow (Figure 7.16) [20]. The snow deforms until the wedge reaches a critical taper, when it locks up and slides over the basal décollement so that deformation is transferred to the front of the wedge, where snow continues to accrete and deform [19, 20]. The model predicts that folds will migrate toward the edge of a fold-and-thrust belt, explaining why these folds and related faults are likely to be active and to present an earthquake hazard [19].

FLEXURAL-SLIP FAULTS

Flexural-slip faulting occurs when the strain resulting from shortening is concentrated directly along the surfaces that are being folded, such as the bedding planes between sedimentary layers (Figure 7.17) [24, 25]. It is important to recognize that flexural-slip displacement along bedding planes reverses across the hinge surface of a fold. Flexural-slip folding is analogous to bending a telephone book so that displacement occurs along the pages.

The central Ventura Basin in Southern California has several good examples of flexural-slip faulting and associated folding. A general tectonic map of the region showing active and potentially active faults and folds is shown in Figure 7.18a. The Devil's Gulch fault (Figure 7.18b) faults Miocene shale over Ventura River terrace gravels (about 40 ka in age) with several meters of displacement along a bedding plane in the shale [24]. Terraces of the Ventura River displaced by the flexural-slip faults are shown in Figures 2.26 and 2.27.

One of the major thrust faults in the central Ventura Basin is the San Cayetano fault (Figure 7.18a), with stratigraphic separation of about 7.5 km in the past 1 M.y. and therefore an average slip rate of about 7.5 mm/yr. Study of Quaternary stratigraphy and tectonic geomorphology suggests a slip rate of 1 mm/yr to 9 mm/yr [26]. Flexural slip is present at one location (Figure 7.18c). The style of deformation at that site is very sim-

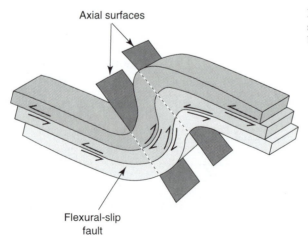

Axial surfaces

Flexural-slip fault

Figure 7.17   Idealized diagram showing flexural-slip faulting on the bedding planes of folded strata.

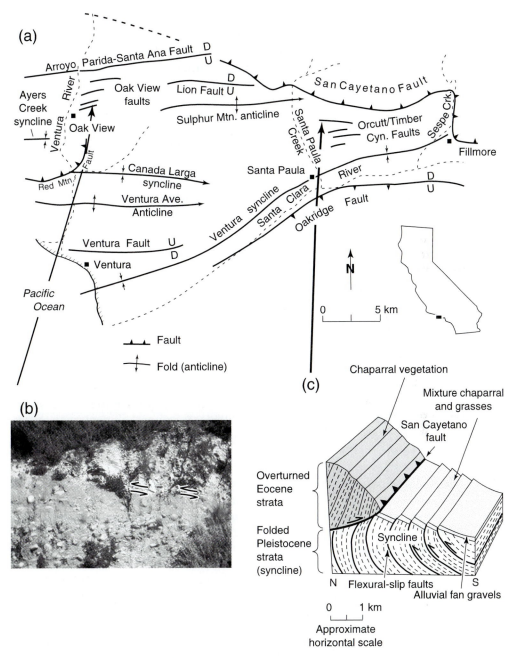

Figure 7.18   (a) Map of the central Ventura Basin, California, showing major geologic structures that deform upper Pleistocene and/or Holocene materials. (b) Devil's Gulch fault, one of the Oak View flexural-slip faults. Distance across the photograph is several meters. (c) Block diagram of flexural-slip and thrust faulting associated with the San Cayetano fault and Ventura syncline. Compare with similar deformation produced by the October 10, 1980, El Asnam earthquake (Figure 7.6b). (Photograph by E. A. Keller.)

ilar to deformation during the 1989 El Asnam event (see Figure 7.6b), suggesting that the flexural slip may be coseismic with large earthquakes on the San Cayetano fault.

Flexural-slip faults, such as those in the central Ventura Basin, can cause a ground-rupture hazard. It is debatable, however, whether flexural-slip faults also present a seismic-shaking hazard. If the fold structures are shallow, then these faults probably do not by themselves produce large earthquakes capable of strong seismic shaking, because the faults may not extend to depths where large earthquakes are generated. On the other hand, the folding is probably related to deeper, buried structures that do produce seismic shaking during earthquakes when the folding occurs. Thus, although the flexural-slip faults themselves may not be capable of producing earthquakes, this does not mean they are not associated with seismic shaking. Most likely, flexural-slip faulting accompanies folding during earthquakes (e.g., during the M_W 7.3 El Asnam event) and causes ground rupture that presents an additional hazard.

FOLDING AND STRIKE-SLIP FAULTING

Where different segments of strike-slip faults overlap, shortening between the segments occurs if the step-over is in the opposite sense of the slip on the fault. For example, two strands of the southern San Andreas fault in the Indio Hills, not far from Palm Springs, California, overlap and form a left-step (Figure 7.19). That is, if you go to the end of either one of the fault segments in the central part of the map, you would have to turn left to get back to another strand of the fault. Within this area of the left-step in the right-lateral San Andreas system, convergence occurs (see Figure 2.11). The cross section in Figure 7.19 shows several folds extending into and through the area of overlap of the fault segments. The folded rocks are geologically young sediments with an age of about 1 m.a [27]. The convergence or shortening that produces the folds is directly related to the strike slip on the two segments.

Folds also can form as a result of **strain partitioning**. Oblique strain in the lower crust can be partitioned into nearly pure tangential and normal strain in the upper seismogenic crust (the upper 10 to 15 km or so) [28]. For example, it has been observed that oblique convergence in subduction zones often results in partitioning into nearly pure thrust faulting and folding as well as nearly pure strike-slip faulting. Similarly, along transform-fault plate boundaries, oblique strain is partitioned into (1) nearly pure strike-slip, and (2) shortening perpendicular to the strike-slip that results in fold-and-thrust belts roughly parallel to the strike-slip faults (Figure 7.20a and b). If the strain partitioning is a local phenomenon that occurs at shallow depths above where large earthquakes nucleate (Figure 7.20b), then the subparallel faults and folds should be considered as a group when evaluating earthquake hazard. However, if the strain partitioning occurs regionally (Figure 7.20b), then individual thrust faults can probably be treated as independent seismic sources [28]. We state "can probably" because we really do not know for sure. The 1957 Gobi-Altay earthquake in China was associated with regional strain partitioning. A zone of about 250 km experienced predominately strike-slip displacement of 3 to 6 m while another 100-km-long zone about 30 km south experienced reverse displacements (vertical components of 1 to 3 m). If a similar event in Southern California were to occur, it could cause simultaneous rupture of the San Andreas fault

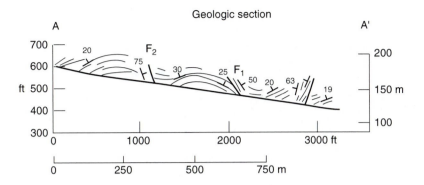

Geologic section

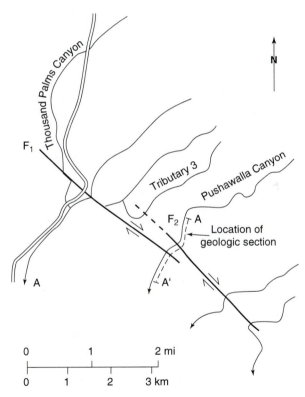

Figure 7.19   Sketch map and geologic section of the Pushawalla Canyon area, near Palm Springs, California. Folds are present in the area of the left-step of the San Andreas fault. (From Keller et al., 1982 [27].)

Figure 7.20   (a) Generalized map of the San Andreas and San Simeon-Hosgri fault systems showing selected historical earthquakes and Quaternary anticlines. Note that the hinge lines are parallel to the strike-slip faults, suggesting that there is shortening perpendicular to the faults.

(b)

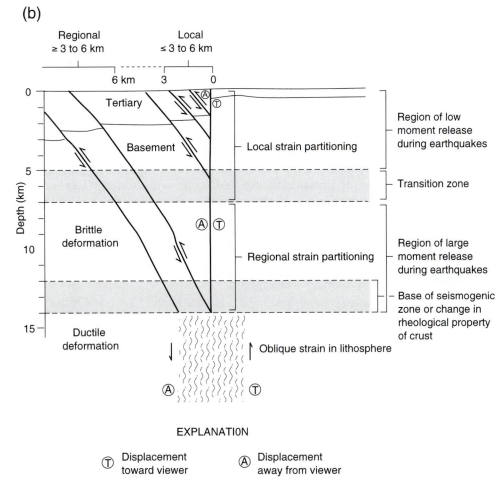

Figure 7.20 (continued) (b) Idealized diagram of how oblique strain in the lithosphere may be partitioned at local or regional scales. The vertical fault is nearly pure strike-slip, whereas the dipping faults are nearly pure reverse (shortening perpendicular to the strike-slip fault). Earthquakes are generated in the brittle zone above the ductile zone. In the ductile zone, oblique strain occurs by processes such as slow flowage without brittle fracture. (After Lettis and Hanson, 1991 [28].)

north of Los Angeles and of one of the reverse faults in the Los Angeles Basin [29]. Certainly we know that the reverse faults in the Los Angeles area do rupture independently of the San Andreas fault, presenting a serious earthquake hazard [30]. What we do not know for sure is what will happen when the San Andrea fault ruptures.

The preceding discussion argues for growth of folds during earthquakes. This is called **coseismic folding**. There is ample evidence that coseismic folding is the dominant process by which folds grow in fold-and-thrust belts of the world. On the other hand, where faults creep, uplift and folding may be a more continuous process, known as **aseismic folding** (that is, without recorded earthquakes). For example, in the southwest

Santa Clara Valley of the San Francisco Bay area, California, some fold growth is evidently associated with "**triggered slip**" and postseismic creep associated with large nearby earthquakes such as the 1989 M_W 7.2 Loma Preita event on the San Andreas fault system [31].

Another example of aseismic folding is Durmid Hill. Resurveying of a level line across and perpendicular to the San Andreas fault at Durmid Hill, located adjacent to the northeast shore of the Salton Sea, suggests that the hill domed upward by 9 mm between September 1985 and December 1991 (about 1.5 mm/yr). The rate of aseismic strike-slip motion (creep) is between 2 mm/yr and 6 mm/yr. These rates are consistent with longer-term rates determined from stratigraphic studies, suggesting that coseismic deformation is not necessary to produce Durmid Hill. Local strain partitioning of oblique creep is sufficient to produce the 8.5 m of relief (elevation difference base to top) of the hill [32]. However, this certainly does not preclude coseismic folding at the site—only that at present creep rates, it is possible to produce the hill since mid-Pleistocene time (last 780 ky) without deformation from large earthquakes.

TECTONIC GEOMORPHOLOGY OF ACTIVE FOLDS

Active folding may occur well below the surface of the Earth, and if the uplift related to that folding is less than the rate of deposition of sediment, then long-term deformation will not be observed at the surface. On the other hand, active folding also may cause uplift of the surface of the Earth that may be observed and measured. In order to evaluate the potential earthquake hazard associated with buried faults that produce folds, it is necessary to evaluate the evidence for active folding at the surface. If a fold stopped growing prior to the Pleistocene (last 1.65 M.y.), then the buried fault producing the fold is probably no longer active.

FAULT AND FOLD GROWTH

Folding in fold-and-thrust belts is intimately related to buried reverse faulting. As total fault displacement from repeated earthquakes increases, so does fault length [33, 34]. In other words, as faults accumulate slip, they propagate and grow laterally. A simple argument, based on two observations, may be constructed to support this statement [34]: (1) The ratio of coseismic fault slip to rupture length during an earthquake generally has a value of approximately 10^{-5} to 10^{-4} (that is, rupture length is generally 10,000 to 100,000 times fault slip as a result of an earthquake); and (2) the ratio of total cumulative displacement on a fault to fault length is approximately 10^{-3} to 10^{-1} (that is, fault length is generally 10 to 1000 times as long as total displacement). It has been pointed out that if these two relationships are to be satisfied simultaneously, then faults are required to increase their lengths as they increase total displacement [10, 12, 33, 34].

LATERAL PROPAGATION OF FOLDS

As faults propagate laterally (grow in length), so must the folds they produce. Although it is difficult to show from evaluation of the landscape that buried reverse faults propagate laterally, the folds these faults produce can provide an indication of the direction rates of lateral propagation. The two processes of faulting and folding occur together, and in many locations **fold scarps** (the scarp produced by active folding) have been mapped as fault scarps. The primary way to demonstrate lateral propagation of folds is to carefully evaluate geomorphic relationships [34]. Geomorphic criteria useful in evaluating rates and direction of lateral propagation of active folds in the direction of lateral propagation include the following [12]:

1. Decrease in drainage density and degree of dissection of a folded surface.
2. Decrease in elevation of wind gaps. A **water gap** is an opening or pass cut through a ridge by streams. If the stream is no longer present, it is called a **wind gap** (see discussion in Chapter 5).
3. Decrease in relief of the topographic profile along the crest of the fold.
4. Development of characteristic drainage patterns.
5. Deformation of progressively younger deposits or landforms.
6. Decrease in rotation and inclination of fold limbs.

These six geomorphic criteria are useful in recognizing lateral propagation of folds but are not proof. This results because the criteria, given specific scenarios, may be consistent with both fold propagation and fold rotation models of fold growth (Figure 7.21). Criteria 4 and 5 are strong evidence of lateral propagation, and if there are at least two wind or water gaps produced by the same stream, then this is very strong evidence of lateral propagation. Assuming that the stream can only be in one place at a time, a model of fold rotation (Figure 7.21) is unlikely to produce two gaps at progressively lower elevation in the direction of fold growth. That is, there could only be one gap and one diversion of drainage. Once a stream is defeated by uplift in the rotation model, it would be deflected to the nose of the fold (which in that model is fixed), and another gap would not form.

Geomorphic methods employed to analyze lateral propagation of folds are keyed to the six criteria just given. Figure 7.22 is an idealized diagram illustrating the basic tectonic geomorphology of a simple fold propagating laterally. Drainage density and degree of dissection decrease in the direction of propagation, as do topographic relief elevation of wind gaps and limb rotation. It is emphasized that Figure 7.22 is a simple idealized model, whereas folds in nature may behave much more complexly. For example, Figure 7.23 shows the development of the El Asnam anticline, which is laterally propagating to the southwest. The Fodda River was diverted twice to the southwest (Figure 7.23b, c), leaving two wind gaps to document its history, only to be then diverted back to the north (Figure 7.23c) to join the Ech Cheliff River. The reversal in direction of diversion is probably due to development of general slope of the topography to the north as a result of subsidence of the Chellif Basin. The Ech Cheliff River is a large drainage, and even the temporary damming of the river as a result of the 1980 M_W 7.3 earthquake, with uplift of 5 m of the anticline and 1 m subsidence of the basin, only temporarily blocked the channel [35].

The first task in evaluating a given fold is to make a detailed drainage map of the fold's fore-limb and back-limb as well as the plunge panel (nose ramp). This analysis may be completed by using aerial photographs and detailed topographic maps. Drainage patterns are then analyzed in order to evaluate **drainage density**, which is a measure of

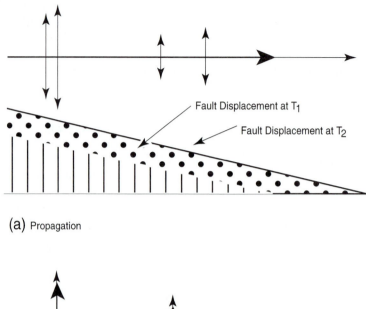

(a) Propagation

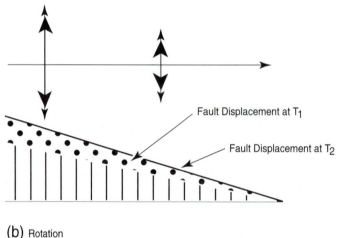

(b) Rotation

Figure 7.21 Two models of fold growth. (a) Lateral propagation where fault length grows. (b) Rotation, where endpoints of buried fault are fixed. (Drawing courtesy of D. Medwedeff.)

degree of dissection. Drainage density (D_d) is defined as the ratio of the sum of total length of channel divided by the drainage basin area.

Construction of topographic profiles along the crest, profiles normal to the fold, and digital elevation models may greatly assist in identifying geomorphic parameters of folds. Elevations of wind gaps along the crest of an anticline may be measured directly, and these elevations are generally lower in the direction of propagation [12, 34]. The topographic profile along the crest of the fold may reveal the direction of fold plunge, and hence the direction of lateral propagation. Drainage parallel to a fold axis will likely be diverted in the direction of propagation [34]. As diversions develop, tributary streams are captured and the size of the upstream drainage basin increases until there

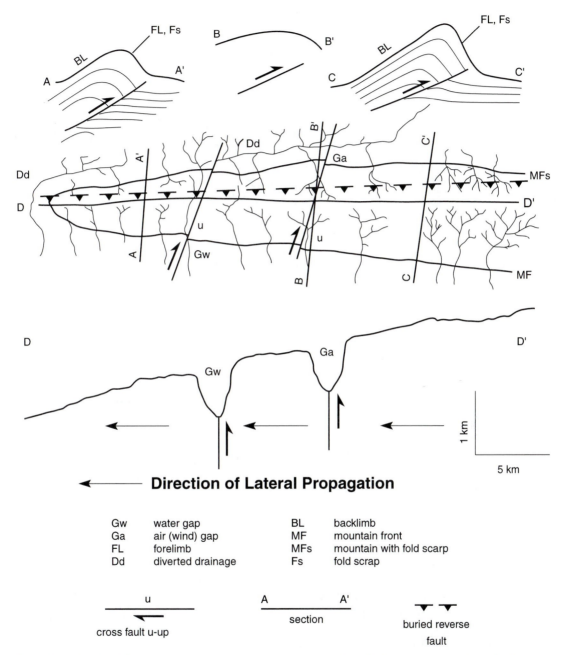

Direction of Lateral Propagation

Gw	water gap	BL	backlimb
Ga	air (wind) gap	MF	mountain front
FL	forelimb	MFs	mountain with fold scarp
Dd	diverted drainage	Fs	fold scrap

cross fault u-up

A ——— A' section

buried reverse fault

Figure 7.22   Idealized diagram showing the tectonic geomorphology of a fold that is propagating laterally. Major features are shown by the three cross sections; A-A' near the terminus or nose of the fold (to the left on the diagram); B-B' which is through a wind gap; C-C' in an older part of the fold. Notice that the degree of dissection decreases from right to left on the diagram, as does the elevation of the topographic profile of the crest of the fold (D-D'). Positions of the wind and water gap coincide with cross-faults that have both a lateral and vertical component of slip (oblique slip). The forelimb of the fold is generally the steepest limb and often forms a straight, prominent fold scarp.

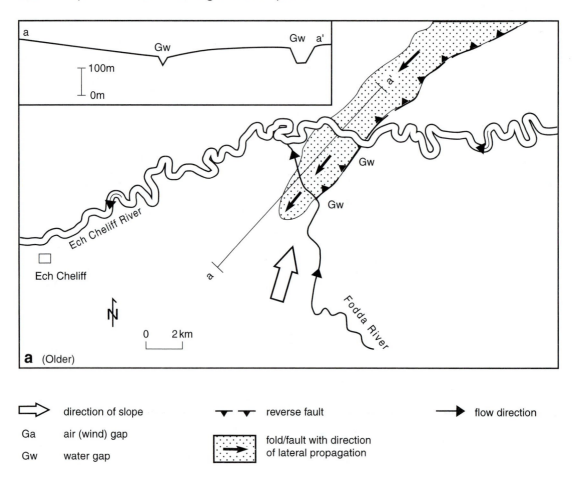

direction of slope   ▼—▼ reverse fault   ——▶ flow direction

Ga air (wind) gap

Gw water gap

fold/fault with direction of lateral propagation

Figure 7.23 (a) Drainage development across the El Asnam anticline. Earlier in the development, the Fodda River was diverted twice as the fold propagated laterally to the southwest. This produced the gaps in panels (a) and(b).

is sufficient stream power to maintain a channel temporarily at the nose of the fold where propagation has not yet occurred. As lateral propagation of the fold continues, this area becomes a water gap and eventually may become a wind gap if the channel is defeated by uplift and/or stream capture. If defeat occurs, the channel may be diverted again in the direction of lateral propagation, and in the course of a fold development, the channel may make several passes around the fold as the drainage develops. For some folds there may be several wind gaps produced in this manner, and the drainage will be repeatedly diverted around the nose of the fold.

Evaluation of drainage patterns for specific folds may suggest that some folds are propagating in two directions, that adjacent folds are propagating toward each other, or that younger folds are propagating parallel toward or against older folds in a fold belt [34].

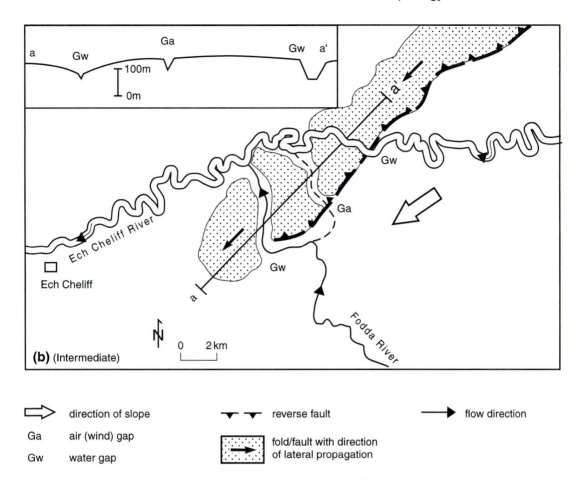

direction of slope

Ga   air (wind) gap

Gw   water gap

reverse fault

fold/fault with direction
of lateral propagation

flow direction

Figure 7.23 (*continued*) (b) The Ech Cheliff River, which is the area's largest river, was not diverted during the growth of the fold and regional slope toward that river eventually resulted in the Fodda River responding to that slope and taking a northerly path to the Ech Cheliff River before crossing the fold.

The processes that transform a water gap into a wind gap are generally complex, and at least two hypotheses are possible [12]: (1) uplift of the fold may block the channel in the water gap, forcing a diversion in the direction of lower topography, which is likely to be in the direction the fold is propagating; and (2) a channel crossing the nose of the fold has a tributary on the mountain side of the fold that erodes headward, parallel to the axis of the fold, toward the water gap (Figure 7.24; also see Figure 5.2). Extension of that tributary captures the drainage feeding the water gap. In many cases a wind gap probably forms by a combination of both tectonic and fluvial processes (uplift diversion and capture of drainage). There is some debate as to whether it is unit stream power (stream power per unit width of channel, proportional to the product of discharge and channel slope) or other factors, such as rate of rock uplift and sediment

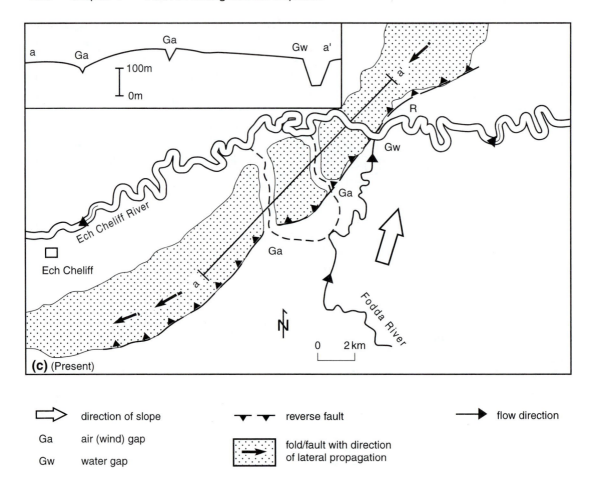

direction of slope

Ga   air (wind) gap

Gw   water gap

reverse fault

fold/fault with direction
of lateral propagation

flow direction

Figure 7.23 *(continued)* (c) This case history illustrates that drainage diversion, which often is in the direction of lateral fold propagation (a and b), may be complex. (Modified after Boudiaf et al., 1998 [35].)

flux, that are most significant in determining if a stream is able to maintain a channel across a rising fold (incision) or is diverted [10, 12, 36, 37].

Consideration of processes related to lateral propagation of folds and stream diversion allows for a hypothetical explanation for the development of drainage across an actively developing fold belt. The process is illustrated in Figure 7.25. A stream establishes a channel across the path of the propagating fold before the fold arrives, and that short stream reach is **antecedent** to the uplift (that is, the stream was present before uplift and rate of incision is about equal to the rate of uplift, discouraging diversion). As fold growth continues, a water gap may form, as discussed previously. With this hypothetical model, drainage is established across a developing fold belt by a series of captures, diversions, and antecedent positioning. The process involves development of relatively short antecedent reaches across the path of propagating folds, followed by defeat and stream deflection along relatively longer reaches parallel to the fold axis [12, 34].

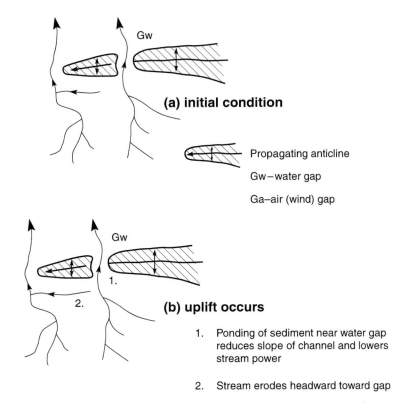

(a) initial condition

Propagating anticline

Gw – water gap

Ga – air (wind) gap

(b) uplift occurs

1. Ponding of sediment near water gap reduces slope of channel and lowers stream power

2. Stream erodes headward toward gap

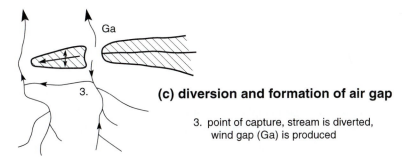

(c) diversion and formation of air gap

3. point of capture, stream is diverted, wind gap (Ga) is produced

Figure 7.24   Idealized diagrams (a–c) showing the transformation of a water gap to a wind gap by uplift, reduction of slope and stream capture.

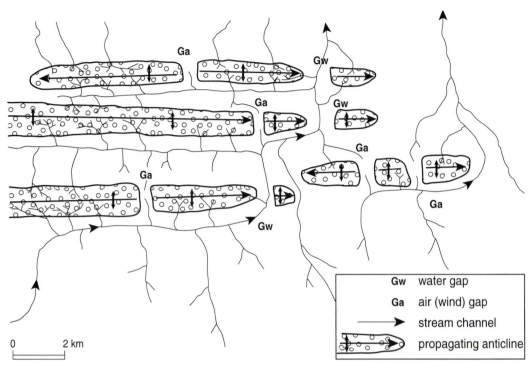

Figure 7.25 Idealized diagram illustrating how drainage transverse to active fold belt deforming surface topography might develop by series of diversions, with short antecedent reaches across folds and longer reaches parallel to fold axes. (Modified after Keller et al., 1999 [12].)

CASE STUDY: WHEELER RIDGE ANTICLINE

The east-west-trending Wheeler Ridge anticline is located on the north flank of the San Emigdio Mountains in the southern San Joaquin Valley of California. Wheeler Ridge (Figure 7.26) is the northernmost topographic expression of the fold-and-thrust belt located on the northern flank of the western Transverse Ranges [11, 38]. The structural geometry of active folding in the western Transverse Ranges and the southern San Joaquin Valley has been described in terms of both fault-bend folding and fault-propagation folding [11, 21, 36]. Several streams have eroded through the fold. The prominent stream valley in the central part of Figure 7.26a is an abandoned valley (wind gap). The stream in the valley was defeated by fold development processes (uplift, stream capture) approximately 30 to 60 ka, as determined by soil analysis and absolute and relative dating [11]. Farther to the east is a water gap, where the stream flowing from north of the ridge into the San Joaquin Valley is still incising through the structure (Figure 7.26). Several geomorphic indicators of lateral propagation at Wheeler ridge are shown in Figure 7.27.

In the central portion of the fold, the subsurface faulting that produced Wheeler Ridge is thought to form a complex fault wedge [36]. Near the eastern end of the structure, the Wheeler Ridge thrust

A

B

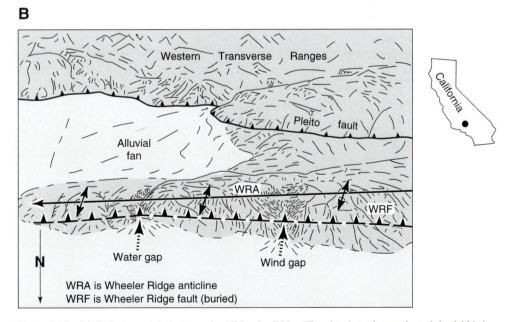

Figure 7.26   (a) Oblique aerial photograph of Wheeler Ridge. The view is to the south, and the fold is in the central portion of the photograph. Note that the fold is strongly asymmetric—the northern limb is more steeply inclined than the southern limb. WRA is the Wheeler Ridge anticline, and WRF indicates the Wheeler Ridge fault. (b) Sketch map of area shown in aerial photograph. (Photograph by J. Shelton.)

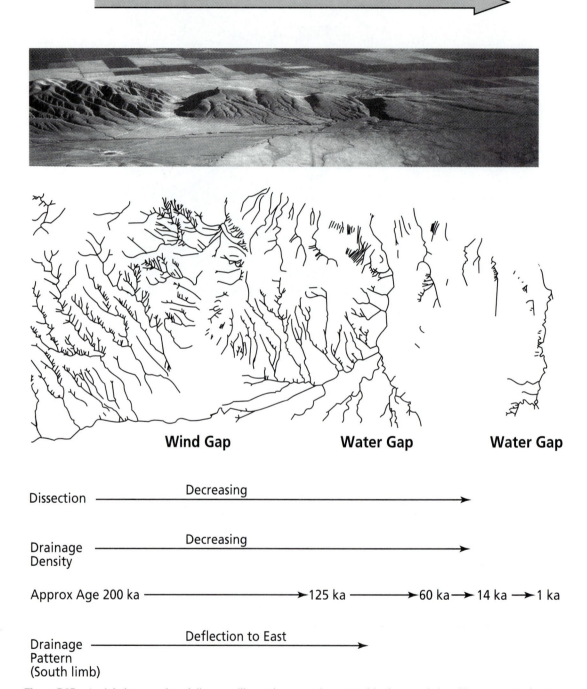

Figure 7.27   Aerial photograph and diagrams illustrating several geomorphic characteristics of later propagation at Wheeler Ridge. Note that this view is to the north, opposite that in Fig. 7.26. (After Keller et al., 1999 [12].)

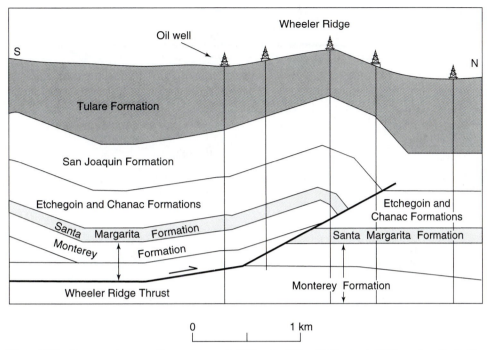

Figure 7.28   Cross section near the eastern end of Wheeler Ridge. To the west of this section the fault geometry is more complex and is described as a fault wedge. (After Medwedeff, 1984. Unpublished report.)

comes closer to the surface, and the fold may be described as a fault-propagation feature (Figure 7.28). A thrust fault near the eastern end of the fold with the expected geometry of the buried Wheeler Ridge fault is exposed in a gravel pit. A sketch of the exposure (Figure 7.29) illustrates that near the surface, a soil horizon is offset approximately 2.5 m. The displacement increases dramatically with depth, and at about 15 m, sedimentary units are folded and dip as steeply as 45°. This fault exposure illustrates an important point—with active folding, the expression of folding and fault displacement may increase significantly with depth. Thus, information obtained from shallow trenches excavated to expose faults in anticlines may be misleading when compared to displacements and dips only a few meters below.

Evaluation of the deformed alluvial-fan deposits at Wheeler Ridge [11] suggests that the rate of uplift of the structure is approximately 3 mm/yr and that the fold is propagating to the east at a rate of about 3 cm/yr. The rates of uplift and propagation of the anticline were determined through careful dating of several parts of the fold. Absolute dating, soil chronology, and topographic data reveal that alluvial-fan segments folded over the structure are lower and younger to the east. Propagation of the fold to the east is also revealed in aerial photographs; notice in Figures 7.26 and 7.27 that the **drainage density** (ratio of total length of streams to area) is lower in the east. Drainage density on the south flank (back-limb) decreases systematically from about 15 to 18 km^{-1} for surfaces older than 100 ka to about 0.4 km^{-1} for surfaces of 17 to 60 ka, and is nearly 0 for Holocene surfaces. Drainage density tends to increase with time and relief, so the decrease in drainage density is additional evidence that the eastern portions of Wheeler Ridge began to fold more recently than the western portions [38].

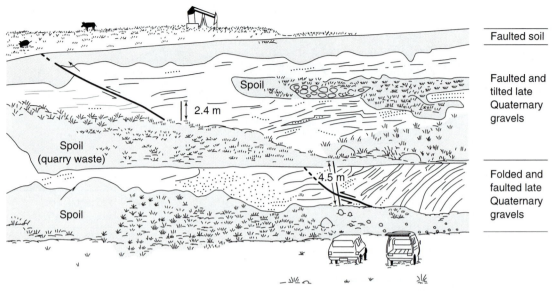

Figure 7.29 Sketch map of exposure of presumed Wheeler Ridge thrust fault in gravel pit at east end of Wheeler Ridge anticline. (From Keller et al., 1998 [11].)

Lateral propagation at Wheeler Ridge evidently is associated with the development of **tear faults** (steeply dipping faults that strike nearly perpendicular to the buried reverse fault system that formed the fold, forming fault segment boundaries, also called **cross-faults**) [36, 39]. Evaluation of subsurface data at Wheeler Ridge suggests that the position of major wind and water gaps coincide with tear faults that produce a scarp facing the direction of lateral propagation, helping establish stream positions at the bases of these scarps where they cross the fold.

Combining the evidence of Holocene and latest Pleistocene folding and faulting near the east end of Wheeler Ridge with subsurface information is useful in evaluating the earthquake hazard. In the Wheeler Ridge area, active deformation is presently partitioned between the Pleito and Wheeler Ridge faults (Figure 7.26). The Pleito fault, located south of Wheeler Ridge, breaks the surface and has an uplift rate of approximately 0.5 mm/yr [40]. The Wheeler Ridge fault is a buried reverse fault with an uplift rate of approximately 3 mm/yr and is well defined by subsurface data [36]. The 1952 M_W 7.5 Kern County earthquake was located beneath Wheeler Ridge on a deeply buried member of the Pleito thrust fault system [36, 41]. In the epicentral region, at Wheeler Ridge, the earthquake produced approximately 1 m of uplift. Assuming that M_W 6 to 7 earthquakes in the Wheeler Ridge area are associated with approximately 1 m of uplift, then the uplift rate of Wheeler Ridge suggests an average recurrence of approximately 300 years for such events (1000 mm divided by an uplift rate of about 3.2 mm/yr = 300 yr). However, the earthquake history is probably much more complex, and as a result an average recurrence interval is probably not very useful. There are several faults present [41], and it is unlikely that each earthquake produces a uniform 1 m of uplift. More likely, both uplift and recurrence intervals are variable. What we can say is that large, damaging earthquakes can be expected to occur in the future along this active fold-and-thrust belt.

Figure 7.30   Photograph of the Ventura Avenue anticline. (Courtesy of A. G. Sylvester.)

CASE HISTORY: VENTURA AVENUE ANTICLINE

The Ventura Avenue anticline is a fold located about 4 km north of Ventura, California (Figure 7.30). Subsurface evidence suggests that the fold is forming above a décollement in Miocene rocks at a depth of approximately 5 km (Figure 7.31) [19]. The anticline is the landwardmost fold of an active fold-and-thrust belt along the Pacific Coast. Rocks in the anticline are of Pliocene and Pleistocene age. Figure 7.32 shows the stratigraphy within the anticline, including the sequence of river terraces that cross the structure and are deformed by it (Figure 7.33). This Pleistocene stratigraphic section is one of the best dated in the world. Table 7.1 shows rates of uplift associated with the Ventura Avenue anticline. The rates are based on absolute ages for the river terraces.

Study of the Ventura Avenue anticline [42] suggests that rates of tectonic activity may vary in time and space as a result of the mechanics of folding. Folding started as recently as about 200 to 400 ka, and the rate of uplift atop the anticline during that time was approximately 7 to 14 mm/yr. The uplift rate today, however, is closer to 4 mm/yr. The shortening that causes this deformation is thought to be constant, at a rate of about 10 mm/yr. The decrease in the uplift rate may be explained by the fact that as the rootless fold buckles, the mechanics of the process dictate that uplift decreases with time (Figure 7.34). Figure 7.34 reveals some surprising information concerning the mechanics of folding above a detachment—notice that the first 4% of the shortening produces approximately 36% of the uplift, and the

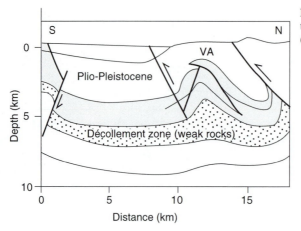

Figure 7.31 Décollement below the Ventura Avenue anticline (VA). (After Yeats, 1986 [19].)

first 8% of the shortening produces one-half of the total uplift! The rate of shortening is constant, but the uplift produced for each increment of shortening decreases with time, so that the uplift rate decreases.

The fact that a constant tectonic driving force can produce different uplift rates over time has important implications for interpretation of rates of active tectonics. It implies that when evaluating folds, the rate of uplift may be somewhat misleading, particularly if the limbs of the fold dip steeply, which would produce relatively low uplift rates. In contrast, if the fold is just beginning to form, uplift rates may be relatively high. You can verify the mechanics of folding that result in a fast initial rate of uplift followed by lower rates with a simple experiment. Use a thin, flexible straight-edge ruler with a length of about 30 cm and shorten the ruler (holding one end fixed) 1 cm at a time while recording the vertical deformation (uplift). Make a graph with uplift (in centimeters) on the vertical axis (x) and amount of shortening (in centimeters) on the horizontal axis (y). It should be noted that the variable rates of uplift on the Ventura Avenue anticline are for one particular type of rootless fold; uplift of a fault-propagation fold that is not rootless would probably be a different story.

In summary, the Pleistocene terraces of the Ventura River are clearly uplifted, tilted, and folded over the Ventura Avenue anticline. Rates of uplift and tilt have decreased since folding began, from approximately 7 to 14 mm/yr to about 4 mm/yr [42]. The limbs of the Ventura Avenue anticline dip as steeply at about 45° on the south flank of the fold, and at this dip additional shortening produces little uplift. As the Ventura Avenue anticline begins to lock up, deformation may be transferred to the Ventura fault, a few kilometers to the south. At present, that fault is a buried feature, with little apparent displacement near the surface, but it is likely to become more active in the future, presenting a potentially serious threat to the Ventura area.

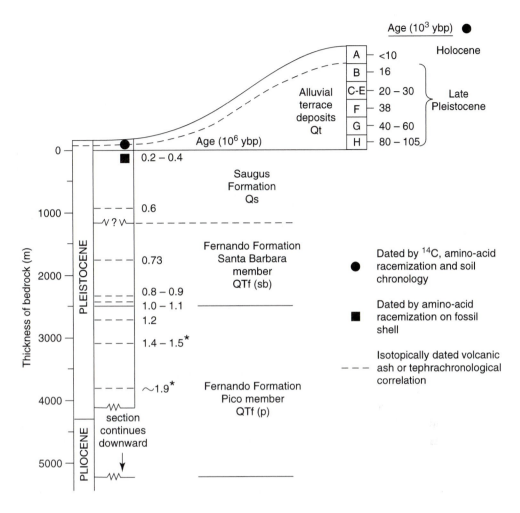

Figure 7.32   Pleistocene stratigraphy and alluvial (river) terraces at the Ventura Avenue anticline. (Pleistocene stratigraphy, in part, from Lajoie et al., 1982. In *Neotectonics in Southern California*, J. D. Cooper (compiler). Cordilleran Section, Geological Society of America Field Trip Guidebook; Rockwell et al., 1988, pp. 43–51 [42].)

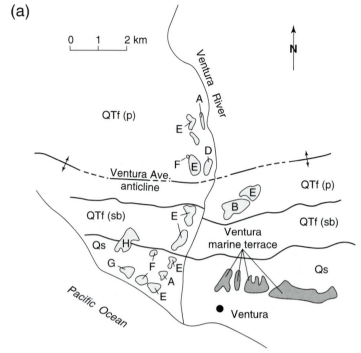

(a)

Note: QTf (p), QTf (sb) and Qs are Quaternary
sedimentary rock formations (see Fig. 7.32)

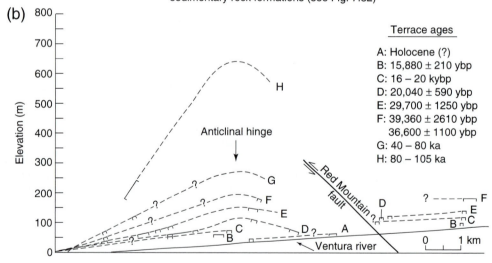

Figure 7.33 (a) Map of Ventura River terraces over Ventura Avenue anticline and (b) terrace profiles
along the Ventura River. (From Rockwell et al., 1988 [42].)

Table 7.1

UPLIFT OF RIVER TERRACES OVER THE VENTURA AVENUE ANTICLINE. LOCATIONS OF TERRACES ARE SHOWN IN FIGURE 7.33.

River Terrace	Age[*]	Present Height Above Ventura River (m)[**]	Best Estimated Uplift Rate (mm/yr)[***]
B.	15,800 ± 210	30.5 ± 10	4.25 ± 0.7
D.	20,040 ± 590	85.3 ± 10	5.65 ± 0.65
E.	29,700 ± 1250	120 ± 10	4.50 ± 0.5
F.	38,000 ± 1900	175 ± 10	4.95 ± 0.5
H.	80,000 or 105,000	625 ± 100	7.10 ± 2.0

[*] Based on ^{14}C and amino-acid racemization chronology
[**] Projected to the hinge of the anticline
[***] Uplift rate determined using the best estimated depths of incision of the Ventura River
(Data from Rockwell et al., 1988 [42])

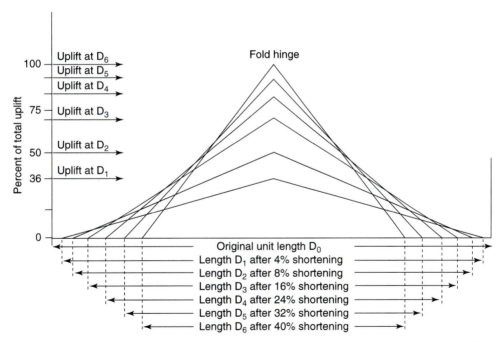

Figure 7.34   Simple model showing that, with folding above a décollement, a large amount of the uplift is produced early in the folding process. Maximum uplift occurs at D_6 with 40% shortening, as the fold locks up and deformation is transferred to another structure. (After Rockwell et al., 1988 [42].)

SUMMARY

Folds are found in a variety of shapes and sizes, and there is a generally accepted terminology for describing them. Construction of geologic cross sections aids in the interpretation of subsurface geological structures such as folds and faults. Interpreting cross sections is facilitated by developing models, such as the fault-propagation and fault-bend fold models, that provide quantitative, geometrically balanced solutions. It is important to understand, however, that the use of these models is controversial, and there are serious questions about the assumptions involved.

Flexural-slip faulting occurs where the strain resulting from shortening is concentrated directly along the surfaces that are being folded. Folding is often associated with strike-slip faulting. Where segments of strike-slip faults overlap or a segment bends, shortening results if the overlap or bend is in the opposite sense of the slip of the fault. Folds can also form as a result of strain partitioning. Along transform-fault plate boundaries, oblique strain may be partitioned into nearly pure strike-slip parallel to the fault and shortening perpendicular to the fault.

Seldom is only a single fold present; rather, a series of folds often defines a fold-and-thrust belt. Some of the thrust faults and high-angle reverse faults may break the surface, but many others may be hidden within the cores of anticlines and are called buried reverse faults. Only recently has it been recognized that these buried active faults present a significant earthquake hazard.

The tectonic geomorphology of active folding is an exciting area of research. We learn that as fault slip on a buried fault with associated fold accumulates, the fault and fold grow longer (that is, propagate laterally). The rate of lateral propagation is often several times the rate of uplift of the fold. Fault growth with earthquakes is difficult to document, but geomorphic criteria are useful in recognizing fold growth and in fact may be the only tool to establish direction and rate of growth.

REFERENCES CITED

1. Dahlstrom, C. D. A., 1969. Balanced cross sections. *Canadian Journal of Earth Sciences*, 6:743–757.

2. Suppe, J., 1983. Geometry and kinematics of fault-bend folding. *American Journal of Science*, 283:684–721.

3. Mitra, S., 1992. Balanced structural interpretations in fold-and-thrust belts. In *Structural Geology of Fold-and-Thrust Belts*, S. Mitra and G. W. Fisher (eds.). Baltimore, MD: Johns Hopkins Press, pp. 53–77.

4. Suppe, J., 1985. *Principles of Structural Geology*. Englewood Cliffs, NJ: Prentice Hall.

5. Stein, R. S., and R. S. Yeats, 1989. Hidden earthquakes. *Scientific American*, 260 (6) : 48–57.

6. Philip, H., and M. Meghraoui, 1983. Structural analysis and interpretation of the surface deformations of the El Asnam earthquake of October 10, 1980. *Tectonics*, 2:17–49.

7. Oskin, M., K. Sieh, T. Rockwell, G. Miller, P. Guptill, M. Curtis, S. McArdle, and P. Elliot, 2000. Active parasitic folds on the Elysian Park anticline: implications for seismic hazard in central Los Angeles, California. *Geological Society of America Bulletin*, 112:693–707.

8. Rivero, C., J. H. Shaw, and K. Mueller, 2000. Oceanside and Thirtymile Bank blind thrusts; implications for earthquake hazards in coastal Southern California. *Geology*, 28:891–894.

9. U.S. Geological Survey Staff, 1996. U.S.G.S. response to an urban earthquake, Northridge 94. *U.S. Geological Survey Open File Report* 96–263.

10. Burbank, D. W., J. K. McLean, M. Bullen, K. Y. Abdrakhmatov., and M. M. Miller, 1999. Partitioning of intermontane basins by thrust-related folding, Tien Shan, Kyrgyzstan. *Basin Research*, 11:75–92.

11. Keller, E. A., R. L. Zepeda, T. K. Rockwell, T. L. Ku, and W. S. Dinklage, 1998. Active tectonics at Wheeler Ridge, southern San Joaquin Valley, California. *Gelogical Society of America Bulletin*, 110:298–310.

12. Keller, E. A., L. Gurrola, and T. E. Tierney, 1999. Geomorphic criteria to determine direction of lateral propagation of reverse faulting and folding. *Geology*, 27:515–518.

13. Rich, J. L., 1934. Mechanics of low-angle overthrust faulting as illustrated by Cumberland thrust block, Virginia, Kentucky and Tennessee. *American Association of Petroleum Geologists Bulletin*, 18:1584–1596.

14. Seeber, L., and C. C. Sorlien, 2000. Listric thrusts in the western Transverse Ranges, California. *Geological Society of America Bulletin*, 112:1067–1079.

15. Hamblin, W. K., 1965. Origin of "reverse drag" on the downthrown side of normal faults. *Geological Society of America Bulletin*, 76:1145–1164.

16. Bruhn, R., 1994. Personal communication.

17. Sanz de Galdeano, C., and López-Garrido, A. C., 1999. Nature and impact of the Neotectonic deformation in the western Sierra Nevada (Spain). *Geomorphology*, 30:259–272.

18. Ikeda, Y., 1983. Thrust-front migration and its mechanism. *Bulletin of the Department of Geography, University of Tokyo*, 15:125–159.

19. Yeats, R. S., 1986. Active faults related to folding. In *Active Tectonics*, R. E. Wallace (ed.) Washington, DC: National Academy Press. pp. 63–79.

20. Davis, D., J. Suppe, and F. A. Dahlen, 1983. Mechanics of fold-and-thrust belts and accretionary wedges. *Journal of Geophysical Research*, 88:1153–1172.

21. Davis, T. L., 1983. Late Cenozoic structure and tectonic history of the western "Big Bend" of the San Andreas fault and adjacent San Emigdio Mountains. Ph.D. dissertation, University of California, Santa Barbara.

22. Keller, E. A., R. L. Zepeda, D. B. Seaver, T. K. Rockwell, D. M. Laduzinsky, and D. L. Johnson, 1987. Active fold-thrust belts and the western Transverse Ranges. *Geological Society of America Abstracts with Programs*, 19(6):394.

23. Chapple, W. M., 1978. Mechanics of thin-skinned fold-and-thrust belts. *Geological Society of America Bulletin*, 89:1189–1198.

24. Rockwell, T. K., E. A. Keller, M. N. Clark, and D. L. Johnson, 1984. Chronology and rates of faulting of Ventura River terraces, California. *Geological Society of America Bulletin*, 95:1466–1474.

25. Cooke, M. L., and D. D. Pollard, 1997. Bedding-plane slip in initial stages of fault-related folding. *Journal of Structural Geology*, 19:567–581.

26. Rockwell, T. K., 1988. Neotectonics of the San Cayetano fault, Transverse Ranges, California. *Geological Society of America Bulletin*, 100:500–513.

27. Keller, E. A., M. S. Bonkowski, R. J. Korsch, and R. J. Shlemmon, 1982. Tectonic geomorphology of the San Andreas fault zone in the southern Indio Hills, Coachella Valley, California. *Geological Society of America Bulletin*, 93:46–56.

28. Lettis, W. R., and K. L. Hanson, 1991. Crustal strain partitioning: implications for seismic-hazard assessment in western California. *Geology*, 19:559–562.

29. Molnar, P., 1995. A review of active deformation of the western Transverse Ranges (and its relevance to the active tectonics and earthquake history of Mongolia). Abstracts from the SCEC Workshop: Thrust Ramps and Detachment Faults in the western Transverse Ranges. January 22–24, University of California, Santa Barbara: 16.

30. Dolan, J. F., K. Sieh, T. K. Rockwell, R. S. Yeats, J. Shaw, J. Suppe, G. J. Huftile, and E. M. Gath, 1995. Prospects for larger or more frequent earthquakes in the Los Angeles metropolitan region. *Science*, 267:199–205.

31. Hitchcock, C. S., and K. I. Kelson, 1999. Growth of late Quaternary folds in southwest Santa Clara Valley, San Francisco Bay area, California: implications of triggered slip for seismic hazard and earthquake recurrence. *Geology*, 27:391–394.

32. Sylvester, A. G., R. Bilham, M. Jackson, and S. Barrientos, 1993. Aseismic growth of Durmid Hill, southeasternmost San Andreas fault, California. *Journal of Geophysical Research*, 98:14,233–14,243.

33. Cowie, P. A., and C. H. Scholz, 1992. Growth of faults by accumulation of seismic slip. *Journal of Geophysical Research*, 97:11,085–11,095.

34. Jackson, J., R. Norris, and J. Youngson, 1996. The structural evolution of active fault and fold systems in central Otago, New Zealand: evidence revealed by drainage patterns. *Journal of Structural Geology*, 18:217–234.

35. Boudiaf, A., J. -F. Ritz, and H. Philip, 1998. Drainage diversions as evidence of propagating active faults: example of the El Asnam and Thenia faults, Algeria. *Terra Nova*, 10:236–244.

36. Medwedeff, D. A., 1992. Geometry and kinematics of an active, laterally propagating wedge thrust, Wheeler Ridge, California. In S. Mitra and G. W. Fisher (eds.), *Structural Geology of Fold-and-Thrust Belts*. Baltimore, MD: Johns Hopkins University Press, pp. 3–28.

37. Humphrey, N. F., and S. K. Konrad, 2000. River incision or diversion in response to bedrock uplift. *Geology*, 28:43–46.

38. Shelton, J. S., 1966. *Geology Illustrated*. San Francisco: W. H. Freeman.

39. Mueller, K., and P. Talling, 1997. Geomorphic evidence for tear faults accommodating lateral propagation of an active fault-bend fold, Wheeler Ridge, California. *Journal of Structural Geology*, 19:397–411.

40. Hall, N. T., 1984. Late Quaternary history of the eastern Pleito thrust fault, northern Transverse Ranges, California. Ph.D. dissertation, Stanford University, California.

41. Davis, T. L., and M. B. Lagoe, 1987. The 1952 Arvin-Tehachapi earthquake (M = 7.7) and its relationship to the White Wolf fault and the Pleito thrust system. *Geological Society of America Abstracts with Programs*, 19(6):370.

42. Rockwell, T. K., E. A. Keller, and G. R. Dembroff, 1988. Quaternary rate of folding of the Ventura Avenue anticline, western Transverse Ranges, Southern California. *Geological Society of America Bulletin*, 100:850–858.

8

Paleoseismology and Earthquake Prediction

PALEOSEISMOLOGY

Our ability to evaluate present and future earthquake hazards is based on understanding the past behavior of seismogenic (earthquake-producing) faults. Basic data for earthquake-hazard evaluation include the location, length, and amount of fault displacement; intensity of shaking; size or magnitude; and dates of previous earthquakes. This information is widely available for earthquakes in recent decades from local, regional, and global seismic networks of recording instruments. The historical record ranges from a few hundred years in the United States to a few thousand years in China. For prehistoric earthquakes, the only source of information is the geologic and geomorphic record [1].

Evaluation of historical earthquakes that predate the use of recording instruments relies on written accounts, coupled with geologic and geomorphic evidence. **Paleoseismology** is defined as the study of the occurrence, size, timing, and frequency of historical earthquakes lacking instrumental seismic records and prehistoric earthquakes [1–3].

Paleoseismology most often utilizes the Holocene (last 10 ky) geologic and geomorphic record, which extends over a much longer time period than the limited instrumental record of seismic activity. When a large damaging earthquake occurs on an unstudied fault (either mapped or unknown), we use paleoseismology methods to investigate previous ruptures as part of the hazards evaluation. This is particularly important because the recurrence intervals for large earthquakes on many faults in tectonically active regions are often a few hundred to a few thousand years. In regions with lesser tectonic activity, recurrence intervals may be tens of thousands of years, requiring the evaluation of the late Pleistocene (last 125 ky) geologic record [1].

Paleoseismology depends on the evaluation of geologic evidence of past earthquakes. This evidence may be [3]

- **geomorphic**; consisting of landform features such as fault scarps, fold scarps, folds, uplifted coastal terraces, deflected or offset streams, and uplifted and/or tilted alluvial fans
- **stratigraphic**; consisting of sedimentary deposits (e.g., stream, alluvial fan, delta, lake, beach) that are deformed (displaced, folded, tilted, etc.)

Often in paleoseismology, investigation begins with recognition of deformed landforms followed by subsurface examination of stratigraphic evidence. That is, geomorphic evaluation guides selection of sites for subsurface examination. Successful completion of a paleoseismic evaluation requires determining dates of past earthquakes and slip rates of faults. Deformation is often recorded as faulted or folded landforms or stratigraphic units. Measurement of such deformation generally is straightforward, but in order to date earthquakes and determine slip rates on faults, it is necessary to obtain numerical dates of deformed landforms and stratigraphy (see Chapter 2). Remember—no dates, no rates!

Earthquakes occur with magnitudes of less than 1 to greater than 9, but earthquakes with M_W less than about 5 to 6 are seldom detected by paleoseismic evaluation.

EVIDENCE FOR PALEOEARTHQUAKES

Large earthquakes in the past have ruptured the surface of the Earth and folded near-surface earth materials in response to subsurface rupture. Surface rupture and folding often can be discerned hundreds, thousands, or even tens of thousands of years later. Evidence for paleoseismic evaluation may be obtained from

- Fault exposures
- Seismic-reflection profiles
- Faulted landforms such as stream terraces, offset streams, and coastal terraces
- Fault scarps, including scarps produced by single earthquakes and composite scarps produced by several earthquakes
- Stratigraphic features such as colluvial wedges, liquefaction features, sand blows, fissure filling, and abrupt burial of deposits
- Folded rocks, sediment deposits, and geomorphic surfaces

FAULT EXPOSURES

The best available scientific information from which to establish paleoseismicity is direct observation of faults. If we are able to study fault exposures directly, we may be able to measure displacements and/or collect material suitable for dating prehistoric earthquakes.

Some ways to examine faults include the following:

- **Natural exposures**. These include landforms such as seacliffs cut by wave erosion and a variety of steep slopes cut by stream and river processes. Natural exposures can be enhanced by clearing recent debris to better expose faulted deposits. A disadvantage of natural exposures is that they may not be in the most favorable orientation to a fault for paleoseismic evaluation.
- **Road cuts and railroad cuts**. Road cuts and railroad cuts are other good locations in which to search for fault exposures. Discovery of flexural-slip faults near Oak View, California (see Figures 2.26 and 7.18 and discussion in Chapter 7) began with one fortuitous observation in a highway cut.
- **Mines and gravel pits**. Gravel pits, in particular, provide important exposures because they typically are excavated in geologically young material, and they may be relatively deep (tens of meters), exposing geologic features that are unlikely to be exposed naturally (see Figure 7.29).
- **Boreholes**. Boreholes may be drilled by a variety of techniques to gain information about a fault. Boreholes vary from small-diameter holes from which a continuous core (cylindrical section of earth material a few centimeters in diameter) is obtained to large-diameter boreholes into which a person can descend, protected by a special cage, for direct observation. Boreholes are effective ways to identify and locate faults, but because of limited exposure, they may not provide sufficient information for detailed paleoseismology investigations. They can; however, provide important information concerning the vertical component of fault displacement and help identify sites suitable for trenching.
- **Trenches**. Where neither naturally occurring nor preexisting exposures of a fault can be found, paleoseismologists may excavate their own exposures using a backhoe or bulldozer (Figure 8.1). Trenches represent one of the principal tools for collecting paleoseismic data for earthquake-hazard assessments.

TRENCHING

Narrow, deep trenches are dangerous to geologists! Numerous construction workers are killed every year when trench walls collapse. Probably more geologists have been killed by trench collapse than by earthquakes. Narrow trenches deeper than 1.5 m are required by law to be stabilized, usually by hydraulic shores (braces) [3]. Even shored trenches in unstable earth materials may fail. Caution, especially in the first few minutes after a trench is opened (when most failures occur), must be exercised. Wide, benched trenches are usually safer and easier to work in but require a greater volume of mater-

Figure 8.1   Excavation of a large fault trench, Point Conception, California. (Photograph by E. A. Keller.)

ial to be removed. For safety, all trenches should have a fence placed around them with signs notifying people of the work and warning them not to enter the site. Before trenching, be sure to consult local and state regulations concerning safety issues. Remember, it's far better to be conservative with respect to safety issues. Always wear a hard hat, and never enter an unshored narrow trench or any trench with evidence of instability.

Sites for trenching are carefully chosen based on preliminary geologic observation and mapping. In order to evaluate both the horizontal and vertical components of displacement on strike-slip or oblique-slip faults (Figure 1.6), it is often useful to excavate two or more trenches. In addition to trenches across (perpendicular) to the fault, which allow the vertical component of displacement to be measured, one or more trenches may be excavated parallel to the strike of the fault to evaluate the horizontal component of displacement. The objective of fault-parallel trenching is to identify a **piercing point**, which is a feature such as a buried channel, a buried pipe, or some other distinctive feature that is offset along the fault (see Figure 8.2). A piercing point can be identified where it enters and exits the opposite sides of a fault. When datable material is recovered from features with measured offset, then slip rates can be calculated as the ratio of displacement to the time over which the displacement has occurred. Of course, if an offset buried pipe is identified after a known earthquake, the only information obtained is the amount of displacement for that event. The best piercing points from which

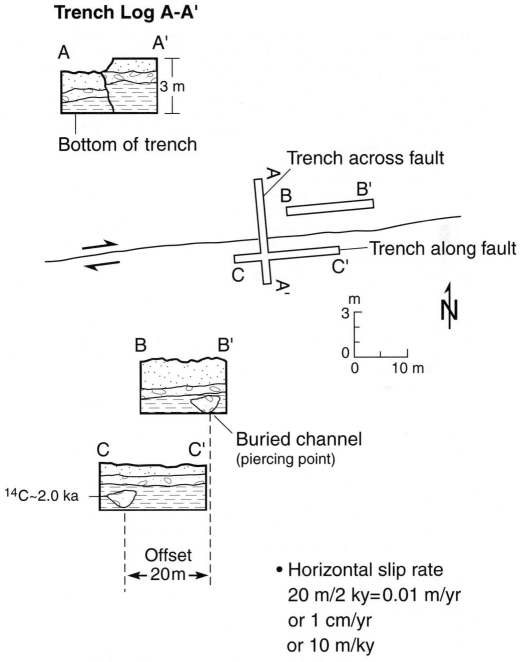

Figure 8.2   Idealized diagram showing how trenches may be excavated across and parallel to a fault zone to evaluate a strike-slip fault.

to calculate a slip rate are features several thousand to several tens of thousands of years old; that is, features that are old enough to have cumulative offset from numerous earthquakes, providing an average slip rate over a longer time period.

Following trench excavation, and taking proper safety precautions, it is necessary to prepare the trench walls for evaluation and logging. Generally only one trench wall is thoroughly cleaned to remove smearing of stratigraphy by the excavation process (the blade of the backhoe smears sediment), but in some instances both walls may be cleaned. Cleaning of trench walls is a tedious job and may involve the removal of several centimeters of material. This may be done with a variety of hand tools, from the relatively wide blade of a mattock to narrower trowels, with work being finished with whisk brooms and even paint brushes [3]. Following cleaning, the trench wall is gridded for mapping, commonly utilizing horizontal and vertical lines of string spaced approximately ~1 m apart. Each grid square is then carefully mapped or logged utilizing a variety of measuring instruments, from simple tape measures to total-station surveying techniques. Following gridding, the investigator often identifies deformation and sites where charcoal or other datable material is present in the trench wall. An important part of logging a trench is determining the various sedimentary units that may be mapped and described. These may be soil horizons or specific units delineated by grain size (for example, gravel from sand or silt) or by compositional differences.

The purpose of the trench logging is to portray the stratigraphy of the trench as carefully as possible in order to get a good understanding of what the subsurface environment looks like [3]. As part of the trench logging process, contacts between various mappable units are marked, perhaps using nails with colored flagging, as are fractures, faults, and other features of interest. In some instances particular sedimentary units (say a gravel or sandy layer) may be clearly truncated by a fault and offset. In other cases, even though there is evidence that a fault is present (for example, a fault scarp present at the surface), faulting in the trench may be difficult to observe. This is sometimes known as the "problem of fault nonvisibility" [3]. In some cases a strand of a fault may be clearly visible above and below a layer in the trench that shows no visible sign of fracturing. Processes that may conceal a fault include a variety of physical and biological processes, including bioturbation (for example, gophers or ground squirrels that dig in the ground), freezing and thawing, shrinking and swelling, and soil formation, among others. Of course, another reason for faults being nonvisible is that they terminate at a particular position in the trench wall. In other words, the fault trace simply dies out where faulting ends or intersected the surface of the earth at some time in the past and is now buried. Identification of where faults terminate in a trench is very important as such termination helps bracket fault activity and thus timing of past earthquakes.

Following careful trench logging, collection of material, and careful observation of faulted and deformed features, the trench is back-filled. This must be done very carefully to ensure that in the future there is no subsidence over the trench (due to compaction) that would cause problems at the surface. The geologist then submits materials to laboratories for dating while drafting trench logs and evaluating and measuring types of displacement. This may also involve construction of a series of drawings that restore stratigraphic units to their original positions prior to deformation (retrodeformation). Retrodeformation is important because it helps delineate displacement (earthquake) events and assist in identifying inconsistencies in the logging that become clear when restoration is attempted. A particular trench log must be able to be restored without the investigator having to depend on unlikely, unreasonable, or physically impossible sequences of events [3].

As an example of retrodeformation, Figure 8.3 depicts deformation that occurred during the past 12 to 25 ky along the Poukawa fault zone on the North Island of New Zealand. This fault zone is part of a fold-and-thrust belt associated with the Hikurangi Subduction Zone. Figure 8.3b to d sequentially retrodeforms the stratigraphy to its pre-deformation condition. A series of numerical ages (^{14}C, from other nearby sites that correlate with trench stratigraphy) help establish the latest Pleistocene-Holocene chronology, which, along with the retrodeformation, suggests that there have been four slip events in the last 12 to 25 ky, with the age of the most recent slip event at approximately 2.4 ka and an average interval between events of 3.5 to 7.5 ka [4]. This study of the Poukawa fault zone demonstrates the utility of trenching, establishment of chronology, and retrodeformation to help evaluate the earthquake hazard.

Some general hints in picking a location for trenching are as follows:

- Careful mapping from aerial photographs, topographic maps, and field work is necessary prior to trenching.
- A series of boreholes excavated prior to trenching will help determine important factors, including depth to water table, types of materials present, and possible displacements of units. Boreholes are less intrusive to the environment and will greatly assist in locating the best place for excavation.
- Fault scarps produced by normal faulting generally will have the fault trace located near the base of the scarp.
- Fault scarps produced by reverse faulting generally will have the fault trace buried approximately one-half the distance up from the base to the top of the scarp.
- Trenching of strike-slip faults is more complicated as relatively long horizontal displacements can be expected on multiple strands of a given fault. As a result, there are several basic approaches necessary for trenching strike-slip faults [5, 6]: (1) Excavate a number of closely spaced trenches at right angles to the fault, one at a time; (2) excavate progressively along the fault zone (generally a wide trench) to expose sequential features that are potential piercing points that enter and exit the fault zone on opposite sides of the trench; and (3) excavate additional trenches perpendicular to the fault, perhaps followed by trenches parallel and on each side of the faults (see Figure 8.2). The first trench excavated is known as the locator trench and spans the entire fault zone to identify the number of fault traces present. Results from this trench will help in determining strategy for subsequent trenching.

SEISMIC REFLECTION

Seismic-reflection profiling is one of a family of geophysical imaging techniques for gaining information about fault offsets without physically excavating or drilling. In this technique, the ground is shaken by a small explosive or a mechanical vibrator at one source (a shotpoint), and the vibration travels beneath the surface as waves and is detected at a series of receivers (geophones) (Figure 8.4). The path and travel time of the vibrations are sensitive to density contrasts of the underlying rocks. Through a computerized process of decoding the seismic vibration, an image of the materials beneath

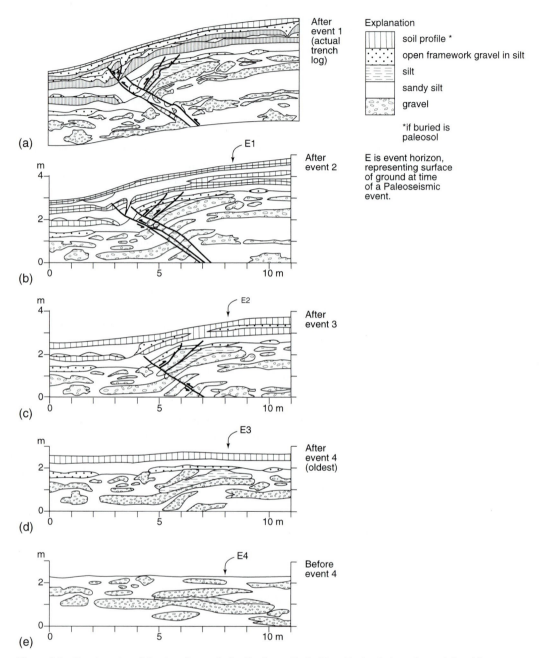

Figure 8.3   Stratigraphy of the Argyll trench site, Poukawa Fault, New Zealand. Actual trench log (a). Sequential retrodeformation back through four earthquakes (b–e) during the past 12 to 25 ky. (Modified after Kelsey et al., 1998 [4].)

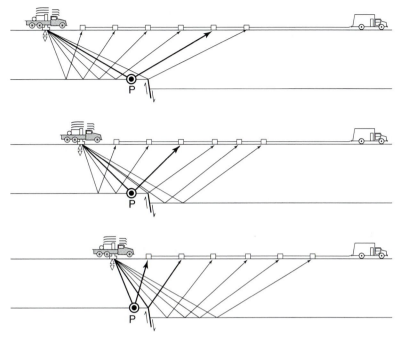

Figure 8.4   Idealized diagram showing how shallow seismic-reflection profiling can be used to identify a fault. (From Cook et al., 1980. *Scientiifc America*, 243(4): 156–157, 159–168. Scientific American, Inc. All rights reserved.)

the surface is created, including any disruptions of the sequence caused by fault offsets. Figure 8.5 illustrates a shallow seismic-reflection profile across the Santa Cruz Island fault, California. The profile was created by a ship towing a vibration source and an array of geophones. Following data processing, the data suggest a 15-m vertical separation between volcanic rocks and recent marine sediments, probably caused by the last two or three ruptures on that fault. Another subsurface imaging technique is **ground-penetrating radar**. Electromagnetic reflections from a radar source are decoded in a manner similar to that for seismic reflection.

FAULTED LANDFORMS

A landform created prior to the last surface-rupturing earthquake at a site may preserve a record of the rupture, such as the following:

- **Stream terraces**. Terraces consist of broad surfaces, sometimes datable, that can be particularly useful in studying paleoearthquakes (see Chapter 2, Figures 2.26 and 2.27, and discussion in Chapter 5). Figure 8.6 is an idealized block diagram that shows two possible scenarios of how active faulting might displace river terraces. These two scenarios illustrate that trying to determine the pattern of displacement of river terraces may be a difficult problem. Sometimes it may be difficult to tell if the scarps present are fault scarps or erosional scarps produced from river processes.

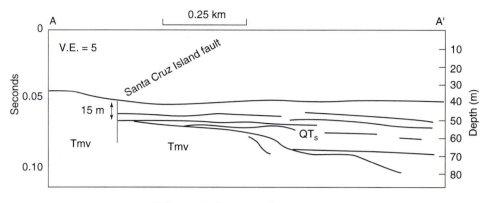

V.E. = vertical exaggeration

Figure 8.5   Interpretative cross section made from shallow seismic data across the Santa Cruz Island fault. This diagram suggests that there are approximately 15 m of vertical displacement of the Tertiary volcanic rock section (Tmv). QTs is Quaternary coastal terrace deposits. (From Pinter and Sorlien, 1991. *Geology*, 19: 909–912.)

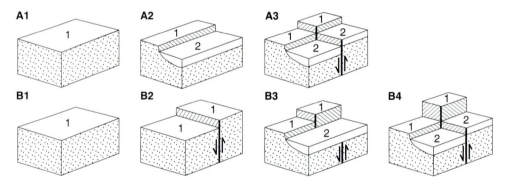

Figure 8.6   Idealized block diagrams illustrating potential complexities of interpreting river terraces that have been faulted. Sequence (a) shows the development of two river terraces (1, 2) that are subsequently faulted (A3). Sequence (B) is more complex. Terrace 1 is faulted (B2). Following faulting, terrace 2 forms (B3), and finally the sequence is faulted again (B4). Because faulting occurred at two specific times, the fault scarp for terrace 1 is higher than that for terrace 2. This illustration (B4) shows a multiple-event scarp on terrace 1. (From McCalpin, 1987 [16].)

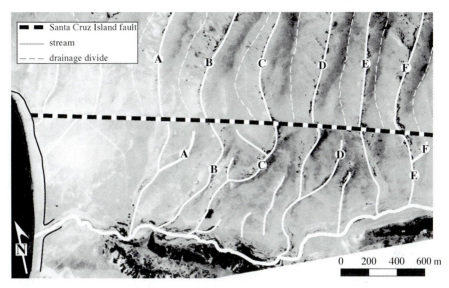

Figure 8.7   Aerial photograph of the Santa Cruz Island fault zone (roughly E to W, right central to left central between arrows) showing several offset and deflected streams (A to F). Each stream consists of a more-or-less straight reach upstream of the fault and the same downstream of the fault, with a ~300 m left deflection at the fault itself. The deflections are smooth because of erosion as the streams attempt to maintain their courses. Some of the streams have found more direct outlets, but the original downstream reaches are preserved as beheaded valleys. (Photograph courtesy of Pacific Western Aerial Survey, Santa Barbara, CA.)

- **Offset streams**. Strike-slip faults with little or no vertical component of motion will not cause large vertical deformation of terraces or other subhorizontal surfaces. However, streams that cross a strike-slip fault may be laterally offset (see Chapters 2 and 5, and Figure 2.5). Figure 8.7 shows several offset or deflected streams along the Santa Cruz Island fault in Southern California. This fault is left-lateral, and so the streams are displaced to the left if you were to follow the stream, either upstream or downstream.

- **Coastal terraces**. Like stream terraces, the broad, subhorizontal surfaces of coastal terraces are useful for measuring vertical fault motions and estimating the age of paleoearthquakes (see Chapter 6). Where a series of coastal terraces of different ages are all displaced by a single fault and the older terraces are displaced more than younger ones, we may infer several ruptures on the fault [7].

Uplift during an earthquake in a coastal area may produce a single coastal terrace. Dating the terrace then establishes when that earthquake occurred. Sometimes, along a tectonically active coastline, a series of uplifted terraces, each produced by a separate earthquake, may be present. This phenomenon has been observed in California, Alaska, New Zealand, Japan, and elsewhere. Figure 8.8a shows a series of five such terraces on Middleton Island, Alaska [8]. All five terraces in a 1947 photograph predate the 1964 M_W 9.2 great earthquake, which uplifted the coast 3.5 m at that site and pro-

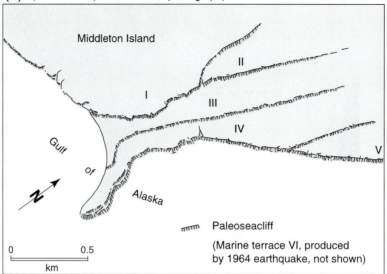

(a) (Drawn from pre-1964 aerial photograph)

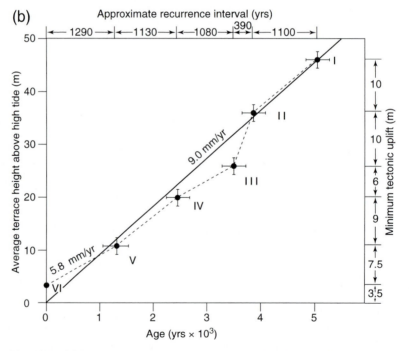

Figure 8.8   (a) Drawing from an aerial photograph (1947) of the southeastern end of Middleton Island, Alaska. The island emerged approximately 4.9 ka, and the terraces were each formed by 5 to 7 m of coseismic uplift. The 1964 earthquake caused additional uplift of approximately 3.4 m, forming a sixth terrace. (b) Graph showing the uplift of all six terraces. Average rate of uplift is 9.0 mm/yr with uplift per event varying from approximately 3.5 to 10 m. Approximate recurrence intervals between earthquakes varied from about 400 to 1300 yr. (From Plafker, 1987 [8].)

duced a sixth terrace. Figure 8.8b shows the uplift history for the six terraces and earthquakes that produced them. Chronology of the terraces is based on carbon-14 dates. Note that the average uplift rate is about 9 mm/yr, but uplift per event varies from about 4 to 10 m, and recurrence intervals from about 400 to 1300 years [8]. All five prehistoric events (I to V, Figure 8.8a) are thought to have been great earthquakes (M > 8) because they produced more uplift than did the 1964 event. Thus, to summarize the paleoseismic activity at Middleton Island, there have been six great earthquakes in this region in the last 5 ky, with average recurrence intervals (with one exception) of about 1 ky.

FAULT SCARPS

Fault scarps are the direct manifestation of surface-rupturing earthquakes. They are produced almost instantaneously as an earthquake rupture propagates to the surface. People who have observed earthquakes firsthand commonly state that the scarps and other fractures form very quickly, racing across the landscape like a giant zipper being undone [9]. Figure 8.9 shows a fault scarp over 2 m high that formed during the 1992 M_W 7.6 earthquake near Landers, California. In normal or reverse faulting, the orientation of the scarps indicates the direction of slip, but in strike-slip faulting, scarps face in different directions (see Chapter 2).

Fault scarps are slopes and, as such, have a basic morphology common to many natural slopes (Figure 8.10). Not all of the slope elements shown on Figure 8.10 may be

Figure 8.9   Fault scarp produced by the 1992 Landers earthquake (M_W 7.6). The scarp is approximately 2 m high. (Photograph by E. A. Keller.)

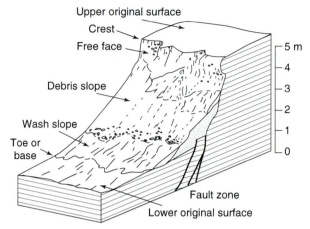

Figure 8.10 Basic slope elements that may be present on a fault scarp. (From Wallace, 1977 [10])

present on a given fault scarp; in fact, the dominance of one element over another will change with time. It is important to recognize that the different elements of a fault scarp are produced by different processes. For example, the free face, produced directly by faulting, may be nearly vertical when it forms. On the other hand, the debris slope and wash slope are related to accumulation of material at the base of the free face, and thus are associated with erosional and sediment-transport processes. **Fault-scarp degradation** proceeds at variable rates, depending on climatic conditions and the types of materials in which the scarp formed [10]. Changes in slope elements through time can be recorded as a percent of the scarp length; for example, a scarp might be composed of a free face over roughly 50% of the profile, and a debris slope over the remaining 50%. Examination of Figure 8.10 suggests some interesting aspects of fault-scarp degradation:

- The **free face** is a steep to vertical slope formed directly by rupture of the surface by an active fault. In loose, unconsolidated material, the free face would instantaneously collapse, but cohesion of the material (in the case of scarps, in sediment) holds the slope together for a period that can range from a few decades to several thousand years.
- The **debris slope** results from material deposited at the base of the free face by gravity. It starts out as a very small portion of the slope, but increases to dominate the scarp within a few thousand years. Later, the dominance of the debris slope wanes in favor of the wash slope.
- The **wash slope** is produced by deposition of material near the toe of the slope, usually by running water. This slope element is less steep than the debris slope, although there often is no abrupt boundary between the two elements. The wash slope begins to form slowly, but eventually dominates the morphology of the scarp. Therefore, very old scarps will have a gentle slope and consist primarily of wash slope.

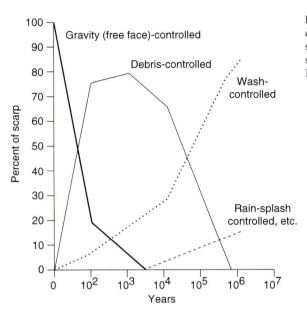

Figure 8.11   Diagram showing change in slope elements (fault-scarp morphology) through time for scarp degradation in the Basin and Range. (From Wallace, 1977 [10].)

Table 8.1

FAULT SCARP-SLOPE MORPHOLOGY.

Slope Element	Morphology	Process (Formation and/or Modification)	Comments and General Chronology
Crest	Top of fault scarp (break in slope) initially sharp becomes rounded with time	Produced by faulting; modified by weathering, mass wasting	Becomes rounded after free face disappears
Free Face	Straight segment; initially 45° vertical to overhanging	Produced by faulting; modified by weathering, gullying, mass wasting; eventually buried from below by accumulation of debris	Dominant element for 100 years or so
Debris Slope	Straight segment; angle of repose of material usually 30° to 38°	Accumulation of material that has fallen down from the free face	Is dominant element after about 100 years,
Wash Slope	Straight to gently concave segment; overlaps the debris slope; slope angle generally 3° to 15°	Erosion and deposition by water; deposition of wedge or fan of alluvium near toe of the slope; some gullying	Is developed by 100 years, significant by 1000 years
Toe	Base of fault scarp (break in slope); may be initially sharp, but with time may become indeterminate as grades into original slope	Erosion and deposition by water; owing to change in process/form from up slope element (free face, debris slope, or wash slope) to original surface below the fault scarp slope	More prominent in young fault scarps or where wash slope is not present

(After Wallace, 1977 [10])

Fault scarps have been studied intensively in the Basin and Range province of Nevada (Figure 8.11; Table 8.1) [10]. The chronology of fault-scarp degradation in the Basin and Range was developed by studying fault scarps that truncate features of known ages such as Pleistocene lake shorelines dated by the carbon-14 method. In other cases, fault scarps may be associated with volcanic ash deposits or are dated by tree rings (dendrochronology) [10].

The processes that govern the morphology and evolution of slopes formed on alluvial materials include erosion, deposition, and sediment transport. It has been recognized that these processes can be modeled quantitatively. Fault scarps are a type of slope that is particularly well suited to numerical models, because scarps form instantaneously during an earthquake and then degrade systematically thereafter. The most common application of scarp-degradation models is to estimate the age of fault scarps cut on loose, unconsolidated material. The technique is known as **morphologic dating**. In particular, models based on the **diffusion equation** are used to answer this question [11, 12]. The diffusion equation is expressed as follows:

$$\frac{\partial z}{\partial t} = \frac{K}{}\frac{\partial^2 z}{\partial x^2}$$

(8.1)

where z is elevation of a point on the slope, x is horizontal distance, t is time, and K is a constant known as **diffusivity** (Figure 8.12), which depends on climate and material (K generally varies from about 0.1 to 15 m^2 ky^{-1}). Equation 8.1 simply states that the rate of erosion or aggradation at a given point is proportional to the curvature of the profile at that point; that is, the sharpest corners on the profile will tend to change, to be smoothed out fastest.

The diffusion equation comes from physics, where it is used to describe the flow of heat across a thermal boundary. The analogy of heat flow (Figure 8.13) makes diffu-

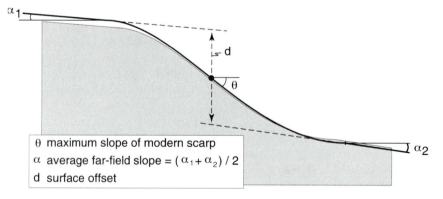

θ maximum slope of modern scarp
α average far-field slope = $(\alpha_1 + \alpha_2)/2$
d surface offset

Figure 8.12   Idealized fault scarp, illustrating parameters that can be directly measured in the field. Numerical values of these parameters are utilized in a solution of the diffusion equation to estimate the age of the scarp, and thus when the earthquake occurred.

sion degradation of fault scarps a bit clearer. Elevation on the scarp profile, z, is equivalent to temperature in the physics problem. The conditions at t = 0 are of an abrupt juxtaposition of high temperature against low temperature at the thermal boundary. Just as heat flows from the region of high temperature to low, sediment is transported from the region of high elevation to low elevation on a degrading slope. The overall rate of flow is determined by the conductive properties of the media in the case of heat flow, and by properties of erosivity (climate) and erodability (texture) in the case of sediment transport. The result in both cases is that the profile becomes increasingly subdued with time.

Applying a diffusion model or a solution of the diffusion equation to morphologically date a fault scarp requires several general prerequisites and assumptions:

- The slope must be transport limited.
- The scarp has formed instantaneously in a single earthquake.
- The sediment must be assumed to be nearly cohesionless.

Transport limited refers to a situation where there is an abundant supply of loose sediment, and degradation of the slope is controlled only by the dynamics of sediment transport. Slopes that are controlled by the rate of weathering (**weathering limited**; for example, bedrock scarps) are fundamentally different. The second requirement is that the scarp formed in one earthquake and was not built by several small offsets on the same fault. Third, diffusion models generally assume that the free face of the scarp crumbled almost instantaneously after the earthquake event and reached the angle of repose.

Given these assumptions, fairly simple solutions to the diffusion equation are possible. For example [11],

$$Kt = \frac{d^2}{4\pi} \frac{1}{(\tan\Theta - \tan\alpha)^2} \qquad (8.2)$$

where d is vertical displacement, θ is maximum scarp angle, and α is far-field slope (Figure 8.12). Note that solutions like Equation 8.2 always include both diffusivity and time; neither of those parameters can be solved independent of the other. On a scarp of known age, a profile measured across it allows us to determine the value of diffusivity. Where two fault scarps occur in approximately the same climate and material, one scarp of known age and the other of unknown age, we can calculate the value of diffusivity on the dated scarp and use it to estimate the age of the undated scarp (Figure 8.14).

Since mass diffusion was introduced as a model for fault-scarp degradation, several complications and refinements have been proposed. For example, in addition to lithology and regional climate, the value of diffusivity can be affected by slope orientation, creating variability between north- and south-facing scarps even in the same vicinity [13]. It has also been found that the gradient of the surface away from the effects of the fault, or "far-field slope", can have first-order effects on age determinations by the diffusion equation [14].

Scarps produced from repeated displacement on strike-slip and normal faults also have been analyzed by morphologic dating based on modeling of change in slope morphology. For example, a study of a fault scarp produced by faulting during the past 30 ky

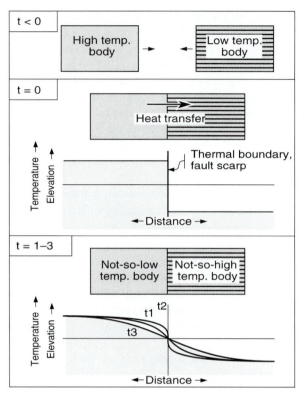

Figure 8.13   Heat transfer from a warm mass to a cold one illustrates the principle of diffusion, which is used to model fault-scarp degradation. Just as heat flows from the hot material (high thermal potential) to the cold material (low thermal potential) in thermal diffusion, sediment erodes from the top of a slope (high gravitational potential) and is deposited at the base (low gravitational potential).

	Heat flow	Scarp degradation
Process	Heat transfer	Gravity-driven mass transfer
Conditions at t = 0	Bodies of unequal temp. come into contact	Faulting creates abrupt step in topography
Boundary	Thermal boundary	Fault scarp
Constant	Thermal diffusivity	Mass diffusivity
Vertical axis of profile	Temperature	Elevation

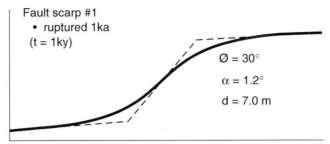

Fault scarp #1
- ruptured 1ka
  (t = 1ky)

$\varnothing = 30°$

$\alpha = 1.2°$

$d = 7.0$ m

Using equation 8.2:

$$K \; (1ky) = \frac{(7.0 \text{ m})^2}{4\pi} \; \frac{1}{[(\tan(30°) - \tan(1.2°)]^2}$$

$$K = 12.62 \text{ m}^2 \, / \text{ ky}$$

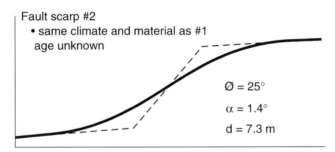

Fault scarp #2
- same climate and material as #1
  age unknown

$\varnothing = 25°$

$\alpha = 1.4°$

$d = 7.3$ m

Using the value of k from scarp #1:

$$(12.6 \text{ m}^2/ \text{ ky}) \, t = \frac{(7.3 \text{ m})^2}{4\pi} \; \frac{1}{[(\tan(25°) - \tan(1.4°)]^2}$$

$$t = \; 1.72 \text{ ky}$$

Figure 8.14   Idealized diagram showing two fault scarps. For fault scarp 1, the date of the earthquake is known and diffusivity, K, is calculated. This value of diffusivity (K) is used for fault scarp 2 to estimate that the earthquake occurred at approximately 1.72 ka.

in the San Andreas fault zone at Wallace Creek in the Carrizo Plain (see Figure 5.4) produced values of diffusivity with weighted mean of 8.6 ± 0.75 m^2 ky^{-1}. This value of K is then applied to normal fault scarps in the same region to estimate age of scarps and slip rates [15]. This study was one of the first successfully to model composite scarps produced by many earthquakes.

STRATIGRAPHIC EVIDENCE FOR EARTHQUAKES

Stratigraphic evidence for paleoseismicity is one type of **event stratigraphy**. This name is used because we are discussing features found in stratigraphic sequences that indicate past and rare events; in this case our concern is past earthquake events. These features include displaced strata, colluvial wedges, sand boils, fissure filling, and evidence of abrupt burial.

Displaced Strata. The clearest evidence of past earthquakes found in fault exposures is displaced strata. Often when a trench is exposed or a natural exposure is examined, there may be several fault strands present. Some of these will displace older material and be buried by younger material, whereas other nearby fault strands may cut the entire sequence. Careful evaluation of the displacement history can help establish the number of faulting events that have occurred. For example, Figure 8.15 shows

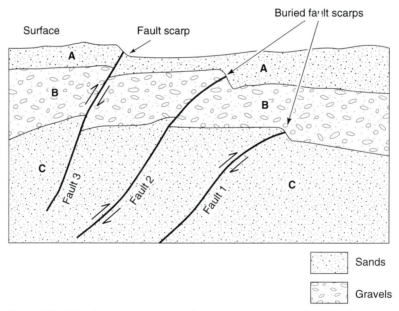

Figure 8.15 Trench exposure showing displacement of sand and gravel deposits, buried fault scarps, and a surface fault scarp. Fault 1 displaces only unit C. Fault 2 displaces B and C, and fault 3 displaces A, B, and C. This stratigraphy, along with buried fault scarps and the surface fault scarp, suggests that three discrete faulting events occurred. The oldest faulting event occurred on fault 1 and the youngest on fault 3.

an idealized diagram of sand and gravel deposits that have been faulted by three splays of a fault system. Fault 1 cuts unit C and has left a buried fault scarp between B and C; fault 2 cuts units B and C and has left a buried fault scarp between A and B; and fault 3 cuts units A, B, and C and has a fault scarp at the surface. One way to interpret this stratigraphy is that there have been three earthquake events: the oldest on fault 1, followed by rupture on fault 2, and the youngest rupture on fault 3. Often the stratigraphy exposed in a fault trench is much more complicated, with sedimentary units folded and faulted in complex ways.

Colluvial Wedges. Unconsolidated material found at the base of steep slopes is known as **colluvium**. Colluvial deposits generally are more angular than stream deposits, not having been transported very far. Following an earthquake that produces a fault scarp, a colluvial wedge may form at the base of the fault scarp as the free face degrades to a debris slope. Sometimes these colluvial wedges may be buried and preserved in the stratigraphic record as evidence of past earthquakes (Figure 8.16). As

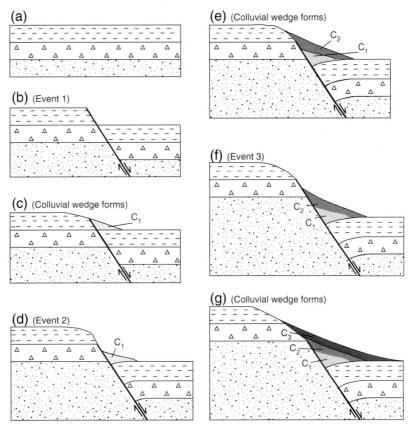

Figure 8.16   Development of a three-event fault scarp. Each faulting event is followed by the generation of a fault-scarp colluvial wedge (C_1, C_2, and C_3). (After McCalpin, 1987 [16].)

Figure 8.16 illustrates, the colluvial wedges may be rotated toward the fault as displacement continues. There may also be some soil development on top of the colluvial wedge, because there often is sufficient time between events for a weak soil to form [16]. If fault-scarp colluvial wedges can be recognized in exposures and dated, this information is valuable in working out the timing of past earthquake events.

Reverse faults also produce colluvial wedges. As a reverse fault ruptures the surface, it has an oversteepened front that quickly collapses, forming a colluvial wedge (Figure 8.17). The hanging wall may be warped, forming **drag folds**, so called because they are dragged (folded) by displacement along the fault plane. Material quickly collapses from the leading edge of a reverse fault, depositing colluvium below the fault as quickly as the area above is uplifted.

Sand Boils. Sand-boil deposits, sometimes also called "sand craters", have been associated with many earthquakes (see discussion of liquefaction in Chapter 1). At the surface, they are characterized by low mounds of sand that have been extruded from fractures. Layered sedimentary deposits beneath the surface may, during an earthquake, experience high fluid pressure and liquefy. Fluidized sand is erupted to the surface, forming a circular apron of sand around the vent (Figure 8.18a). Evidence of sand boils in a stratigraphic sequence (Figure 8.18b) represents liquefied sand extruded onto the surface and later buried by other materials. When such deposits are identified in fault exposures, they are evidence of a past earthquake event, but they are not con-

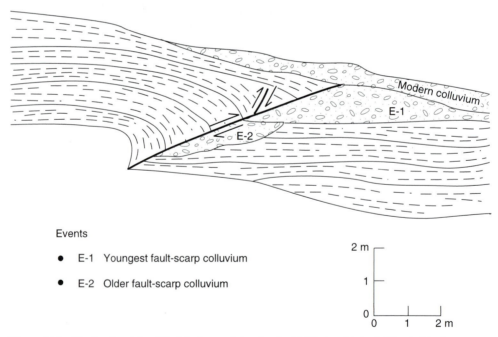

Events

●   E-1   Youngest fault-scarp colluvium

●   E-2   Older fault-scarp colluvium

Figure 8.17   Idealized and simplified fault-trench log for the McKinleyville fault, Humboldt County, California, showing two events (E-1 and E-2) of faulting and formation of fault-scarp colluvium. (From Carver, 1987. In A. J. Crone and E. M. Omdahl (eds.), *Directions in Paleoseismology*, U.S. Geological Survey Open-File Report 87–673, 115–128.)

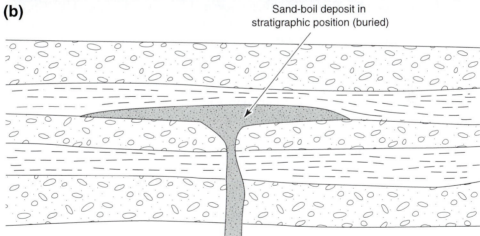

Figure 8.18   (a) Sand boil produced by the 1989 Loma Prieta (M_W 7.2) earthquake. Note street curb for scale. (b) Idealized diagram showing how a sand boil may appear in the stratigraphic record following burial. It is important to keep in mind, however, that sand boils are also produced by processes other than earthquakes. (Photograph courtesy of D. Laduzinsky.)

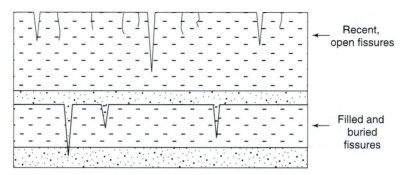

Figure 8.19   Recent, open fissures and older, filled and buried fissures. As with sand boils, fissures are not absolute proof of earthquakes because they may be produced by several other processes as well as earthquakes.

clusive evidence because sand boils also can form without earthquakes (for example, by discharge of sediment-laden runoff routed from preexisting fissures) [17].

Fissure Filling. Large earthquakes may form numerous fissures and cracks (Figure 8.19). These are unlikely to remain open for very long because the sides of the fissures and cracks are very steep. Material from the surface and from the sides of the fissures soon fills them. Eventually new material may be deposited over the filled fracture. When these features are recognized in fault exposures, they may indicate past earthquakes. As with sand boils, fissures can form without earthquakes (for example, due to landsliding or groundwater withdrawal).

Abrupt Burial. Figure 8.20 shows a "ghost forest", the trees of which were killed suddenly by a subsidence event approximately 300 years ago. It has been suggested recently that abrupt burial of forest or of salt-marsh deposits is characteristic of large subduction-zone earthquakes along the Cascadia Subduction Zone (see discussion in Chapter 6) [18]. The hypothesis is that during great subduction-zone earthquakes, parts of the coast may subside by a meter or so. Until recently it was thought that no great earthquakes occurred along the Washington and Oregon coasts; however, the discovery of the abrupt burial of salt-marsh and forest environments suggests that this is not the case. In several locations from Northern California to Washington there is evidence of two or more episodes of abrupt apparently coseismic burial.

FAULT-ZONE SEGMENTATION

The phenomenon of **fault-zone segmentation** has been recognized for over 20 years [19]. The basic idea is that for long faults, an earthquake seldom ruptures along the entire length of the fault. More commonly, only one, or perhaps two, segments rupture during a large event. For example, on a regional scale, the San Andreas fault of California is subdivided into four segments (Figure 8.21) [10]. Segment boundaries are based on the historical behavior of the fault. The south-central segment ruptured in 1857, producing a great earthquake. The northern segment ruptured in 1906. The southern segment has not produced a major earthquake ($M_W > 7$) in historical time.

Figure 8.20   Photograph of the "ghost forest" on the coast of western Washington. It is hypothesized that these trees were killed approximately 300 years ago by a giant earthquake that caused subsidence, submerging the trees below sea level. (Photograph courtesy of United States Geological Survey and B. Atwater.)

The simple regional segmentation of the San Andreas fault (Figure 8.21) might lead you to believe that the concept of fault segmentation is simple. Unfortunately, this is not the case. There is general agreement that faults may be segmented at a variety of scales (from a few meters to several tens of kilometers in length) [20]. The most basic approach to fault-zone segmentation is to define **earthquake segments** based on rupture behavior during earthquakes. Earthquake segmentation may be determined from historical earthquakes or by paleoseismic evaluation. An earthquake segment is defined as those parts of a fault zone that rupture as a unit during an earthquake [21]. In addition, segmentation based on earthquake activity is only one approach. **Structural segmentation** can be recognized by changes in fault-zone geomorphology or fault-trace orientation, such as bends, step-overs, and separations or gaps [22]. Structural segmentation also occurs where segments intersect with other faults or folds. The end of a segment is a structural discontinuity. Finally, structural segmentation can also result from changes in geologic materials along a fault zone or local heterogeneities along a fault plane [20, 22]. An improved understanding of all types of fault segmentation is necessary to better understand earthquake mechanics, including why and where ruptures start and terminate [20]. Of particular importance for major earthquakes ($M_W > 7$) is the need to know how rupture of multiple fault segments occurs during earthquakes [22].

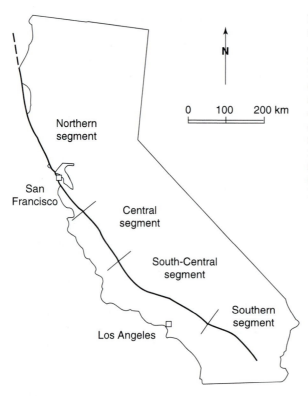

Figure 8.21   Simplified segmentation model for the San Andreas fault zone, California. The northern segment ruptured in the 1906 San Francisco earthquake; the central segment is characterized by relatively frequent modern earthquakes and creep; the south-central segment most recently ruptured in a great earthquake in 1857; and the southern segment has not ruptured in historical times. (From Schwartz and Coppersmith, 1986 [24].)

The concept of fault segmentation is important because it has implications for the following:

- Long-term earthquake forecasting involving probabilistic assessment of seismic hazard (see Chapter 1). Conditional-probability analysis requires information on slip rate, slip per event, and recurrence interval of earthquakes for specific fault segments.
- Estimating the maximum earthquake likely to occur on a particular fault [19] (see Figure 1.32).
- Estimating ground motion produced by an earthquake. It is believed that rupture propagation is related to fault-zone structure, and thus to segmentation [20].
- Identifying areas along a fault zone where earthquakes nucleate (rupture starts), as well as areas that may act as barriers to earthquake rupture (where rupture ends) [21].
- Better understanding the fundamental mechanisms associated with earthquake generation and faulting.
- Better understanding the complexities of large, damaging earthquakes that produce ruptures along several geometric and structural segments of a fault zone. This is particularly important because it is a common practice to map a particular fault zone by linking two or more segments together. As total length of a fault zone increases, the magnitude of the largest probable earthquake increases.

CASE STUDY: SEGMENTATION AND PALEOSEISMICITY OF THE WASATCH FAULT ZONE, UTAH

The Wasatch fault zone (WFZ) is one of the most-studied normal fault zones in the western United States. Extending nearly 350 km from southern Idaho southward past Great Salt Lake and Salt Lake City, it is also one of the longest. Paleoseismic evaluation of the WFZ resulted in the development of the concept of characteristic earthquakes (see Chapter 1) as well as an improved understanding of fault-zone segmentation [21, 23, 24].

Although there have not been historical earthquakes on the Wasatch fault zone (Figure 8.22), there is paleoseismic evidence that it has ruptured several times in the Holocene (last 10 ky); the most recent event was about 400 years ago [21, 23, 24]. The WFZ is generally coincident with the boundary between the Wasatch Mountains and the adjacent basin in which Great Salt Lake and Utah Lake are located (see Figure 8.22). Proposed fault segments for the WFZ (Figure 8.22) have boundaries that generally coincide with geometric or structural changes, such as changes in surface trend of the fault zone; major salients (a salient is defined structurally as a block of bedrock that extends into the basin, or geomorphically as a landform that extends outward from surrounding topography); and other geologic and structural discontinuities, such as older thrust faults or cross faults [24]. There has been discussion of how many segments are present in the WFZ [21, 23], but there is general agreement that segmentation is very important in characterizing the past earthquake activity and evaluating the seismic hazard.

During the past 30 years, over 40 trenches have been excavated to gather paleoseismic data at 18 sites along the segments of the WFZ. Table 8.2 summarizes some of the information obtained from trench studies and studies of natural fault exposures. These data [21, 23, 24] for the WFZ suggest the following:

- The northern (Collinston) segment has not been active in the Holocene.
- Assuming that each prehistoric event ruptured most of a particular segment (30 to 70 km), past earthquakes range from strong to major (M_W 6.5 to M_W 7.7). These magnitudes are consistent with historical earthquakes in the Basin and Range.
- Segments 2 to 5 have experienced multiple strong to major (M_W 6.5 to M_W 7.7) earthquakes during the Holocene, some in the past few hundred years.
- There is no evidence for a major earthquake on the WFZ in the past 400 years.
- During the past 5.5 ky, there has been a strong to major earthquake (M_W 6.5 to M_W 7.7) about every 400 years somewhere on segments 2 to 6.
- The central portion of the fault zone (segments 2 to 5) is more active (more events, higher slip rates, and lower recurrence interval) than the northern or southern ends (segments 1 and 6).
- Approximate Holocene recurrence intervals on each of the more active center segments (segments 2 to 5) vary from about 1 to 4 ky.

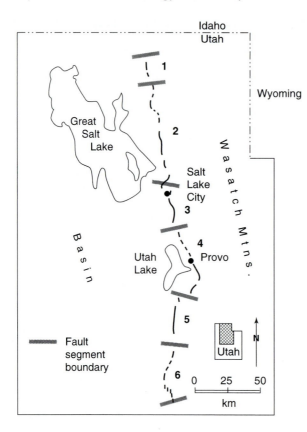

Figure 8.22 Map of the Wasatch fault zone, Utah, showing six fault segments. The fault system has normal displacement, with the upthrown side being the Wasatch Mountains. Dark bands define boundaries to fault segments and generally correspond to geometric and structural discontinuities along the fault zone. (After Schwartz and Coppersmith, 1986 [24].)

Table 8.2

PALEOSEISMIC EVALUATION FOR THE SIX SEGMENTS OF THE WASATCH FAULT ZONE SHOWN IN FIGURE 8.22.

Segment	Length (km)	Approximate displacement per event (m)	Approximate slip rate (mm/yr)	Approximate recurrence interval (yr)	Comment
1	30	—	—	>10,000	No known surface displacement past 13,500 yr
2	70	1.6	1.3 (+0.5, -0.2)	1000 to 1500	4 Holocene events. Most recent about 500 yr ago, oldest 4000 yr
3	35	2.0	0.76 (+0.6, -0.2)	1500 to 3500	2 Holocene events. Most recent about 1500 yr ago, oldest 5000 yr
4	55	1.6 to 2.3	0.85 to 1.0	150 to 3000	3 Holocene events. Most recent about 500 yr ago, oldest 5000 yr ago
5	35	2.3	1.27 to 1.36	1500 to 2000	3 Holocene events. Most recent about 400 yr ago
6	40	2.5	less than 0.35	7000	1 Holocene event about 1000 yr ago

(Data from Schwartz and Coppersmith, 1984 [23]; Schwartz and Coppersmith, 1986 [24]; and Machette et al., 1989 [21])

Evaluation using detailed age control of faulting on the WFZ [21] suggests:

- During the past 1 ky, there has been one major earthquake about every 200 years somewhere on segments 2 to 6.
- There is no evidence for strong clustering of earthquakes between about 1 and 5.5 ka.

In summary, paleoseismic evaluation of WFZ suggests that large, damaging earthquakes have occurred on several fault segments in the past few thousand years. The most recent events occurred 400 to 500 years ago. In the past 1 ky, there has been a temporal clustering of events on the central segments (2 to 5), with strong to major earthquakes (M_W 6.5 to M_W 7.7) about every 200 years. This paleoseismic information is critical for developing strategies to minimize future earthquake damage to urban areas such as Salt Lake City and Provo, Utah.

Study of the Wasatch fault zone suggests that, for this fault zone, each fault segment has particular geomechanical properties that lead to repeated earthquakes on that segment through time with approximately equal magnitudes and amounts of displacement [23]. Some investigators have used this to infer that the recurrence time or recurrence interval is approximately the same as well; this would follow if (1) the slip rate of the fault, (2) the magnitude of the earthquake, and (3) the displacement per event were all constant. Under these circumstances, the average recurrence time is equal to the ratio of displacement per event to the slip rate [25].

Figure 8.23 relates the three parameters above, based on the assumption that a M_W 7 earthquake will produce a 1-m characteristic displacement, a M_W 8 will produce 5 m, and a M_W 9 will produce 20 m. Figure 8.23 does appear to provide "ballpark" figures for recurrence intervals of earthquakes. However, when we examine the history of a particular fault and have sufficient data to estimate slip rates and recurrence intervals, it is not uncommon to find that slip rate and time between events vary considerably for different fault segments, as illustrated by the WFZ example. The more we learn about earthquakes, the more it becomes apparent that there is a general lack of uniformity in earthquake frequency for a particular fault zone composed of several segments. Nevertheless, although average slip rates and recurrence intervals have limitations in predicting the magnitude and time of the next rupture, they do provide useful guidelines for land-use planning, building codes, and engineering design [26].

MODELS OF EARTHQUAKE RECURRENCE

Several models have been proposed to describe earthquakes and slip that recur on fault zones [27, 28]. If a particular fault or fault segment tends to generate earthquakes with about the same maximum magnitude, then that fault is said to generate **characteristic earthquakes**. In the characteristic earthquake model, (1) the displacement per event at a point on the fault is constant, (2) the slip rate along the length of the fault or fault segment may be variable, and (3) the size of large earthquakes is nearly constant (the range in magnitude for an event is narrow and near the maximum, and moderate events are infrequent) [27]. A competing concept is the **uniform-slip model** [28]. In the uniform-slip

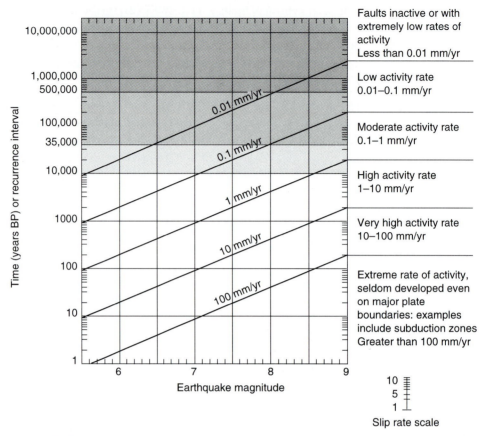

Figure 8.23   Relationships between recurrence interval, slip rate, and earthquake magnitude. (After Slemmons and Depolo, 1986. In *Active Tectonics*,. Washington, DC: National Academy Press , 45–62.)

model there is (1) constant displacement per event at a point along the fault, (2) constant slip rate along the length of the fault or fault segment, and (3) constant size of earthquakes (more frequent moderate earthquakes may occur). The main difference between the characteristic-earthquake and uniform-slip models is that the characteristic-earthquake model predicts the recurrence of large earthquakes with infrequent, moderate events, whereas the uniform-slip model allows for more frequent moderate events. A third model is known as the **variable-slip model**, in which the amount of slip and the length of rupture may both vary from one earthquake to the next, producing variability in earthquake size [27, 29].

As yet, we do not have sufficient paleoseismic data on individual faults to determine which of the models (characteristic earthquake, uniform slip, or variable slip) best characterizes recurrent earthquakes and slip on faults. In fact, all three models are apparently present in nature. The characteristic earthquake model seems to fit for the Wasatch fault and perhaps for some segments of the San Andreas fault [27]. The uniform-slip model may also be invoked for some segments of the San Andreas fault [28]. Still other faults are likely to have displacement histories and earthquakes that are best represented by a variable-slip model. Variable slip need not be a random pattern of earthquake size and displacement, as sometimes suggested [27]. Rather, it is hypothesized that the more we learn about the behavior of particular faults (especially reverse faults; see Figure 1.6), the more variable (but perhaps systematically variable) the pattern of earthquakes is likely to be. For example, three subduction-zone earthquakes (reverse faulting) in 1707, 1854, and 1946, which uplifted the coastal area near Kyoto, Japan (Nankaido earthquakes), produced rupture lengths that varied from 300 to 500 km and uplift that varied from 1.2 to 1.8 m. Similarly, a series of earthquakes off the Colombia-Ecuador coast in 1906, 1942, 1958, and 1979 ruptured the same 500-km length of subduction zone; the first event had a seismic moment over five times greater than the moments of the other three events, suggesting that the fault slip in the 1906 event was much greater than the other three [29]. The Japan and Colombia-Ecuador events support the variable-slip model.

EARTHQUAKE CLUSTERING

Some faults may be characterized by clusters of earthquakes over periods of a 1000 years or so, followed by long, quiet periods before new clusters of events occur. This pattern of activity may fit the Oued Fodda fault, which produced the 1980 Algeria earthquake (M_W 7.3). Paleoseismic investigation suggests that the three most recent earthquakes occurred during the last 900 years (including the 1980 event) and provides a recurrence interval of about 450 years. However, these recent earthquakes represent a clustering of events large enough to produce surface rupture. During the late Pleistocene, the earthquake activity on the fault was concentrated in relatively short periods, with earthquakes occurring every few hundred years, separated by much longer periods (thousands to tens of thousands of years) with no large earthquakes [30].

Earthquakes on a long, segmented fault system sometimes occur in a rapid, progressive sequence. For example, in the twentieth century, subsequent to a 1912 event, a remarkable series of M_W > 6.7 earthquakes occurred generally from east to west on the North Anatolian fault in Turkey (Figure 8.24), resulting in surface ruptures along a 1000-km section of the fault. Two events in 1999—M_W 7.4, Izmit (September 14) and M_W 7.1 Duzce (November 12)—produced right-lateral strike slip of 1.5 to 5 m [31]. The sequence has been described as a "falling-domino" scenario (the 1939, 1942, 1943, 1944, 1957, 1967, and two 1999 events), where one earthquake sets up the next, eventually rupturing nearly the entire length of the fault in a cluster of events [32]. Clusters of earthquakes that form the progressive sequence, with time between events a few months to a decade or so, are apparently separated by seismic quiescence of several hundred years. The process that facilitates the sequence may be a stress transfer, where an earthquake alters the stress field on surrounding faults or fault segments, increasing the probability of subsequent earthquakes. As this process continues, a cluster of events occurs. For example, during the 75-year time period prior to the M_W 7.7 1906 earthquake on the San Andreas fault in the San Francisco Bay region, there were 14 earthquakes equal or exceeding M_W 6. In the 75 years following the 1906 event, only one M_W 6 occurred. The inference is that earthquake interactions occur, and as a result the rate of seismicity is

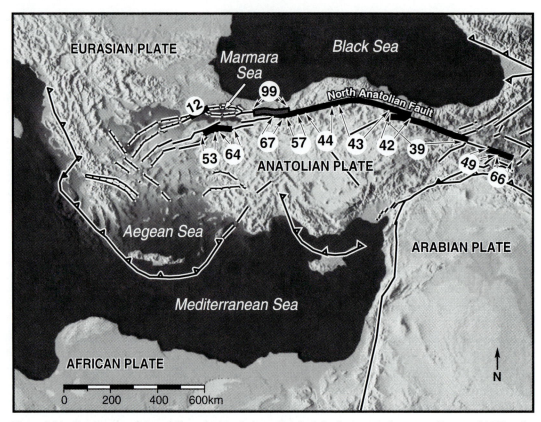

Figure 8.24 Earthquakes ($M_W > 6.7$) on the North Anatolian fault in the twentieth century. Events of 1992 and 1951 are not shown. Arrows show approximate limits of rupture in year indicated (i.e., 57 = 1957). Year 1999 shows rupture length of two events combined (Izmit, August; and Duzce, November). (Modified after Reilinger, et al., 2000. *GSA Today*, 10(1):1–6.)

not constant. In terms of understanding the earthquake hazard, this means that the probability of an earthquake on one fault is not independent of the probability of an event on another nearby fault or another segment of the same fault [32].

CASE STUDY: TWELVE CENTURIES OF EARTHQUAKES ON THE SAN ANDREAS FAULT, SOUTHERN CALIFORNIA

Undoubtedly one of the most remarkable sites for paleoseismic studies in the world is Pallett Creek, located approximately 55 km northeast of Los Angeles, on the San Andreas fault. The site contains evidence of ten large earthquakes, two of which occurred in historical time (1812 and 1857). High-precision radiocarbon analyses provide accurate dating of most of the eight prehistoric events, extending back to approximately 671 A.D. (Figure 8.25) [33]. When the dates with their accompanying error bars are plotted, it is apparent that the earthquakes are not evenly distributed through time, but tend to be

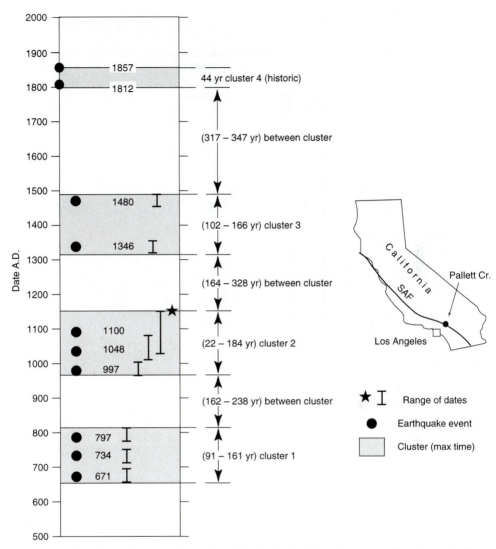

Figure 8.25   Graph showing chronology of two historical and eight prehistoric earthquakes on the San Andreas fault at Pallett Creek. (Data from Sieh et al., 1989 [33].)

clustered (Figure 8.25). Most of the earthquakes within clusters are separated by a few decades, but the time between clusters varies from approximately 160 to 350 years. The earthquakes were identified through careful evaluation of natural exposures and fault trenches at Pallett Creek [33]. Figure 8.26 shows part of the trench wall and the offset of organic-rich sedimentary layers that correspond to an earthquake that occurred approximately A.D. 1100.

Of course, the question is, When will the next big earthquake occur on the San Andreas fault northeast of Los Angeles? Studies using conditional probability suggest a 26% probability for the 30-year period from 1994 to the year 2024 (see Figure 1.22). This seems reasonable in light of the history of clustering

Figure 8.26   Part of a trench wall crossing the San Andreas fault zone at Pallett Creek showing a 30-cm offset. Notice that the strata above the 22-cm scale (central part of photograph) are not offset. The offset sedimentary white beds layers and dark organic layers were produced by an earthquake that occurred approximately 900 years ago during cluster 2 (see Figure 8.25). (Photograph courtesy of K. Sieh.)

of events on the fault. Assuming that the two historical events are indeed a cluster, and that clusters can be separated by approximately 160 to 350 years, then the next large earthquake could be expected between the years 2017 and 2207. Thus, a 26% probability for an event by approximately the year 2024 seems reasonable. On the other hand, the data set at Pallett Creek, although the most complete record for any fault in the world, is still a study in small numbers. For example, during cluster 3, the time period between events could have been as long as 166 years. Thus we might argue that another earthquake in cluster 4 could happen as late as the year 2023 without violating the history at Pallett Creek. However, this seems unlikely given the fact that almost all the time intervals during clusters are closer to 40 to 50 years.

CONDITIONAL PROBABILITIES FOR FUTURE EARTHQUAKES

Geologists in California have estimated conditional probabilities for major earthquakes along segments of the San Andreas fault for the 30-year periods from 1994 to 2024 for Southern California, and 2000 to 2030 for Northern California (Figures 1.22 and 1.23). These probabilities are based on a synthesis of historical records and geologic evaluation of prehistoric earthquakes (see Chapter 1). Results of an earlier study [34] suggested that the Parkfield segment is almost certain to rupture by the year 2018. This estimate is based in part on the observation that moderate to large earthquakes have occurred on the Parkfield segment in the years 1857, 1881, 1902, 1922, 1934, and 1966, or on average every 21 to 22 years. The southern segment of the fault in the Coachella Valley has been assigned a probability of approximately 22% to produce a major earthquake by the year 2024. A study, compiled in early 1989, assigned a probability of about 30% for a major event on the San Andreas fault in the Santa Cruz Mountains, where the M_W 7.2 Loma Prieta earthquake occurred on October 17, 1989. The earthquake is believed to support the validity of the conditional probability approach. On the other hand, the 1992 M_W 7.6 Landers earthquake occurred east of the San Bernardino Mountains. That event caused ground rupture over nearly 100 km, with maximum right-lateral displacement of about 6 m. Because the displacement is almost entirely pure strike-slip, the fault that ruptured is probably part of the right-lateral plate boundary system between the Pacific and North American Plates. Although in retrospect there is evidence for moderate to large earthquakes in the past on the fault that ruptured in 1992, it was not generally recognized that the next right-lateral earthquake in southern California would be so far to the east of the San Andreas fault. There is speculation now that the San Andreas fault, and thus the boundary between the North American and Pacific Plates, is either wider than we thought or is in the process of migrating eastward. Of course, this does not mean that large earthquakes will not continue to occur on the main San Andreas fault. The conditional probabilities shown in Figures 1.22 and 1.23 are our best estimate of the probabilities of large earthquakes occurring in the next few decades on the San Andreas fault.

EARTHQUAKE PREDICTION

Short-term **prediction**, or forecasting, of earthquakes is an area of serious research. The term **forecasting** is preferred by some scientists because, like forecasting the weather, an earthquake forecast specifies a time period for the event to occur and a probability of occurrence. The Japanese made the first attempts at earthquake prediction, with some success, using the frequency of microearthquakes, repetitive geodetic surveys, water-tube tiltmeters, and geomagnetic observations. They found that earthquakes in the areas they studied were nearly always accompanied by swarms of microearthquakes that occurred several months before the major shocks. Furthermore, ground tilt correlated strongly with earthquake activity. Anomalous magnetic fluctuations were also reported [35].

Chinese scientists made the first successful prediction of a major earthquake in 1975. The M_W 7.3 Haicheng earthquake of February 4, 1975, destroyed or damaged about 90% of structures in a city of 9000 people. The short-term prediction of that event was based primarily on a series of foreshocks that began four days prior to the main event. On February 1 and 2, there were several shocks with a M < 1. On February 3, less than 24 hours before the main shock, a foreshock of M 2.4 occurred, and in the next 17 hours, eight shocks with M > 3 occurred. Then, as suddenly as it began, the foreshock activity became relatively quiet for 6 hours until the main earthquake struck [36, 37]. The lives of thousands of people were saved by massive evacuation from potentially unsafe housing just before the earthquake.

Unfortunately, foreshocks do not always precede large earthquakes. In 1976, a catastrophic earthquake—one of the deadliest in recorded history—struck near the mining town of Tangshan, China, killing several hundred thousand people. There were no foreshocks!

Optimistic scientists around the world today believe that eventually we will be able to make consistent, long-range forecasts (tens to a few thousand years), medium-range predictions (a few years to a few months), and short-range predictions (a few days or hours) for the general locations and magnitudes of large, damaging earthquakes. Unfortunately, earthquake prediction is still a complex problem and it will probably be many years before dependable short-range prediction is possible. Such predictions most likely will be based on precursory phenomena such as the following:

- Deformation of the ground surface
- Seismic gaps along faults
- Patterns and frequency of earthquakes
- Changes in electrical resistivity of the Earth
- Changes in the amount of dissolved radioactive gases (radon) in groundwater
- Perhaps, anomalous behavior of animals

Preseismic Uplift and Subsidence. Rates of uplift and subsidence, especially when rapid or anomalous, may be significant in predicting earthquakes. For example, for more than ten years before the 1964 earthquake near Niigata, Japan (M 7.5), there was broad uplift of the Earth's crust of several centimeters (Figure 8.27) [38]. Similarly, broad, slow uplift of several centimeters occurred over a five-year period prior to the 1983 M_W 7.7 Sea of Japan earthquake (Figure 8.27, inset map). The mechanism responsible for the uplift is thought to be deep, stable fault slip prior to the earthquakes [39]. Less well understood, but possibly important observations include preinstrument measurements of 1 to 2 m of uplift preceding large Japanese earthquakes in 1793, 1802, 1872, and 1927. The uplift was recognized by sudden oceanward shifts of the coastline of as much as several hundred meters. For example, on the morning of the 1802 earthquake (about 10 A.M.), the sea suddenly withdrew about 300 m from a harbor in response to a seismic (preseismic) uplift of about 1 m. Four hours later, at 2 P.M., the earthquake struck, destroying many houses and uplifting the land another meter, causing the sea to withdraw farther [39].

Seismic Gaps. Seismic gaps are defined as areas along active fault zones that are capable of producing large earthquakes but have not produced one recently. These areas are thought to store tectonic strain and thus are candidates for future large earthquakes [40].

Seismic gaps have been useful in medium-range earthquake prediction. At least ten large, plate-boundary earthquakes have been successfully forecast since 1965, including one in Alaska, three in Mexico, one in South America, and three in Japan. In the United States, seismic gaps are currently recognized along the San Andreas fault near Fort Tejon, California, where the fault last ruptured in 1857, and in the Coachella Valley, a segment of the fault that has not produced a great earthquake for several hundred years. Both sections are likely candidates to produce a great earthquake in the next few decades [40, 41].

As earth scientists examine patterns of seismicity, two ideas are emerging. First, sometimes there are reductions in small to moderate earthquakes prior to a larger event. For example, prior to the 1978 (M 7.8) Oaxaca, Mexico, earthquake there was a 10-year period (1963 to 1973) of relatively high seismicity of earthquakes (mostly M 3 to M 6.5),

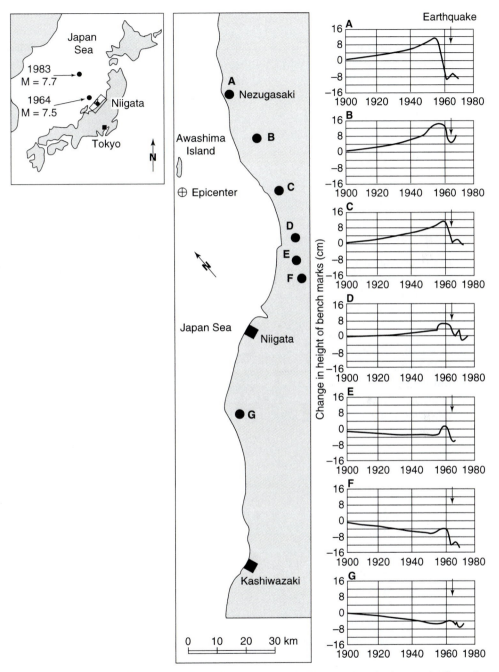

Figure 8.27   Anomalous uplift that occurred during the years before the M 7.5 earthquake that struck Niigata, Japan, in 1964. Uplift was measured by plotting changes in the benchmarks (points of known elevation) with time. The black dots are the locations of benchmarks, and the graphs show the uplift at those locations. The data suggest that there was both uplift and subsidence for several decades until the mid-1950s, when an episode of rapid uplift occurred at all stations. This activity stabilized by about 1960 and was followed by several years of subsidence prior to the 1964 earthquake. (From Press, 1975 [38]. [copyright] May, 1975 by Scientific American, Inc. All rights reserved.)

followed by a quiet period of 5 years. Renewed activity began ten months before the M 7.8 event, which was the basis of a successful prediction [39]. Second, small earthquakes may tend to ring an area where a larger event will eventually occur. Such a ring (or "donut") was noticed in the 16 months prior to the 1983, M 6.2 earthquake in Coalinga, California. Unfortunately, hindsight is clearer than foresight, and this ring was not noticed or identified until after the event.

Electrical Resistivity. Electrical resistivity is a measure of a material's ability to conduct an electrical current. Conductors, such as copper and aluminum, have a very low resistance, whereas other materials, such as quartz (an insulator), have a very high resistance to electrical current. In general, the Earth is a good conductor of electricity, but its resistivity varies with the amount of groundwater present and many other factors. If a current is fed into the Earth between two points separated by several kilometers, changes in the voltages will be noticed if any change in electrical resistivity takes place in the rocks. Changes before earthquakes have been reported in the United States, Eastern Europe, and China (Figure 8.28) [38]. The mechanisms responsible for the decrease in resistivity are related to physical changes that accompany the earthquake cycle (see Figure 1.28). As rocks dilate, there is an influx of water prior to an earthquake and the electrical resistance decreases. Following the earthquake, fluids drain and resistivity increases.

Amount of Radon. The radioactive gas radon occurs naturally dissolved in water of deep wells. In some cases, the concentration of the gas increases significantly in the hours to days before an earthquake (Figure 8.29). The technique of monitoring radon gas has been studied in Eastern Europe, China, and the United States and appears promising. The increase in radon probably results from precursory cracking and dilation of rocks before an earthquake, allowing more radon to dissolve into the groundwater.

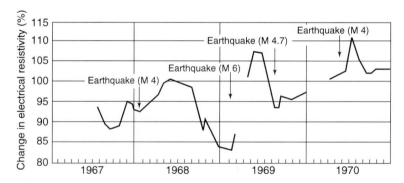

Figure 8.28 Changes in electrical resistivity of the Earth's crust before earthquakes were measured in the former USSR, China, and the United States. The data for this graph represent earthquakes monitored in the former USSR between 1967 and 1970. The measurements are made by feeding an electric current into the ground and recording voltage changes a few kilometers away. In general, it has been observed that earthquakes are preceded by a decrease in resistivity. (After Press, 1975 [38]. © Scientific American, Inc. All rights reserved.)

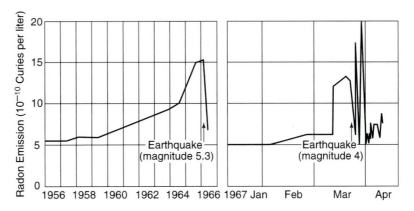

Figure 8.29   The amount of radon dissolved in the water of deep wells may increase significantly before an earthquake. The two examples shown here were recorded before earthquakes in the former USSR. The 1966 event (left) was M 5.3; the 1967 aftershock (right) was M 4. The technique is used in both the former USSR and China and is being tried in the United States. (From Press, 1975 [38]. © by Scientific American, Inc. All rights reserved.)

Anomalous Animal Behavior. Anomalous animal behavior often has been reported prior to large earthquakes; everything from unusual barking of dogs to chickens that refuse to lay eggs, horses or cattle that run in circles, rats that come out of the ground and perch on power lines, and snakes that crawl out of the ground in the winter and freeze. Anomalous behavior reportedly was common before the Haicheng earthquake [36]. Unusual animal behavior also was observed prior to the 1971 San Fernando earthquake. Unfortunately, the significance and reliability of animal behavior is difficult to evaluate. Nevertheless, there has been considerable interest in the topic, and it remains an interesting mystery.

PROGRESS TOWARD EARTHQUAKE PREDICTION

We are still a long way from a working, practical methodology to predict earthquakes reliably. On the other hand, a good deal of data currently is being gathered concerning possible precursor phenomena associated with earthquakes.

Although progress on short-range earthquake prediction has not matched expectations, medium- to long-range forecasting, including hazard evaluation and probabilistic analysis of areas along active faults, has progressed faster than expected [26]. The Borah Peak earthquake of October 28, 1983, in central Idaho has been lauded as a success story for medium-range earthquake-hazard evaluation. Previous evaluation of the Lost River fault suggested that the fault was active [42]. The earthquake, which was about M 7, killed two people and did about $15 million damage, producing fault scarps several meters high and numerous ground fractures along the 36-km rupture zone of the fault. The important fact was that the scarp and faults produced during the earthquake were superimposed on previously existing fault scarps, validating the usefulness of careful mapping of scarps produced by prehistoric earthquakes. The principle is that where the ground has broken before, it may break again!

EARTHQUAKE-HAZARD REDUCTION

The United States has developed a **National Earthquake Hazard Reduction Program** in cooperation with the United States Geological Survey and other scientists. The major goals of the program [43] are to:

- Develop an understanding of earthquake sources. This involves an understanding of the physical properties and mechanical behavior of faults as well as development of quantitative models of the physics of the earthquake process (see Figure 1.28).
- Determine earthquake potential. This involves characterizing seismically active regions, including determining the rates of crustal deformation, identifying active faults, determining characteristics of paleoseismicity, calculating long-term probabilistic forecasts, and, finally, developing methods for intermediate- and short-term prediction of earthquakes.
- Predict effects of earthquakes. This includes gathering data necessary for predicting ground rupture and shaking (see Chapter 1), and the response of buildings and other structures in areas likely to be affected by earthquakes. This goal also involves evaluating the losses associated with earthquake hazards.
- Apply research results. At this level, the program is interested in the transfer of knowledge about earthquake hazards to people, communities, states, and the nation. This knowledge concerns what can be done to better plan for earthquakes and reduce potential losses of life and property.

EARTHQUAKES AND CRITICAL FACILITIES

Critical facilities are those that, if damaged or destroyed, might cause significant to catastrophic loss of life, property damage, or disruption of society. In the urban environment, examples include schools, medical facilities, police stations, and fire stations. Other examples include dams, power plants, and other necessary facilities. Three aspects of the decision-making process concerning critical facilities and earthquake hazard [44] are as follows:

- Evaluation of the hazard
- Evaluation of whether the facility may be designed or modified to accommodate the hazard
- Subjective evaluation of an "acceptable risk"

The first two factors have a strong scientific component, while risk assessment is a public-safety issue, since no facility can be rendered absolutely safe.

In most cases regarding siting of critical facilities, the main scientific challenge is in estimating the activity of a particular fault system. For example, the seismic evaluation for the site of Auburn Dam near Sacramento, California, was very controversial. Millions of dollars were spent and hundreds of meters of fault trenches were excavated during geologic studies, and there was still no agreement about the hazard associated with the system of faults in the vicinity of the proposed site. The general area was considered to have a relatively low seismic risk until a M 5.7 earthquake occurred in 1975 near Oroville, about 80 km from the Auburn site. This earthquake rekindled concern and initiated a new round of seismic-risk evaluation and ultimately led to the virtual abandonment of the site. It was determined from the evaluation that the faults on and near the site would produce only a very small displacement, but this was enough to cause major concern because the concrete dam being planned could not accommodate even such small deformation.

The main problem remains: It is often very difficult to numerically date prehistoric earthquakes. However, we are making progress in understanding earthquakes, identifying active faults, and estimating the maximum credible earthquake for a particular area.

ADJUSTMENTS TO EARTHQUAKE ACTIVITY

The mechanisms that trigger earthquakes are still poorly understood, and therefore earthquake prevention and warning systems are not yet viable alternatives. There are, however, reliable protective measures we can take:

- Structural protection, including the construction of large buildings able to accommodate at least moderate shaking. This has been relatively successful in the United States. The 1988 Armenia earthquake (M 6.8) was slightly larger than the 1994 Northridge, California, event (M_W 6.7), but the loss of life and destruction in Armenia was staggering—at least 45,000 killed (compared with 61 in California) and near-total destruction in some towns near the epicenter. Most buildings in Armenia were constructed of unreinforced concrete and instantly crumbled, crushing or trapping their occupants.
- Land-use planning, or earthquake zonation (also called microzonation), includes the siting of important structures such as schools, hospitals, and police stations in areas away from active faults or sensitive earth materials likely to accentuate seismic shaking.
- Increased insurance and relief measures to help recovery following earthquakes. Following the 1994 Northridge earthquake, total insurance claims were very large, and some insurance companies stopped issuing earthquake insurance. Earthquake zonation could help insurance companies identify high-hazard areas and assign premiums accordingly.

A fourth possible measure is to take little or no action in advance and to pay the consequences when an earthquake occurs. This philosophy is not what we espouse, but it is, in fact, what we often do.

At the same time that we are adjusting to existing hazards, we are increasing the need for adjustment by creating new hazards where none existed before. When any new building or structure is built on sensitive earth materials or on active faults, we are creating a new problem where none existed before. We know from studying earthquake damage that building on unconsolidated deposits that become unstable (subject to liquefaction) or amplify shaking during an earthquake (see Chapter 1) is more hazardous than building on bedrock. We also know that building astride active faults is unwise.

We hope, eventually, to be able to predict earthquakes. The federal plan for issuing prediction and warning is shown in Figure 8.30. The general flow of information is from scientists to a prediction council for verification. A prediction that a damaging earthquake of a specific magnitude will occur at a particular location over a specified

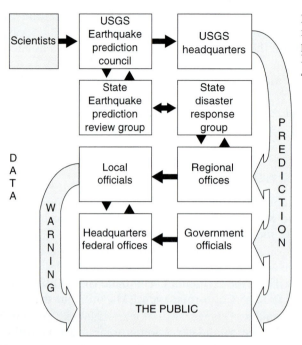

Figure 8.30   Proposed information flow for issuing earthquake predictions and warnings. (From McKelvey, 1976. *U.S. Geological Survey Circular*, 729:10–12.)

time span then may be issued to state and local officials, who will be responsible for issuing a warning to the public to take defensive action (that has, one hopes, been planned in advance). Potential responses to a prediction depend on lead time (Table 8.3), but even a short time (as little as a few days) would be sufficient to mobilize emergency services, shut down important machinery, and evacuate particularly hazardous areas.

Table 8.3

POTENTIAL RESPONSE TO AN EARTHQUAKE PREDICTION WITH GIVEN LEAD TIME.

Lead Time	Buildings	Contents	Lifelines	Special Structures
3 Days	Evacuate previously identified hazards	Remove selected contents	Deploy emergency materials	Shut down reactors, petroleum products pipelines
30 Days	Inspect and Identify potential hazards	Selectively harden (brace and strengthen) contents	Shift hospital patients; alter use of facilities	Draw down reservoirs, remove toxic materials
300 Days	Selectively reinforce		Develop response capability	Replace hazardous storage
3000 Days		Revise building codes and land-use regulations; enforce condemnation and reinforcement		Remove hazardous dams from service

(From Thiel, 1976. *U.S. Geological Survey Circular* 729)

SUMMARY

Paleoseismology is defined as the study of the occurrence, size, timing, and frequency of preinstrumental and prehistoric earthquakes. Paleoseismic data may be gathered from fault exposures, faulted landforms, fault scarps, stratigraphic features, and folded deposits, rocks, and geomorphic surfaces. Of particular importance is the excavation of trenches across and along fault zones to examine, date, measure, and evaluate evidence of past earthquakes.

The concept of fault-zone segmentation is important in evaluation of earthquake hazards. Earthquake segments may be identified in terms of rupture length, determined from historical earthquakes or paleoseismic evaluation. Faults may also be structurally segmented, based on changes in the surface expression of subsurface fault-zone structure or direct structural evidence. The concept of fault segmentation is particularly important as it relates to long-term earthquake forecasting involving probabilistic assessment of seismic hazard. The concept also is useful in constraining estimates of the maximum earthquake likely to occur along a particular fault and in constraining estimates of seismic ground motion.

Short-term prediction of earthquakes is an area of serious research. Optimistic scientists believe that eventually we will be able to make consistent long-range forecasts, medium-range predictions, and short-range predictions for the general locations and magnitudes of damaging earthquakes. Such predictions will be based on factors such as deformation of the ground surface, seismic gaps along faults, pattern and frequency of earthquakes, changes in electrical resistivity of the Earth, changes in the amount of dissolved radon gas and groundwater, and, perhaps, anomalous behavior of animals.

The United States, in cooperation with the U.S. Geological Survey and other scientists, is developing programs for earthquake-hazard reduction. Major goals are to develop an understanding of earthquake source, to determine earthquake potential, to predict effects of earthquakes, and to apply research results to minimize the earthquake hazard to people.

REFERENCES CITED

1. Crone, A. J., 1987. Introduction to directions in paleoseismology. In A. J. Crone and E. M. Omdahl (eds.), *Directions in Paleoseismology*, U.S. Geological Survey Open-File Report 87–673, 1–6.

2. Wallace, R. E., 1987. A perspective of paleoseismology. In A. J. Crone and E. M. Omdahl (eds.), Directions in Paleoseismology, U.S. Geological Survey Open-File Report 87–673, 7–16.

3. McCalpin, J. P., and A. R. Nelson, 1996. Introduction to paleoseismology, In *Paleoseismology*, J. P. McCalpin (ed.). San Diego: Academic Press, 1–32.

4. Kelsey, H. M., A. G. Hull, S. M. Cashman, K. R. Berryman, P. H. Cashman, J. H. Trexler Jr., and J. G. Begg, 1998. Paleoseismology of an active reverse fault in a forearc setting: The Poukawa fault zone, Hikurangi forearc, New Zealand. *Geologic Society of America Bulletin*, 110:1123–1148.

5. Weldon, R. J. II, J. P. McCalpin, and T. K. Rockwell, 1996. Paleoseismology of strike-slip tectonic environments, In *Paleoseismology*, J. P. McCalpin (ed.). San Diego: Academic Press, 271–329.

6. Yeats, R. S., K. Sieh, and C. R. Allen, 1997. *The Geology of Earthquakes*. New York: Oxford University Press.

7. Carver, G. A., and R. M. Burke, 1992. Late Cenozoic deformation on the Cascadia Subduction Zone in the region of the Mendocino Triple Junction. In *Friends of the Pleistocene, Pacific Cell Guidebook*. Arcata, CA: Humboldt State University.

8. Plafker, G., 1987. Application of marine-terrace data to paleoseismic studies. In A. J. Crone and E. M. Omdahl (eds.), *Directions in Paleoseismology*, U.S. Geological Survey Open-File Report 87–673, 146–156.

9. Geophysics Study Committee, 1986. Overview and recommendations. In *Active Tectonics*, R. E. Wallace (ed.). Washington, DC: National Academy Press, 3–19.

10. Wallace, R. E., 1977. Profiles and ages of young fault scarps, north-central Nevada. *Geological Society of America Bulletin*, 88:1267–1281.

11. Colman, S. M., and K. Watson, 1983. Ages estimated from a diffusion equation model for scarp degradation. *Science*, 221:263–265.

12. Andrews, D. J., and R. C. Bucknam, 1987. Fitting degradation of shoreline scarps by a nonlinear diffusion model. *Journal of Geophysical Research*, 92:12, 857–12, 867.

13. Pierce, K. L., and S. M. Colman, 1986. Effect of height and orientation (microclimate) on geomorphic degradation rates and processes, late-glacial terrace scarps in central Idaho. *Geological Society of America Bulletin*, 97:869–885.

14. Hanks, T. C., and D. J. Andrews, 1989. Effect of far-field slope on morphological dating of scarp-like landforms. *Journal of Geophysical Research*, 94:565–573.

15. Arrowsmith, J. R., D. D. Rhodes, and D. D. Pollard, 1998. Morphologic dating of scarps formed by repeated slip events along the San Andreas Fault, Carrizo Plain, California. *Journal of Geophysical Research*, 103:10, 141–10,160.

16. McCalpin, J., 1987. Geologic criteria for recognition of individual paleoseismic events in extensional environments. In A. J. Crone and E. M. Omdahl (eds.), *Directions in Paleoseismology*, U.S. Geological Survey Open-File Report 87–673, 102–114.

17. Holzer, T. L., and M. M. Clark, 1993. Sand boils without earthquakes. *Geology*, 21:873–876.

18. Atwater, B. F., 1987. Evidence for great Holocene earthquakes along the outer coast of Washington State. *Science*, 236:942–944.

19. Allen, C. R., 1968. The tectonic environments of seismically active and inactive areas along the San Andreas fault system. In W. R. Dickinson and A. Grantz (eds.), *Proceedings of the Conference on Geologic Problems of the San Andreas Fault System*, Stanford University Publications, Geological Sciences, 11:70–80.

20. Schwartz, D. P., and R. H. Sibson, 1989. Introduction to workshop on fault segmentation and controls of rupture initiation and termination. In D. P. Schwartz and R. H. Sibson (eds.), *Fault Segmentation and Controls of Rupture Initiation and Termination*, U.S. Geological Survey Open-File Report 89–315, i–iv.

21. Machette, M. N., S. F. Personius, A. R. Nelson, D. P. Schwartz, and W. R. Lund, 1989. Segmentation models and Holocene movement history of the Wasatch Fault Zone. In D. P. Schwartz and R. H. Sibson (eds.), *Fault Segmentation and Controls of Rupture Initiation and Termination*, U.S. Geological Survey Open-File Report 89–315, 229–245.

22. DePolo, C. M., D. G. Clark, D. B. Slemmons, and W. H. Aymard, 1989. Historical Basin and Range province surface faulting and fault segmentation. In D. P. Schwartz and R. H. Sibson

(eds.), *Fault Segmentation and Controls of Rupture Initiation and Termination*, U.S. Geological Survey Open-File Report 89–315, 131–162.

23. Schwartz, D. P., and K. J. Coppersmith, 1984. Fault behavior and characteristic earthquakes: examples from the Wasatch and San Andreas Fault Zones. *Journal of Geophysical Research*, 89:5681–5698.

24. Schwartz, D. P., and K. J. Coppersmith, 1986. Seismic hazards; new trends in analysis using geologic data. In *Active Tectonics*. Washington; DC: National Academy Press, 215–230.

25. Wallace, R. E., 1970. Earthquake recurrence intervals on the San Andreas fault. *Geological Society of America Bulletin*, 81:2875–2889.

26. Allen, C. R., 1983. Earthquake prediction. *Geology*, 11:682.

27. Schwartz, D. P., and K. J. Coppersmith, 1984. Fault behavior and characteristic earthquakes: examples from the Wasatch and San Andreas fault zones. *Journal of Geophysical Research*, 89:5681–5698.

28. Sieh, K. E., 1981. A review of geological evidence for recurrence times of large earthquakes. In D. W. Simpson and P. G. Richards (eds.), *Earthquake Prediction, An International Review*, M. Ewing Series 4. Washington, DC: American Geophysical Union, 181–207.

29. Scholz, C. H., 1990. The mechanics of earthquakes and faulting. New York: Cambridge University Press.

30. Swan, F. H., 1988. Temporal clustering of paleoseismic events on the Oued Fodda fault, Algeria. *Geology*, 16:1092–1095.

31. Reilinger, R., N. Toksoz, S. McClusky, and A. Barka, 2000. 1999 Izmit, Turkey earthquake was no surprise. *GSA Today*, 10:1–6.

32. Stein, R. S., 1999. The role of stress transfer in earthquake occurrence. *Nature*, 402:605–609.

33. Sieh, K., M. Stuiver, and D. Brillinger, 1989. A more precise chronology of earthquakes produced by the San Andreas fault in Southern California. *Journal of Geophysical Research*, 94:603–623.

34. Heaton, T. H., D. L. Anderson, W. J. Arabasz, R. Buland, W. L. Ellsworth, S. H. Hartzell, T. Lay, and P. Spudich, 1989. National seismic system science plan. *U.S. Geological Survey Circular* 1031.

35. Pakiser, L. C., J. P. Eaton, J. H. Healy, and C. B. Raleigh, 1969. Earthquake prediction and control. *Science*, 166:1467–1474.

36. Raleigh, B., et al., 1977. Prediction of the Haicheng earthquake. *Eos: Transactions of the American Geophysical Union*, 58:236–272.

37. Simons, R. S., 1977. Earthquake prediction, prevention and San Diego. In P. L. Abbott and J. K. Victoria (eds.), *Geologic Hazards in San Diego: Earthquakes, Landslides and Floods*, San Diego, CA: San Diego Society of Natural History, 13–22.

38. Press, F., 1975. Earthquake prediction. *Scientific American*, 232(May):14–23.

39. Scholz, C. H., 1990. *The Mechanics of Earthquakes and Faulting*. New York: Cambridge University Press.

40. Rikitakr, T., 1983. *Earthquake Forecasting and Warning*. London: D. Reidel.

41. Hanks, T. C., 1985. The National Earthquake Hazards Reduction Program: scientific status. *U.S. Geological Survey Bulletin* 1659.

42. Hait, M. H., and W. E. Scott, 1978. Holocene faulting, Lost River Range, Idaho. *Geological Society of America Abstracts with Programs*, 10(5):217.

43. Page, R. A., D. M. Boore, R. C. Bucknam, and W. R. Thatcher, 1992. Goals, opportunities, and priorities for the USGS Earthquake Hazards Reduction Program. *U.S. Geological Survey Circular* 1079.

44. Cluff, L. S. 1983. The impact of tectonics on the siting of critical facilities. *Eos: Transactions, American Geophysical Union*, 64:860.

Mountain Building

INTRODUCTION

LANDSCAPE SCALE

If the Earth were as smooth as a cue ball, it would not be a very scenic place, and there would be no such science as geomorphology. The Earth's surface is varied and interesting because it has relief—from individual hills and valleys to great mountain chains. The difference between a hill and a mountain is a difference in **scale**, both in terms of the heights of the landforms and their lateral dimensions. The importance of landscape scale has been recognized for a long time, but geomorphologists have not yet agreed on a single classification system. Joseph Le Conte, an eminent geologist during the late nineteenth and early twentieth centuries, classified relief as either greater or lesser [1]. Greater relief would include the continents, ocean basins, and great mountain chains and is created by processes occurring in the Earth's interior. Lesser relief would include all smaller landforms shaped principally by erosion. Le Conte did not recognize that tectonic processes are responsible for many small landforms. Furthermore, as this chapter will discuss, we are increasingly aware that erosional processes play a part in the development of large-scale topography. Classifications of geomorphic scale since the

Table 9.1

A CLASSIFICATION OF LANDFORMS BY SCALE.

Order	Lateral dimensions (km)		Examples	Le Conte classification
1	>10,000 km		Continents, ocean basins	
2	1000–10,000 km	**Tectonic**	Physiographic provinces, mountain belts	**Greater**
3	100–1000 km	**Units**	Sedimentary basins, mountain ranges	**Relief**
4	10–100 km		Volcanoes, fault-block mountains, troughs	
5	1–10 km	**Erosional-**	Deltas, piedmonts, major valleys	
6	100 m–1 km	**depositional**	Floodplains, alluvial fans, moraines	
7	10–100 m	**units**	Terraces, sand dunes, gullies	**Lesser**
8	1–10 m	**Geomorphic**	Hillslopes, sections of stream channels	**Relief**
9	10 cm–1 m	**process**	River bars, pools and riffles, beach cusps	
10	1–10 cm	**units**	Fluvial or eolian ripples, glacial striations	

(After Baker, 1986 [2]; Le Conte, 1909 [1])

discovery of tectonics (for example, Table 9.1) tend to focus more on quantification and description than the older schemes [2]. The previous eight chapters of this book focused on local features caused directly or indirectly by tectonic activity. With that background now complete, this final chapter integrates those local tectonic processes and features over larger scales.

THE SHAPE OF THE EARTH

The concept of **hypsometry**—the distribution of different elevations across a landscape—was introduced in Chapter 4. Hypsometry can be evaluated over small areas, such as a single drainage basin (see Chapter 4), or over areas as large as the entire planet (Figure 9.1). The total elevation-frequency distribution of the Earth reveals much about the structure of the planet and the processes operating in the interior and on the surface.

The mean elevation of the Earth is 2.3 km below sea level [3], but the distribution is strongly bimodal (Figure 9.1). One mode is between about 3.0 and 5.0 km below sea level, and the other is a sharp spike between about 0.2 km below and 0.5 km above sea level. This bimodal distribution primarily represents the profound differences between oceanic crust and continental crust (see Chapter 1). Oceanic crust is thin (<10 km), dense (3000 to 3100 kg/m^3), and composed largely of basaltic rock [4]. In contrast, continental crust is thick (35 to 40 km, on average), predominantly granitic, and relatively buoyant (2700 to 2800 kg/m^3), and it therefore stands higher than the ocean basins. The abrupt transition from continental crust to oceanic crust, at the edge of the continental shelves at depths of 100 to 200 m below sea level, contributes to the distinct separation of elevations in Figure 9.1.

The Earth's surface also has been profoundly shaped by the presence of its hydrosphere (oceans, lakes, rivers, glaciers, groundwater, etc.). The oceans weigh down on oceanic crust and accentuate the contrasts in Figure 9.1. Furthermore, the concentration of the Earth's surface elevation near sea level is a result of running water—areas above

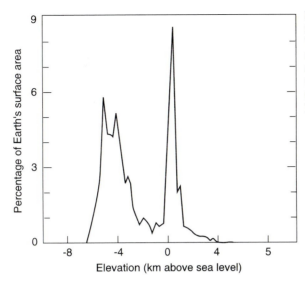

Figure 9.1   The elevation-frequency distribution of the Earth. The curve indicates the percent of area for each 200-m interval. Data are averages from 1° by 1° areas; therefore, the range indicated is less than the total range of elevation of Earth. (After Head et al., 1981 [5].)

sea level erode to base level, and the resulting detritus is deposited at and in proximity to the coastline.

Lest anyone become complacent about the peculiar topography of the Earth, compare Figure 9.1 with Figure 9.2, which illustrates the hypsometry of the planet Venus [5]. Venus has only one type of crust—basaltic—and hence a unimodal height distribution. With an average surface temperature of about 200°C and atmospheric pressure up to 90 times that of Earth [6], Venus is shaped by processes completely alien to us. At these surface conditions, Venus has only a thin, brittle crust over the ductile mantle that forms a lithosphere without distinct plates like Earth's but which rather deforms by countless small-scale rips and tears across zones up to several hundred kilometers wide [7]. However, in spite of profound differences, the tectonic processes on both planets have led to the formation of mountain ranges (the extremely high elevations seen as small "tails" at the right sides of Figures 9.1 and 9.2). For example, on Venus, a broad zone of crustal convergence has uplifted a tall mountain chain (Freyja Montes) and a high-standing plateau (Lakshmi Planum) (Figure 9.3) that are directly comparable to the Himalaya and Tibetan Plateau [8, 9].

MODELS OF LANDSCAPE AND MOUNTAIN DEVELOPMENT

There may be as many different theories of landscape development as there have been geologists who have studied the subject, but no model has shaped thinking about landscape and landscape change more than William Morris Davis's 'Cycle of Erosion' [10]. This model was introduced in Chapter 2 and illustrated in Figure 2.1. In brief, Davis proposed that landscapes are subject to brief periods of tectonism and mountain building, followed by long periods of erosion. In this cycle, a landscape goes from a stage of "youth", characterized by deep incision, to a "mature" stage, with a fully developed

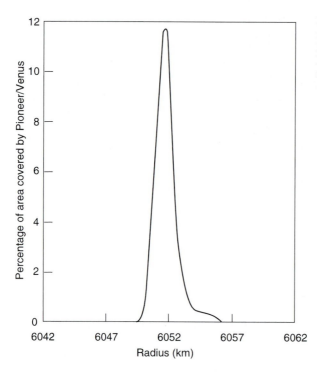

Figure 9.2 The elevation-frequency distribution of Venus. Data is radar altimetry from the Pioneer/Venus spacecraft, with a resolution between 100 and 200 km. (After Head et al., 1981 [5].)

drainage network and moderate relief, to "old age" [10]. Landscapes in old age are characterized by peneplains, which are regional surfaces of low relief, produced by long-continued erosion in the Davis model.

Davis's model was geomorphology's governing paradigm through much of the twentieth century. At its heart, the Cycle of Erosion is based on two fundamental assumptions: (1) that when tectonic activity occurs, it is much more powerful and faster than erosional processes; and (2) that the most important variable in long-term landscape evolution is **time**. Through the 1960s and 1970s, however, a growing number of geomorphologists came to the conclusion that both of these assumptions are wrong. As a result, the science made a fairly abrupt shift away from its previous evolutionary focus and toward measurement and observation of tangible geomorphic processes [11, 12].

A NEW "CYBERNETIC" MODEL OF OROGENESIS

Within the past few years, there has been renewed interest in examining large-scale surficial processes and landscape change over long time periods [11, 13]. A growing number of researchers have been looking at **orogenesis**—large-scale regional mountain building—as an integrated system [14]. Technically speaking, a **system** can be defined as a number of components and component processes that function together or act as a whole. Orogenesis is widely seen as the net effect of three major sets of processes: (1)

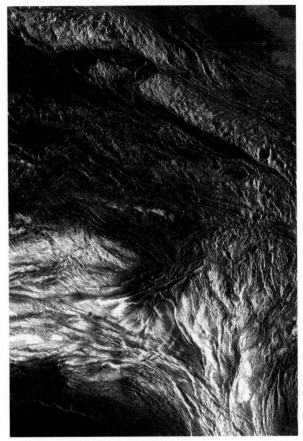

Figure 9.3   Radar image of central Freyja Montes region of Venus from *Magellan* spacecraft. The region shown is interpreted to be a collision zone, very similar to the Himalaya of Earth. (Image courtesy of Jet Propulsion Laboratories.)

tectonics, (2) erosion, and (3) climate (Figure 9.4). At the broadest possible scale, tectonic processes add excess mass to the crust, driving uplift of the surface. Surficial processes generally act in the opposite direction: disaggregating and removing that mass. Climate controls the distribution of precipitation and temperatures at the Earth's surface, thereby determining the types and the rates of geomorphic processes at work.

It has been suggested that a requirement for any sweeping model is a concise (and preferably a catchy) name. This new systems-based model of mountain building has been called the "cybernetic" model of orogenesis [15, 16]. Given the popular connotation of the word cybernetic, most people associate it with robotics, but the term more generally refers to the *command-and-control processes* in electronic, mechanical, and biological systems. In referring to mountain building as cybernetic, we are just adding geologic systems to that list. The new system is called cybernetic because it emphasizes the interconnectedness and the importance of feedback between the elements of the

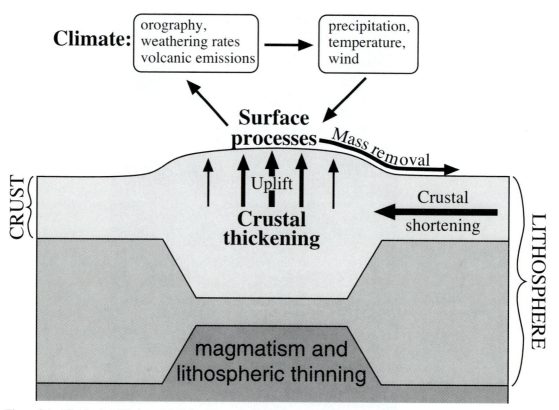

Figure 9.4   Highly simplified model of the mountain-building system. Tectonic processes thicken the crust, driving uplift of the surface. Climate-driven surficial processes act to remove that high-standing mass. Each set of processes influences the other sets in a number of ways. (After Isacks, 1992 [85].)

orogenic system. Many of the principles outlined in this chapter illustrate this systems-based model, focusing on the dominant processes that form and shape regional topography, the linkages between the different sets of processes, and some of the implications and applications of a dynamic, interconnected model of orogenesis.

A reasonable question at this point is, what makes this new model better than the Cycle of Erosion or other sweeping models that have been proposed in the past? The first part of the answer is that the predictions of the new model seem to fit empirical observations of the world much better, and this will be outlined in detail throughout the rest of this chapter. Even more fundamentally, the data about mountain ranges and other regional topography available today are much broader and far more quantitative than available previously. For example, W. M. Davis predicated his model on the central importance of time, and yet he and his contemporaries lacked the numerical dating tools available today. Recent studies have utilized a broad and widening range of geochronometric tools, including techniques that illustrate rates and patterns of exhumation [e.g., 17, 18] and surface-exposure age [e.g., 19, 20]. Other quantitative evidence has come from analysis of regional landscape using digital elevation models (DEMs), which allows quantitative characterization of large-scale topography and measurement of orogenic dynamics [e.g., 21, 22]. Finally, our quantitative understanding of tectonic, geologic, and geomorphic processes in mountain ranges is now such that numerical models of orogenic behavior and evolution over time are yielding highly realistic results [e.g., 23, 24].

DYNAMICS OF OROGENESIS

DRIVING MECHANISMS

The principal driving mechanism for orogenesis is plate tectonics. Each of the three types of plate boundary—divergent, convergent, and transform—can form high-standing topography. Divergent plate boundaries (midocean ridges and continental rifts) stand higher than stable plate interiors because of heating of the crust. Transform plate boundaries also may be the site of significant uplift, where strike-slip motion includes a component of convergence. However, true orogenesis, uplift of the largest areas to the greatest elevations, occurs at convergent plate boundaries, meaning at subduction zones and continental collisions.

Convergence between oceanic lithosphere and continental or oceanic lithosphere (subduction) causes compression and thickening of the overriding plate. Partial melting of the downgoing plate also generates magma, which is added to the area above. The classic example of a mountain chain formed by these processes is the Andes of South America. The modern Andes began forming in the early Jurassic (around 200 Ma), when subduction began along the western margin of Pangea (the supercontinent of which South America was a part at the time) [25, 26]. Up to 240 km of horizontal shortening has accompanied subduction [27]. The topography and the geology of the modern Andes can be subdivided into several distinct domains from west to east across the range (Figure 9.5). The Western Cordillera rises 5000 m from the Pacific and the narrow coastal zone. The Eastern Cordillera rises even further, to 6000 m, and is separated from the western ranges by the Altiplano ("high plain") in the Peruvian and Bolivian Andes. The Altiplano did not rise above sea level until about 60 Ma and was still less than half of its present-day elevation at 10 Ma, indicating rapid recent uplift [26]. East of the Eastern Cordillera is the Subandean zone, foothills of the Andes up to 2000 m high that constitute a Pliocene-age belt of folding and thrusting [28]. The morphology of the Andes today also varies from north to south, apparently in response to variations in the geometry of the subducting plate. Where the plate subducts at a low angle (~10°, northern and central Peru), there is no history of recent volcanism, but there is widespread seismicity; where the plate subducts at a steeper angle (~30°, southern Peru), there has been volcanism, and seismicity has been concentrated along the downgoing plate [29].

Uplift, volcanism, and faulting in subduction-zone mountain ranges can occur over long periods of geologic time; indeed, they will occur as long as the plates converge and oceanic lithosphere is subducted. However, a different kind of mountain building occurs when plate convergence can no longer be accommodated by subduction. For example, 80 Ma ago, the Indian continent and Asia were separated by 4000 to 5000 km of ocean (the Tethys Sea) (Figure 9.6). Oceanic crust of the Indian Plate was subducted beneath the southern margin of the Eurasian Plate, which formed a volcanic mountain chain along the Asian coast similar to the modern Andes. A belt of granite 2500 km long is the

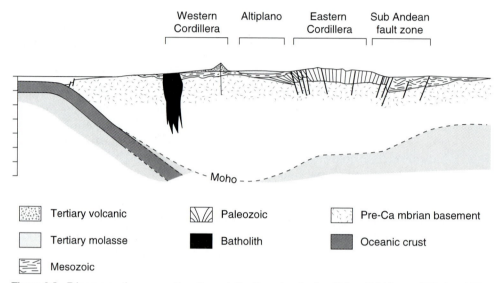

Figure 9.5 Diagrammatic cross section through the Peruvian Andes. (After Cobbing and Pitcher, 1972. *Nature Physical Science*, 240:51–53.)

eroded remnant of that mountain chain today. At this time, India was moving northward, narrowing the Tethys Sea at a rate of 15–20 cm/yr [30]. By 40 to 50 Ma, the ocean was closed and continental crust of the Indian Plate had impinged on the boundary. As discussed in Chapter 1, continental crust is thick and buoyant and will not descend into the underlying mantle. In the 40 M.y. since India ran into Asia, the collision has reshaped the face of Asia, uplifted the Himalaya and the Tibetan Plateau, and altered the climate of the region and perhaps the entire world.

After the Indian continent ran into Asia, the rate of convergence slowed to about 5 cm/yr [30], but that rate has been maintained to the present—a total of over 2000 km of continued convergence. In the process, India has been partially thrust beneath Asia [31]. Convergence between the two continents appears to have been accommodated by three mechanisms:

1. Movement on regional thrust faults
2. Continental extrusion
3. Uplift

Several hundred kilometers of motion is confirmed on several crustal-scale thrust faults (the Himalayan Frontal thrust, the Main Boundary thrust, and the Main Central thrust).

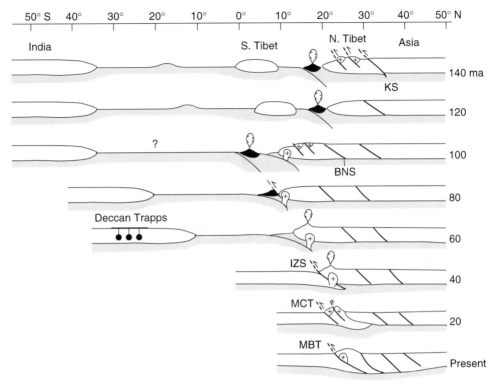

Figure 9.6   The closure of the Tethys Sea and collision of India and Asia. Bangong Nujiang suture (BNS); Indus-Zangbo suture (IZS); Kokoxili suture (KS); Main Boundary thrust (MBT); Main Central thrust (MCT). (From Allègre et al., 1984. *Nature*, 307:17–22.)

A large portion of the convergence was accommodated by lateral extrusion of portions of Asia along regional strike-slip fault systems (Figure 9.7) [32]. The timing of uplift across the Himalaya and Tibet is the subject of continuing discussion, with some workers suggesting that most occurred during the middle to late Tertiary and others suggesting that most of the uplift did not occur until the Pliocene to Pleistocene [33]. Different areas seem to yield different ages of activity, probably reflecting distinct faulting and uplift histories associated with the different regional thrust faults [e.g., 33–36].

Uplift of the Himalaya and the Tibetan Plateau has had several dramatic effects: topography that cross-cuts the modern rivers, voluminous discharge of sediment, and initiation of the monsoonal weather system. The major rivers of the region are **antecedent** (see Chapter 5), meaning that they predate the collision, and the Himalaya and Tibet have been lifted up beneath the rivers. These rivers now flow through deep gorges—for example, the Indus River flows less than 21 km in distance from Nanga Parbat peak, but over 7000 m lower in elevation [34] (Figure 9.8). Another effect of uplift has been the erosion and discharge of tremendous volumes of sediment, much of which has accumulated in the vast submarine fans at the outlets of the Indus and Ganges Rivers [37]. Uplift of so large an area has caused region-wide climate changes that have intensified

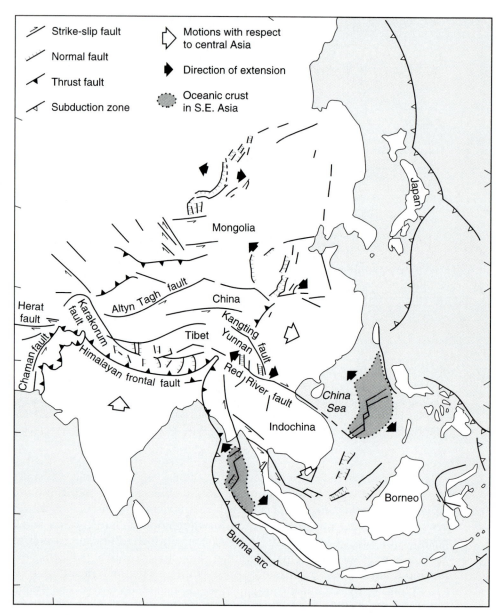

Figure 9.7 Extrusion of eastern and southeastern Asia caused by the India-Asia collision. India is interpreted to have penetrated into Asia, causing widespread strike-slip faulting and regional reshaping of the Asian continent. (From Tapponier et al., 1982. *Geology*, 10:611–616.)

Figure 9.8   View of Nanga Parbat massif, with the Indus River at its base. Relief between the river channel and the Nanga Parbat summit is more than 7000 m. (Photo courtesy of P. Zeitler.)

erosion of the mountain range. About 8 Ma, uplift initiated, or at least dramatically strengthened, the seasonal storm system known as the Asian **monsoon** [32]. Uplift of the Himalaya and Tibet is also associated with increasing aridity in central Asia, north of the collision zone [38]. Even more broadly, it has been suggested that uplift of the Himalaya and Tibet altered the global climate and intensified erosion worldwide (see the section titled "Long-Term Geochemical Cycling" later in this chapter).

STRUCTURAL SUPPORT OF MOUNTAINS

Tremendous forces are required not only to build a mountain range, but also to maintain one. Like a waiter trying to carry a great load of dishes on his tray, at some point the question of how to add another plate to the stack becomes less important than how to keep the entire pile from collapsing to the floor. Indeed, the Tibetan Plateau is now cut by north-south-trending normal faults, indicating that the landmass may have begun to slide to the east and west under its own weight [39, 40, 41].

The primary mechanism of mountain-range support is **isostasy**. Remember that both oceanic crust (density of about 3000 kg/m^3) and continental crust (density of about 2700 kg/m^3) are less dense than the mantle (density of about 3300 kg/m^3 at the base of the crust). For most purposes, the buoyant crust can be thought of as floating on the mantle below. According to the principle of isostasy, the weight of the crust is borne by the fluidlike properties of the mantle. Two different hypotheses for isostatic support of mountains were proposed in 1855, one by George Biddell Airy, Astronomer Royal of England, and the other by John Henry Pratt, Archdeacon of Calcutta.

According to Airy, any high-standing landmass at the surface is supported by a deep root at the base of the crust (Figure 9.9a). To understand this, visualize two icebergs floating in the ocean, one standing twice as far above the water as the other, we know that the larger iceberg also has correspondingly more ice beneath the surface. The density of ice is about nine-tenths that of water, and thus an iceberg must displace nine kilograms of water to support one kilogram of ice above the waterline. In contrast, according to the Pratt hypothesis, the crust is generally of uniform thickness but is characterized by regional variations in density. According to this hypothesis, high mountainous topography can be supported by low-density material beneath the mountains (Figure 9.9b). To give an analogy, a block of styrofoam floats much higher in water than does a block of dense wood. We now know that (as is usually the case) there are elements of truth in both hypotheses. The crust beneath mountains is indeed much thicker than in low-lying areas, but it is also true that density differences, such as those caused by heating of the crust, do cause regional uplift in some areas [42].

A third, nonisostatic mechanism exists to support loads upon the surface—**flexural support**. When a skyscraper is built, it does not slowly sink beneath the surface until nine-tenths of it is below ground level, as pure isostasy might suggest. Skyscrapers can be built because the Earth's lithosphere has strength. Any local load is supported not only by the buoyancy of the crust directly under it, but also by the rigidity of the lithosphere in a large area around it. Another way to say this is that isostatic support is regional—a skyscraper built in Dallas will depress the lithosphere all over Texas, but not by a measurable amount. A geologic example of flexural support is the Himalaya, which are neither thick enough (only 55 km thick; they would need to be >80 km for Airy support) nor buoyant enough to support the great mass of the range [43]. Instead, the Himalaya partially rest on the edge of the Indian Plate, flexing India downward and forming the Ganga basin, a deep, sediment-filled trough at the boundary of the two plates [30].

Flexural support of loads on the lithosphere depends on the **wavelength** (the lateral dimensions) of the load. A single skyscraper has a wavelength of no more than 100 m, whereas the Himalaya have a wavelength of hundreds to thousands of kilometers. Flexural support also depends on the strength of the crust. In general terms, thick crust is strong crust; rigidity is measured by the **elastic thickness** (T_e) of the crust, which is the depth to which it will act as a rigid solid. In North America, the elastic thickness of the crust ranges from about 4 km in the Basin and Range province to about 128 km in the cratonic core of the continent [44]. The shorter the wavelength of the load and the greater the rigidity of the crust, the more that flexure will support a load rather than isostasy (Figure 9.10).

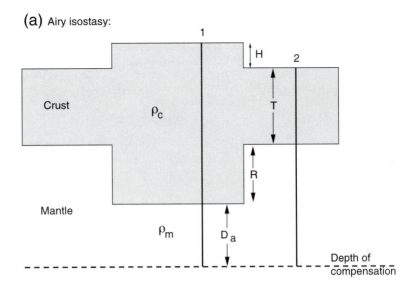

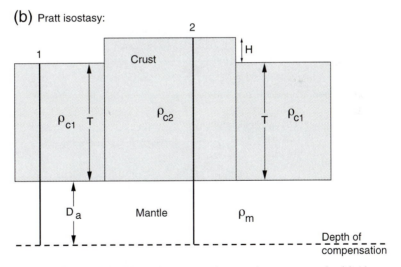

Figure 9.9   Two models of isostatic support of mountainous topography. (a) Airy isostasy is based on the premise of deep crustal roots beneath mountains. (b) Pratt isostasy is based on a column made of less-dense material. Variables are explained in the Box titled "Calculating Isostatic Support of Mountains." that follows.

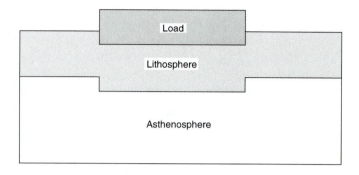

(b) Finite rigidity

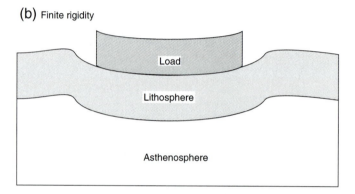

(c) Infinite rigidity

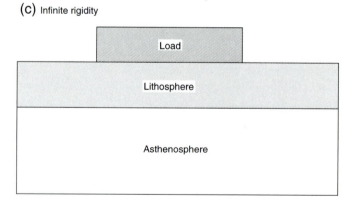

Figure 9.10 Support of loads on the surface (such as mountain ranges) by flexure of the lithosphere. Flexure depends on the strength (rigidity) of the lithosphere. (a) If zero rigidity is assumed, support is purely isostatic. (b) If there is some finite rigidity, then regional flexure occurs. (c) If infinite rigidity is assumed, no isostasy occurs, and the load is carried exclusively by the strength of the lithosphere.

CALCULATING ISOSTATIC SUPPORT OF MOUNTAINS

Isostasy is based upon **Archimedes principle**—that the weight of a floating solid is supported by the weight of the fluid that it displaces. Archimedes was the ancient Greek scientist who is said to have made his insights about fluid displacements getting into the bathtub. The reason that an iceberg with a mass of 100,000 kg does not sink to the bottom of the ocean is that it displaces exactly 100,000 kg of water. The reason that nine-tenths of the iceberg is below the waterline is that ice is only nine-tenths as dense as water—100,000 kg of water are displaced by 90,000 kg of ice, and the remaining 10,000 kg of the iceberg stand above the surface.

The calculation of isostatic support is based on the principle that at any given depth within a fluid, called the **depth of compensation**, the pressures generated by overlying material are everywhere equal (see Figure 9.9). This means that the weights of any two columns of material, measured down to the depth of compensation, are equal. Given an area of uniform crustal density (ρ_c) and thickness (T), a mountain block with elevation H above the surface, and an arbitrary distance down to the depth of compensation (D_a), we can calculate the mountain-root depth (R) beneath the base of the crust for Airy isostasy (Figure 9.9a). Weight (W) equals mass (volume • density) times gravity (g). For column 1, through the mountain block

$$W_1 = (H\rho_c + T\rho_c + R\rho_c + D\rho_m)g. \tag{9.1}$$

The weight of column 2 equals

$$W_2 = (T\rho_c + R\rho_m + D\rho_m)g. \tag{9.2}$$

The weights of the two columns equal each other; therefore,

$$W_1 = W_2 \tag{9.3}$$

$$(H\rho_c + T\rho_c + R\rho_c + D\rho_m)\, g = (T\rho_c + R\rho_m + D\rho_m)\, g \tag{9.4}$$

$$H\rho_c + T\rho_c + R\rho_c + D\rho_m = T\rho_c + R\rho_m + D\rho_m \tag{9.5}$$

$$H\rho_c + T\rho_c + R\rho_c = T\rho_c + R\rho_m \tag{9.6}$$

$$H\rho_c + R\rho_c = R\rho_m \tag{9.7}$$

$$R\rho_m - R\rho_c = H\rho_c \tag{9.8}$$

$$R = H\rho_c\, /\, (\rho_m - \rho_c). \tag{9.9}$$

This expression is easily solved to give the depth of the crustal root beneath the Tibetan Plateau, which has an average elevation of 5000 m. We know that the density of continental crust is about 2700 kg/m^3 and the density of the mantle is about 3300 kg/m^3; therefore,

$$R = \frac{5000\,\text{m} \cdot 2700\ \text{kg/m}^3}{(3300\,\text{kg/m}^3 - 2700\,\text{kg/m}^3)} \tag{9.10}$$

$$R = \frac{5000\,\text{m} \cdot 2700\ \text{kg/m}^3}{(600\,\text{kg/m}^3)} \tag{9.11}$$

$$R = 22{,}500\ \text{m}. \tag{9.12}$$

In Pratt isostasy (Figure 9.9b), the base of the crust is uniform, and the height of the mountain (H) is supported by a lesser density (ρ_{c2}) than the density in the surrounding region (ρ_{c1}). In this problem, we will solve for the value of ρ_{c2}. Once again, we set the weights of the two columns equal to each other:

$$(T\rho_{c1} + D\rho_m)g = (H\rho_{c2} + T\rho_{c2} + D\rho_m)g \tag{9.13}$$

$$T\rho_{c1} + D\rho_m = H\rho_{c2} + T\rho_{c2} + D\rho_m \tag{9.14}$$

$$T\rho_{c1} = H\rho_{c2} + T\rho_{c2} \tag{9.15}$$

$$T\rho_{c1} = \rho_{c2}(H + T) \tag{9.16}$$

$$\rho_{c2} = T\rho_{c1} / (H + T). \tag{9.17}$$

If the Tibetan Plateau were of normal thickness (T $\cong$ 40 km), Equation 9.17 states that it would require a crustal root with an average density of 2400 kg/m^3 to support the additional 5000 m of elevation.

SURFACE PROCESSES IN MOUNTAINS

When discussing mountainous landscapes, uplift processes are, at most, only half of the story. Earthquakes, deformation, and orogeny are impressive, but it is the persistent, sculpting hand of erosion that should be given most of the credit for the rugged topography of the Alps, the Rocky Mountains, and the other ranges of the world. Analysis of regional topography using DEMs shows that broad characteristics of the landscape in some active mountain ranges seem to depend more on the package of geomorphic processes at work than on tectonic or lithological controls [36, 45].

It is worth reviewing some of the vocabulary of erosion before discussing the process in greater detail. **Weathering** is the sum of physical and chemical processes that act to break down bedrock at and near the Earth's surface. Weathering is the process that disaggregates and decomposes rock, providing soil, sediment, and solutes to transport processes. **Erosion** is the most general of the terms used; it includes both processes of weathering and of transport. Erosion typically is used to describe both small-scale processes, such as soil creep on a single hillside, and large-scale behavior, such as the removal of material from an entire continent over geologic time periods. Finally, **exhumation** refers to the movement of rock bodies deep in the crust up to shallower levels as a result of erosion at the surface.

Rates of erosion are not uniform within a single mountain range, nor from one range to another [46] (Table 9.2). The major control on erosion rates is climate, both regional and local. Climatic conditions control the amount and timing of precipitation, the temperature at the surface, and the range and effectiveness of the geomorphic processes at work. In certain hyperarid settings, erosion rates may be almost indistinguishable from zero. For example, in the Dry Valleys region of Antarctica, erosion rates of 0.2 m/M.y. have been documented, and even minor surficial features appear to be

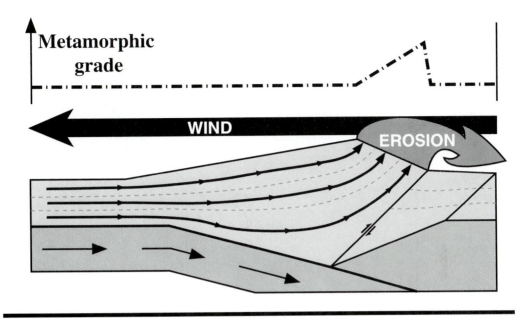

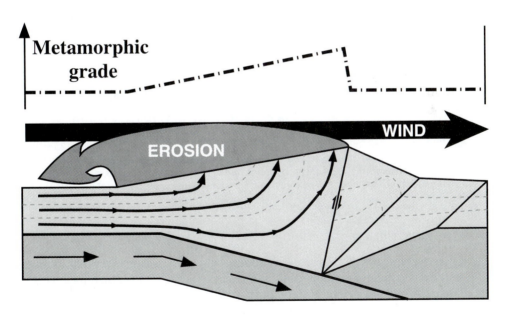

Figure 9.11   The direction of prevailing winds over a mountain range may have profound effects on the basement geology and the tectonic behavior of the orogen. Because erosion is concentrated on the windward side of the range, exhumation depths and strain gradients are much higher there. (After Willett et al., 1993. *Geology*, 21:371–374.)

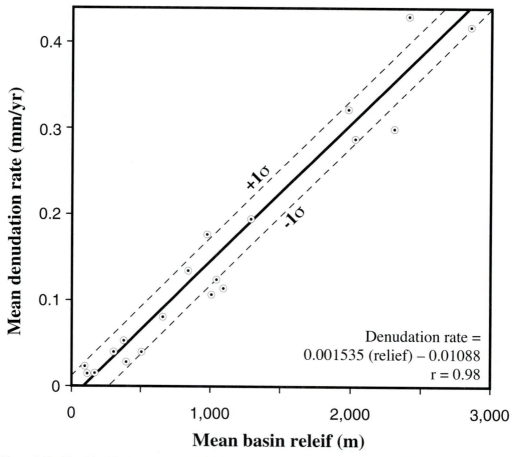

Figure 9.12   Humid midlatitude basins exhibit a surprisingly linear relationship between basin relief, which is the elevation change between the highest and lowest points, and the mean erosion rate in each basin. This is inferred to be the result of increasing precipitation at higher elevations as well as higher slope gradients and gravitational potential energy that are commonly associated with higher topographic relief. (After Ahnert, 1970 [57].)

ISOSTATIC UPLIFT

On a dynamic Earth, things are sometimes not as they seem. Denudation is the process that wears a landscape down, but as a result of isostasy, it is also a mechanism for uplift [14, 62, 63]. To understand this, consider the example of a 100,000-kg iceberg (Figure 9.13). Imagine that Admiral Frost, intrepid explorer of the Arctic, is adrift on the iceberg and is intent on shoveling away the 10,000 kg of ice above the waterline (Figure 9.13b). After removing that quantity of ice, is the Admiral standing at the water's edge (Figure 9.13c)? No—he stands on 9000 kg of ice (Figure 9.13d), because isostasy (or, in this case, "buoyancy") always readjusts the position of the iceberg so that nine-tenths of its mass is above the waterline. By digging downward (eroding the iceberg), Admiral Frost has caused the entire iceberg to move upward. The same counterintuitive process appears to operate in mountain belts.

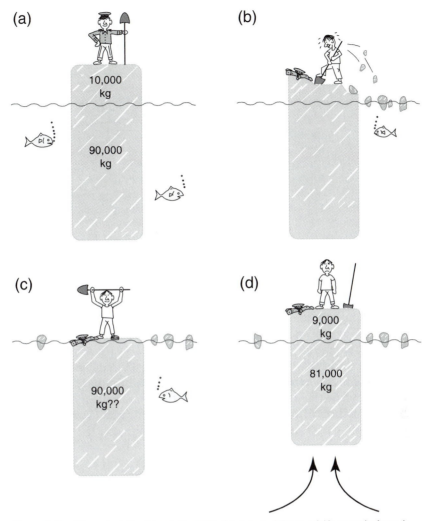

Figure 9.13   The principle of isostatic uplift. (a) Admiral Frost, adrift on an iceberg, is uncomfortable so far above the surface of the water. (b) He attempts to remove the 10,000 kg of ice above the waterline. (c) Without isostatic (buoyant) uplift, he would have reached his goal, but in the real world (d), isostasy (buoyancy) always keeps one-tenth of the iceberg above the water.

It is important to note that the average elevation of an area can only decrease as a result of isostatic uplift acting alone—isostatic uplift may mask the effects of down-wearing, but it cannot reverse those effects. However, consider the case where erosion is not uniform across a mountain range, but rather is concentrated in the valley bottoms, increasing total relief. The high peaks of the range are isolated from most of the erosion below, but because of the rigidity of the lithosphere, they experience all of the isostatic uplift of the mountain block. In the absence of tectonic uplift, *average* elevation decreases through time, but the *maximum* elevation of the region will increase (Figure

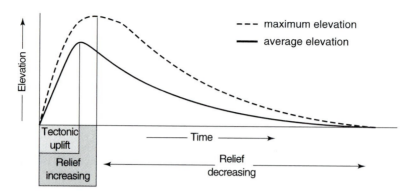

Figure 9.14　A brief pulse of tectonic uplift of a mountain range is combined with the long-term effects of erosion. The maximum elevation of the range continues to increase for a time after tectonic uplift ceases because, as long as relief is increasing, it triggers isostatic uplift. (After Isacks, 1992 [85])

9.14). In the Sierra Nevada of California, it has been estimated the mass removed by Quaternary glaciers has been enough for isostatic compensation to have caused at least 40%, and perhaps all, of the measured uplift of the range as a whole [64].

It has also been suggested that deep river incision can create a positive feedback loop, weakening the crust and actively driving upward advection of crustal material. Where the Indus and Tsangpo Rivers cross the Himalayas, exhumation rates are exceptionally high, thermal gradients are steep, and the base of the crust appears to be very shallow. Zeitler et al. [65] suggest that the rivers have formed "tectonic aneurysms", whereby incision of the deep gorges has concentrated erosion and drawn crustal and subcrustal material rapidly upward, further weakening the crust and accelerating uplift, erosion, and exhumation.

It is important to note that not every aspect of mountain ranges can be explained by isostasy [66]. Remember that the Earth's crust has strength, and isostatic compensation is always regional. For this reason, uplift of small geomorphic features, especially small warps such as described in Chapter 7, cannot be driven by isostasy. Furthermore, a number of authors have tried to invoke isostasy as a mechanism to explain episodic uplift and rejuvenation, but Gilchrist and Summerfield [67] have pointed out that isostasy operates continuously, without significant thresholds, and should not trigger episodic activity. However, Stüwe [68] outlines a credible exception to the rule whereby large-scale retreat of escarpments on the African coast may drive a repeating pattern of crustal flexure and renewed scarp retreat.

LONG-TERM GEOCHEMICAL CYCLING

In addition to orography and isostatic uplift, another potentially important linkage in the orogenic system is the role that surficial processes play in mobilizing and sequestering terrestrial chemicals that influence global climate. In particular, attention has focused on the sources and sinks of carbon dioxide in the Earth's atmosphere. Any process that either (1) decreases the rate of release of CO_2 into the atmosphere or (2) accelerates the rate of CO_2 consumption would tend to cool the climate; changes in the opposite direction would tend to warm the climate. It has been noted that during the late Cenozoic, from at least the Miocene to the present, several mountain ranges around the world (and the Indo-Eurasian collision zone in particular) seem to have experienced substantially faster tectonic uplift than during earlier periods [69]. Over the same period of time, however, there has also been dramatic global cooling leading up to and through the Pleistocene glaciations.

In what has been called a geological "chicken and egg" debate, these two global trends (accelerated uplift and a cooling climate) have led to two diametrically opposed models. The first model assumes that tectonics are the major driving force and that the late Cenozoic surge of orogenesis caused the climate change [69–71]. The second model traces causation in the opposite direction, assuming that changes in climate-driven erosion have caused the apparent increases in late Cenozoic mountain building [72]. In the first, tectonic-driven model, it has been suggested that growing mountain masses increased the average rates of siliclastic weathering worldwide and caused a corresponding increase in the consumption of CO_2 during weathering reactions [69]. Early support for this model came from paleooceanographic records of seawater strontium, which suggested that the growth of the Himalaya and Tibet did indeed cause significant increases in chemical weathering [73, 74], but more recent research has questioned the link between the isotopic data and the inferred weathering history [75–77].

The second model assumes that climate is the independent variable in the equation—that late Cenozoic cooling caused increased erosion rates in mountain ranges worldwide, accelerating isostatic uplift in response to the erosional removal of mass, and thereby resulting in what has been called an "illusion of uplift" [72]. The higher late Cenozoic uplift rates are illusory, according to this model, because they do not measure **surface uplift**, but rather **uplift of rocks**. "Surface uplift" refers to an actual increase in altitude of the topographic surface, whereas "uplift of rocks" is any kind of upward vertical motion, including isostatic uplift, whether or not the elevation of the surface increases as a result [62]. It has been suggested that most or all geomorphic and geodetic uplift measurements are actually measurements of uplift of rocks, incorporating both tectonic and isostatic driving forces [62]. The late Cenozoic climate shift—involving cooling, local increases in precipitation, and lowering of glacial ELAs—could have dramatically increased the overall pace of erosion in mountain ranges. In addition, as discussed previously, fluvial or glacial erosion concentrated in deep valleys would trigger isostatic uplift over broad regions that theoretically could lift isolated peaks up to elevations higher than they would be without the accelerated erosion [14, 64, 78].

So which of the models just outlined correctly characterizes the causation between late Cenozoic climate change and orogenesis? Like the original "chicken and egg" question, no quick answer presents itself. However, it should be noted that in feedback-rich systems, like mountain building appears to be, this kind of circularity is not a paradox, but rather it is the rule. Multiple positive feedback mechanisms can operate simultaneously and together promote and amplify any change regardless of the initial trigger.

To further cloud the question, the reader should also note that one of the central pieces of evidence, that rates of tectonic uplift seem to have accelerated, is not a recent or isolated discovery. In 1949, James Gilluly, then the retiring president of the Geological Society of America, extolled the membership to beware of "the pull of the present"

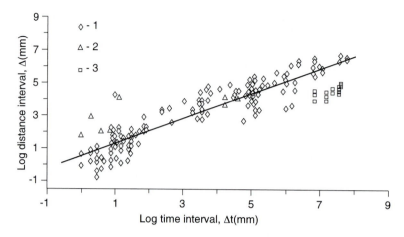

Figure 9.15   Log-log plot of uplift versus time interval for a large number of measured uplift rates. Different symbols indicate specific types of process: 1— vertical crustal uplift, 2—glacial crustal rebound and fault movement, 3—salt or magma doming. (From Gardner et al., 1987 [81].)

[79]. Sedimentation [80], erosion [81], and evolution [82] all appear to have accelerated through the Cenozoic. Examining a large number of rates calculated from different intervals of time, the bias turns out to be quite systematic (Figure 9.15)—the longer the interval over which a rate is measured, the slower the apparent rate [81]. Gilluly argued that the older geologic record is less complete, and the less complete the record, the more it will incorporate quiet times along with active periods. The "pull of the present" remains an odd phenomenon, the implications of which may not be fully appreciated across a range of disciplines.

LANDSCAPE EVOLUTION

The concept of landscape evolution implies how large-scale regional topography changes over long periods of geologic time. As discussed earlier in this chapter, the prevailing model of landscape evolution during most of the past century was the Cycle of Erosion of W. M. Davis. Davis's framework was subsequently abandoned for good reasons, but its abandonment left something of a vacuum as researchers steered well clear of questions of long-term evolution. The heart of the Cycle of Erosion was the idea that regional landscape passes through a series of predictable stages as a function of time. It is ironic that now, as researchers begin to return to studying the evolution of regional topography, it seems that orogens may indeed pass through certain common morphological and geologic stages. For example, it has been suggested that, in spite of quite different settings and geologic and tectonic differences, the Tibetan Plateau, the Turkish-Iranian Plateau, and the U.S. Great Basin have all experienced broadly similar episodes at different times in their tectonic histories [83]. At a smaller scale, a study of the response of drainage networks in Japan to a tectonic forcing points out systematic morphological changes that cross-cut local lithological and structural variations [84].

Recent advances in our understanding of orogenesis seem to bring time back into the equation while simultaneously keeping a solid footing on tangible geomorphic and tectonic evidence. Various researchers have begun considering large-scale landscape evolution as a dynamic balance between (1) the tectonic processes that add mass to orogens and (2) the erosional processes that remove that mass. Viewed this way, the evolutionary stage of a mountain range depends on the relative balance between these two sets of processes, and it is suggested that over time ranges pass through the following stages [85, 86] (Figure 9.16):

1. uplift > erosion
2. uplift = erosion
3. uplift < erosion

During the initial stage, such as shortly after the initiation of convergence in an area, uplift rates exceed erosion rates, and average elevations would increase over time. As elevations grew, erosion rates would tend to increase as a result of orography. In Japan, progression through this stage of evolution has been characterized by the gradual disappearance on an antecedent low-relief surface of Tertiary age, increases in elevation and topographic relief, and a corresponding increase in erosion rates [87]. Given tectonic uplift of long enough duration, an orogen may pass into the second stage in the model, characterized by regional denudation rates equaling average uplift rates and theoretically resulting in a regionally averaged steady-state condition. This possibility of a dynamic equilibrium between constructive and destructive forces in a mountain range

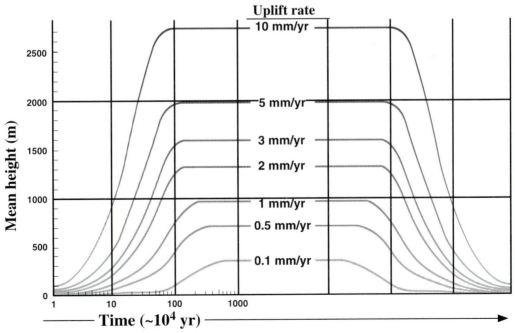

Figure 9.16   Development of mean elevation of a mountain range by the simultaneous action of uplift and erosion. Although developed with data from Japan, this model illustrates the principle of coupled tectonic-erosional orogenesis in general. Note that a stable maximum elevation is predicted when the erosion rate becomes equal to the tectonic uplift rate; the higher the uplift rate in a particular area, the higher the equilibrium elevation. (After Yoshikawa, 1985 [86].)

represents one of the most intriguing and controversial implications of recent research, and it is the focus of the next section of this chapter. The third and final stage of evolution in this three-stage model is a period of waning topography that might follow a reduction in tectonic uplift rates (or possibly an increase in climate-driven erosion). A waning landscape would be characterized by reductions over time in average elevation and lithospheric thickness—rapid at first but with rates declining over time—as both high-standing topography and the associated isostatic root are gradually consumed by erosion. For example, the average elevation of the Appalachian Mountains during the culmination of the last collision with east Africa has been estimated at 3500 to 4500 m [88]. Given any reasonable erosion rate for this setting (see Table 9.2), the original topography should have been removed long ago; why then are the Appalachians still a region of relatively high relief? The answer seems to be that erosion must consume not only the topography of the original mountain range, but also the underlying crustal root that will be fed to the surface by isostasy over time.

EQUILIBRIUM LANDSCAPES

The possibility of some sort of geomorphic equilibrium is not altogether a new concept. In 1960, John Hack proposed that some portions of the landscape may reach a dynamic equilibrium where landforms remain stable over time [89]. Some recent studies, however, have gone much further, suggesting that tectonic input into an orogen may eventually be balanced by erosional removal, with no net change in regional topography after that time. For example, Figure 9.16 was developed based on uplift and erosion measurements in Japan [86], and it illustrates some of the implications of regional equilibrium. For a given climate and uplift rate, there may be a maximum elevation that the topography can reach. A range with a low uplift rate would reach its dynamic balance at a lower elevation than a range with more rapid uplift. According to this model, no amount of additional uplift will increase the elevation above the dynamic maximum, but any increase in the *rate* of uplift could lead to a new, higher equilibrium elevation. It is important to note, however, that equilibrium conditions do not imply that geologic processes would cease; on the contrary, equilibrium appears most likely in areas of very robust uplift and erosion. With rates of tectonic uplift and erosional removal equal, rock mass would continually move upward and be exhumed, without any change in average elevation of the surface. Numerical simulations of combined tectonic and erosional processes illustrate the tremendous volumes of material that may be moved up and out of an orogen over spans of geologic time [54].

Studies have suggested that steady-state conditions may exist in a number of mountain ranges around the world, including the Southern Alps of New Zealand [47], Taiwan [90, 91], the northern Andes [55], Japan [86], the Santa Monica Mountains of California [92], the Olympic peninsula of Washington [17], and other areas. Although research into landscape steady state remains in its infancy, it is worth looking for characteristics shared in areas for which equilibrium has been suggested. Most of these mountain ranges seem have very high rates of both uplift and precipitation. In the Southern Alps, convergence across the Alpine fault results in uplift rates as high as 22 mm/yr [47]. It appears that variations in rock type and structure, Quaternary climate change, vegetation, and human influence in the Southern Alps are all of secondary importance, swamped by the overwhelming influence of upheaval and erosion [47]. Almost all of the areas of purported steady-state topography are quite wet, but local equilibrium conditions have been suggested even in the arid Basin and Range of the United States [24]. In addition, most areas of steady-state topography noted to date, with some notable exceptions, seem to occur in relatively small landmasses. This correlation may exist because smaller landmasses are more likely have fully integrated drainage networks. The presence, locations, and geometry of rivers seems to play a surprisingly important role in the evolution of orogens [93, 94], and effective removal of eroded mass by fluvial systems seems to be a prerequisite for regional equilibrium. Even stronger connections between rivers and tectonic processes have also been proposed (see the previous discussion of "tectonic aneurysms").

A number of studies have focused on the importance of variations in rock strength in regional topography [92, 95–97]. Because areas cut on resistant rocks stand higher (all other factors being equal), there should be a pervasive relationship between topography and rock strength. In its simplest form, this means that the landscape should be higher and steeper over a durable substrate than over weak rock. The Appalachian Mountains are a candidate for at least quasi-equilibrium ("quasi" because the only uplift is isostatic, and therefore elevations can only be reduced). In the Valley and Ridge Province of the Appalachians (Figure 9.17), the highest topography is on fold limbs because erosion is controlled by the most resistant rock layers. The vertical thickness, and therefore the overall resistance, of a layer is greatest where the layer dips the most (Figure 9.18). Understanding rock strength is also important for understanding the potential for local change within a regional equilibrium. Although dynamic equilibrium is

Figure 9.17   Satellite image of Valley and Ridge province of the Appalachians. (Photo by Earth Satellite Corp./SPL/Photo Researchers, Inc.)

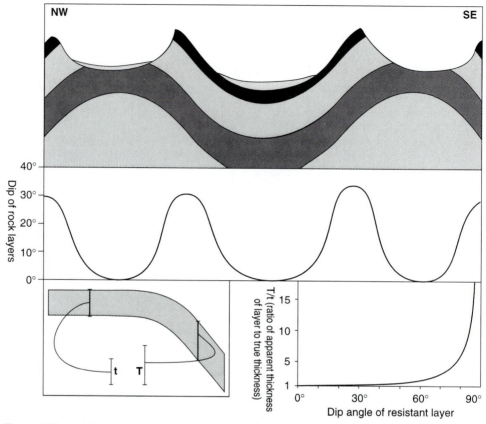

Figure 9.18   (a) Schematic illustration of relationship between structure and topography in the Valley and Ridge province of the Appalachians. (b) Note that, for any warped stratum, the vertical thickness is greatest where the layer dips the most steeply (inset). For a strong layer interbedded with weaker ones, the resistance to erosion is greatest where its thickness is greatest. (c) This fact explains why elevations are highest on the limbs of the folds.

defined by no change in *average* elevation over time, this does not require that *local* topography remain static. Over the life span of an orogen, local topography should continuously change as heterogeneous rock masses are exhumed and exposed at the surface.

SUMMARY

Mountain ranges are the product of tectonic forces that cause uplift and erosional forces that sculpt the landscape. The effects of these processes can be seen at a variety of scales, from a single mountain peak to the entire planet. Regional mountain building can be modeled as an integrated system, reflecting the strongly interconnected effects of three major subsystems: tectonics, erosion, and climate. Tectonic construction can be viewed as the sum of all processes that add excess mass to the crust. Once a mountain range has been constructed, it must also be supported. Possible support mechanisms include the buoyancy of the crust (isostasy) and the rigidity of the lithosphere (flexure). Climate-driven erosion also plays an important role during and after construction of mountain ranges. Erosion rates tend to increase with increasing elevation and relief, meaning that growing mountain ranges are self-destructive over time—the taller they grow, the more potent become the erosional processes acting the wear them away. In some locations, rates of uplift and erosion may be approximately equal, creating mountain ranges in dynamic equilibrium between creation and destruction.

REFERENCES CITED

1. Le Conte, J., 1909. *Elements of Geology*, 5th ed. New York: D. Appleton and Company.

2. Baker, V. R., 1986. Regional landform analysis. In N. M. Short and R. W. Blair, Jr. (eds), *Geomorphology from Space: A Global Overview of Regional Landforms*. Washington, DC: National Aeronautics and Space Administration, pp. 1–11.

3. Balmino, G., K. Lambeck, and W. M. Kaula, 1973. A spherical harmonic analysis of Earth's topography. *Journal of Geophysical Research*, 78:478–481.

4. Burchfiel, B. C., 1983. The continental crust. *Scientific American*, 249(3):130–142.

5. Head, J. W., S. E. Yuter, and S. C. Soloman, 1981. Topography of Venus and Earth: a test for the presence of plate tectonics. *American Scientist*, 69:614–623.

6. Saunders, R. S., R. E. Arvidson, J. W. Head III, G. G. Schaber, E. R. Stofan, and S. C. Soloman, 1991. An overview of Venus geology. *Science*, 252:249–252.

7. Soloman, S. C., J. W. Head, W. M. Kaula, D. McKenzie, B. Parsons, R. J. Phillips, G. Schubert, and M. Talwani, 1991. Venus tectonics: analysis from Magellan. *Science*, 252:297–312.

8. Head, J. W., 1990. Formation of mountain belts on Venus: evidence for large-scale convergence, underthrusting, and crustal imbrication in Freyja Montes, Ishtar Terra. *Geology*, 18:99–102.

9. Vorder Bruegge, R. W., and J. W. Head III, 1991. Processes of formation and evolution of mountain belts on Venus. *Geology*, 19:885–888.

10. Davis, W. M., 1899. The geographical cycle. *Geographical Journal*, 14:481–504.

11. Burbank, D. W., and N. Pinter, 1999. Landscape evolution: the interactions of tectonics and surface processes. *Basin Research*, 11:1–6.

12. Smith, B. J., W. B. Whalley, P. A. Warke, and A. Ruffell, 1999. Introduction and background: interpretations of landscape change. In B. J. Smith, W. B. Whalley, and P. A. Warke (eds.), *Uplift, Erosion and Stability: Perspectives on Long-Term Landscape Development*. Geological Society Special Publication, 162:vii–xi.

13. Merritts, D., and M. Ellis, 1994. Introduction to special section on tectonics and topography. *Journal of Geophysical Research*, 99:12,135–12,141.

14. Pinter, N., and M. T. Brandon, 1997. How erosion builds mountains. *Scientific American*, 276(4):74–79.

15. Pinter, N., 1997. The "cybernetic" model of orogenesis. *Geological Society of America Abstracts with Programs*, 29:6.

16. Pinter, N., 2000. Global geomorphology. In P. L. Hancock and B. J. Skinner (ed.), *Oxford Companion to the Earth*. Oxford University Press, 456–458.

17. Brandon, M. T., M. K. Roden-Tice, and J. I. Garver, 1998. Late Cenozoic exhumation of the Cascadia accretionary wedge in the Olympic Mountains, northwest Washington State. *Geological Society of America Bulletin*, 110:985–1009.

18. House, M. A., B. P. Wernicke, and K. A., Farley, 1998. Dating topography of the Sierra Nevada, California, using apatite (U-Th)/He ages. *Nature*, 396:66–69.

19. Small, E. E., and R. S. Anderson, 1998. Pleistocene relief production in Laramide mountain ranges, western United States. *Geology*, 26:123–126.

20. Summerfield, M. A., D. E. Sugden, G. H. Densmore, D. R. Marchant, H. A. P. Cockburn, and F. M. Stuart, 1999. Cosmogenic isotope data support previous evidence of extremely low rates of denudation in the Dry Valleys region, southern Victoria Land, Antactica. In B. J. Smith, W. B. Whalley, and P. A. Warke (eds.), *Uplift, Erosion and Stability: Perspectives on Long-Term Landscape Development*. Geological Society, London, Special Publications, 162:255–267.

21. Abbott, L. D., E. A. Silver, R. S. Anderson, R. Smith, J. C. Ingle, S. A. King, D. Haig, E. Small, J. Galewsky, and W. Sliter, 1997. Measurement of tectonic surface uplift rate in a young collisional mountain belt. *Nature*, 385:501–507.

22. Fielding, E., B. Isacks, M. Barazangi, and C. Duncan, 1994. How flat is Tibet? *Geology*, 22:163–167.

23. Kooi, H., and C. Beaumont, 1996. Large-scale geomorphology: classical concepts reconciled and integrated with contemporary ideas via a surface processes model. *Journal of Geophysical Research*, 101:3361–3386.

24. Ellis, M. A., A. L. Densmore, and R. S. Anderson, 1999. Development of mountainous topography in the Basin Ranges, USA. *Basin Research*, 11:21–41.

25. James, D. E., 1973. The evolution of the Andes. Scientific America, 229(2): 60–69.

26. Gregory-Wodzicki, K. M., 2000. Uplift history of the Central and Northern Andes: a review. *Geological Society of America Bulletin*, 112:1091–1105.

27. Ramos, V. A., 1989. The birth of southern South America. *American Scientist*, 77:444–450.

28. Mògard, F., 1987. Structure and evolution of the Peruvian Andes. In J. P. Scaer and J. Rodgers (eds.), *The Anatomy of Mountain Ranges*. Princeton, NJ: Princeton University Press, 179–210.

29. Isacks, B. L., 1988. Uplift of the Central Andean Plateau and bending of the Bolivian orocline. *Journal of Geophysical Research*, 93:3211–3231.

30. Molnar, P., 1986. The geologic history and structure of the Himalaya. *American Scientist*, 74:144–154.

31. Molnar, P., 1989. The geological evolution of the Tibetan Plateau. *American Scientist*, 77:350–360.

32. Harrison, T. M., P. Copeland, W. S. F. Kidd, and A. Yin, 1992. Raising Tibet. *Science*, 255:1663–1670.

33. Zheng, H., C. M. Powell, Z. An, J. Zhou, and G. Dong, 2000. Pliocene uplift of ther northern Tibetan Plateau. *Geology*, 28:715–718.

34. Sorkhabi, R. B., and E. Stump, 1993. Rise of the Himalaya: a geochronologic approach. *GSA Today*, 3:85, 88–92.

35. Yin, A., T. M. Harrison, M. A. Murphy, M. Grove, S. Nie, F. J. Ryerson, W. X. Feng, and C. Z. Le, 1999. Tertiary deformation history of southeastern and southwestern Tibet during the Indo-Asian collision. *Geological Society of America Bulletin*, 111:1644–1664.

36. Brozovíc, N., D. W. Burbank, and A. J. Meigs, 1997. Climatic limits on landscape development in the northwestern Himalaya. *Science*, 276:571–574.

37. Clift, P., N. Simizu, G. Layne, C. Gaedicke, H. U. Schlüter, M. Clark, and S. Amjad, 2000. Fifty-five million years of Tibetan evolution recorded in the Indus Fan. *EOS: Transactions of the American Geophysical Union*, 81:279–281.

38. Rea, D. K., H. Snoeckx, and L. H. Joseph, 1998. Late Cenozoic eolian deposition in the North Pacific: Asian drying, Tibetan uplift, and cooling of the northern hemisphere. *Paleoceanography*, 13:215–224.

39. England, P., and G. Houseman, 1989. Extension during continental convergence with application to the Tibetan Plateau. *Journal of Geophysical Research*, 94:17,561–17,579.

40. Yin, A., P. A. Kapp, M. A. Murphy, C. E. Manning, T. M. Harrison, M. Grove, D. Lin, D. Xi-Guang, and W. Cun-Ming, 1999. Significant late Neogene east-west extension in northern Tibet. *Geology*, 27:787–790.

41. Garzione, C. N., D. L. Dettman, J. Quade, P. G. DeCelles, and R. F. Butler, 2000. High times on the Tibetan Plateau: paleoelevation of the Thakkhola graben, Nepal. *Geology*, 28:339–342.

42. Gurnis, M., 1992. Long-term controls on eustatic and epeirogenic motions by mantle convection. *GSA Today*, 2:141, 144–145, 156–157.

43. Molnar, P., 1986. The structure of mountain ranges. *Scientific American*, 255(1):70–79.

44. Bechtel, T. D., D. W. Forsyth, V. L. Sharpton, and R. A. F. Grieve, 1990. Variations in effective elastic thickness of the North American lithosphere. *Nature*, 343:636–638.

45. Burbank, D. W., J. Leland, E. Fielding, R. S. Anderson, N. Brozovíc, M. R. Reid, and C. Duncan, 1996. Bedrock incision, rock uplift and threshold hillslopes in the northwestern Himalayas. *Nature*, 379:505–510.

46. Summerfield, M. A., and N. J. Hulton, 1994. Natural controls of fluvial denudation rates in major world drainage basin. *Journal of Geophysical Research*, 99:13,871–13,884.

47. Adams, J., 1985. Large-scale tectonic geomorphology of the Southern Alps, New Zealand. In M. Morisawa and J. T. Hack (eds.), *Tectonic Geomorphology*. Boston: Allen & Unwin, 105–128.

48. Holeman, J. N., 1968. Sediment yield of major rivers of the world. *Water Resources Research*, 4:787–797.

49. Bull, W. B., 1991. *Geomorphic Responses to Climate Change*. New York: Oxford University Press.

50. Tucker, G. E., and R. Slingerland, 1997. Drainage basin responses to climate change. *Water Resources Research*, 33:2031–2047.

51. Rinaldo, A., W. E. Dietrich, R. Rigon, G. K. Vogel, and I. Rodriguez-Iturbe, 1995. Geomorphological signatures of varying climate. *Nature*, 374:632–635.

52. Whipple, K. X., E. Kirby, and S. H. Brocklehurst, 1999. Geomorphic limits to climate-induced increases in topographic relief. *Nature*, 401:39–43.

53. Whipple, K. X., and G. E. Tucker, 1999. Dynamics of the stream-power river incision model: implications for height limits of mountain ranges, landscale response timescales, and research needs. *Journal of Geophysical Research*, 104:17,661–17,674.

54. Willett, S. D., 1999. Orogeny and orography: The effects of erosion on the structure of mountain belts. *Journal of Geophysical Research*, 104:28,957–28,981.

55. Masek, J. G., B. L. Isacks, T. L. Gubbels, and E. J. Fielding, 1994. Erosion and tectonics at the margins of continental plateaus. *Journal of Geophysical Research*, 99:13,941–13,956.

56. Hoffman, P. F., and J. P. Grotzinger, 1993. Orographic precipitation, erosional unloading, and tectonic style. *Geology*, 21:195–198.

57. Ahnert, F., 1970. Functional relationships between denudation, relief, and uplift in large mid-latitude drainage basins. *American Journal of Science*, 268:243–263.

58. Hallet, B., L. Hunter, and J. Bogen, 1996. Rates of erosion and sediment evacuation by glaciers: a review of field data and their implications. *Global and Planetary Change*, 12:213–235.

59. Anderson, S. P. J. I. Drever, and N. F. Humphrey, 1997. Chemical weathering in glacial environments. *Geology*, 25:399–402.

60. Burbank, D. W., 1992. Characteristic size of relief. *Nature*, 359:483–485.

61. Densmore, A. L., R. S. Anderson, B. G. McAdoo, and M. A. Ellis, 1997. Hillslope evolution by bedrock landslides. *Science*, 275:369–372.

62. England, P., and P. Molnar, 1990. Surface uplift, uplift of rocks, and exhumation of rocks. *Geology*, 18:1173–1177.

63. Bishop, P., and R. Brown, 1992. Denudational isostatic rebound of intraplate highlands: the Lachlan River valley, Australia. *Earth Surface Processes and Landforms*, 17:345–360.

64. Small, E. E., and R. D. Anderson, 1995. Geomorphically driven Late Cenozoic rock uplift in the Sierra Nevada, California. *Science*, 270:277–280.

65. Zeitler, P. K., A. S. Meltzer, P. O. Koons, D. Craw, B. Hallet, C. P. Chamerlain, W. S. F. Kidd, S. K. Park, L. Seeber, M. Bishop, and J. Shroder, 2001. Erosion, Himalayan geodynamics, and the geomorphology of metamorphism, 2001. *GSA Today*, 11(1):4–8.

66. Pinter, N., and E. A. Keller, 1991. Comment on "Surface uplift, uplift of rocks, and exhumation of rocks" (England and Molnar, 1990). *Geology*, 19:1053.

67. Gilchrist, A. R., and M. A. Summerfield, 1991. Denudation, isostasy, and landscape evolution. *Earth Surface Processes and Landforms*, 16:555–562.

68. Stüwe, K., 1991. Flexural constraints on the denudation of asymmetric mountain belts. *Journal of Geophysical Research*, 96:10,401–10,408.

69. Raymo, M. E., W. F. Ruddiman, and P. N. Froelich, 1988. Influence of late Cenozoic mountain building on ocean geochemical cycles. *Geology*, 16:649–653.

70. Ruddiman, W. F., and M. E. Raymo, 1988. Northern Hemisphere climate regimes during the past 3 Ma: Possible tectonic connections. *Philosophical Transactions of the Royal Society of London*, Series B, 318:411–430.

71. Ruddiman, W. F., and J. E. Kutzbach, 1989. Forcing of Late Cenozoic northern hemisphere climate by plateau uplift in southern Asia and the American West. *Journal of Geophysical Research*, 94:18,409–18,427.

72. Molnar, P., and P. England, 1990. Late Cenozoic uplift of mountain ranges and global climate change: chicken or egg? *Nature*, 346:29–34.

73. Edmond, J. M., 1992. Himalayan tectonics, weathering processes, and the strontium isotope record in marine limestones. *Science*, 258:1594–1597.

74. Krishnaswami, S., J. R. Trivedi, M. M. Sarin, R. Ramesh, and K. K. Sharma, 1992. Strontium isotopes and rubidium in the Ganga-Brahmaputra River system: weathering in the Himalaya, fluxes to the Bay of Bengal and contributions to the evolution of oceanic $^{87}Sr/^{86}Sr$. *Earth and Planetary Sciences Letters*, 109:243–253.

75. Derry, L. A., and L. C. France, 1996. Neogene Himalayan weathering history and river $^{87}Sr/^{86}Sr$; impact on the marine Sr record. *Earth and Planetary Sciences Letters*, 142:59–74.

76. Jacobson, A. D., and J. D. Blum, 2000. Ca/Sr and $^{87}Sr/^{86}Sr$ geochemistry of disseminated calcite in Himalayan silicate rocks from Nanga Parbat; influence on river-water chemistry. *Geology*, 28:463–466.

77. Rea, D. K., 1992. Delivery of Himalayan sediment to the northern Indian Ocean and its relation to global climate, sea level, uplift, and seawater strontium. *Synthesis of Results from Scientific Drilling in the Indian Ocean*, Geophysical Monograph, 70:387–402.

78. Montgomery, D. R., 1994. Valley incision and the uplift of mountain peaks. *Journal of Geophysical Research*, 99:13,913–13,921.

79. Gilluly, J., 1949. Distribution of mountain building in geologic time. *Geological Society of America Bulletin*, 60:561–590.

80. Sadler, P. M., 1981. Sediment accumulation and the completeness of stratigraphic sections. *Journal of Geology*, 89:569–584.

81. Gardner, T. W., D. W. Jorgensen, C. Shuman, and C. R. Lemieux, 1987. Geomorphic and tectonic process rates: effects of measured time interval. *Geology*, 15:259–261.

82. Gingerich, P. D., 1983. Rates of evolution: effects of time and temporal spacing. *Science*, 222:159–161.

83. Dilek, Y., and E. M. Moores, 1999. A Tibetan model for the early Tertiary western United States. *Journal of the Geological Society, London*, 156:929–941.

84. Sugai, T., and H. Ohmori, 1999. A model of relief forming by tectonic uplift and valley incision in orogenesis. *Basin Research*, 11:43–57.

85. Isacks, B. L., 1992. 'Long-term' land surface processes: erosion, tectonics and climatic history in mountain belts. In P. M. Mather (ed.), *TERRA-1: Understanding the Terrestrial Environment. The Role of Earth Observations from Space*. London: Taylor & Francis, 21–36.

86. Yoshikawa, T., 1985. Landform development by tectonics and denudation. In A. Pitty (ed.), *Themes in Geomorphology*. London: Croom Helm, 194–210.

87. Ohmori, H., 1985. A comparison of the Davisian scheme and landform development by concurrent tectonics and denudation. *Bulletin of the Department of Geography, University of Tokyo*, 17:19–28.

88. Slingerland, R., and K. P. Furlong, 1989. Geodynamics and geomorphic evolution of the Permo-Triassic Appalachian Mountains. *Geomorphology*, 2:23–37.

89. Hack, J. T., 1960. Interpretation of erosional topography in humid temperate regions. *American Journal of Science*, 258-A:80–97.

90. Lundberg, N., and R. J. Dorsey, 1990. Rapid Quaternary emergence, uplift, and denudation of the Coastal Range, eastern Taiwan. *Geology*, 18:638–641.

91. Petley, D. N., and S. Reid, 1999. Uplift and landscape stability at Taroko, eastern Taiwan. In B. J. Smith, W. B. Whalley, and P. A. Warke (eds.), *Uplift, Erosion and Stability: Perspectives on Long-Term Landscape Development*. Geological Society Special Publication, 162:169–182.

92. Meigs, A., N. Brozovíc, and M. L. Johnson, 1999. Steady, balanced rates of uplift and erosion of the Santa Monica Mountains, California. *Basin Research*, 11:59–73.

93. Pavlis, T. L., M. W. Hamburger, and G. L. Pavlis, 1997. Erosional processes as a control on the structural evolution of an actively deforming fold and thrust belt: An example from the Pamir-Tien Shan region, central Asia. *Tectonics*, 16:810–822.

94. Burbank, D. W., 1992. Causes of recent Himalayan uplift deduced from deposited patterns in the Ganges basin. *Nature*, 357:680–683.

95. Schmidt, K. M., and D. R. Montgomery, 1995. Limits to relief. *Science*, 270:617–620.

96. Weissel, J. K., and M. A. Seidl, 1997. Influence of rock strength properties on escarpment retreat across passive continental margins. *Geology*, 25:631–634.

97. Battiau-Queney, Y., 1999. Crustal anisotropy and differential uplift: their role in long-term landform development. In B. J. Smith, W. B. Whalley, and P. A. Warke (eds.), *Uplift, Erosion and Stability: Perspectives on Long-Term Landscape Development*. Geological Society Special Publication, 162:65–74.

Appendix A:
Selected Dating Methods

			Age Range and Optimum Resolution					
	Method	Applicability	10^2	10^3	10^4	10^5	10^6	Basis of Method and Remarks
Numerical Dates — Annual	Historical records	X to XXX	====+===					Requires preservation of pertinent records; applicability depends on quality and detail of records. Limited to several hundred years in western hemisphere.
	Dendrochronology	XX	=======					Requires either direct counting of annual rings back from present or construction of a chronology based on variations in annual ring growth. Restricted to areas where trees of the required age and (or) environmental sensitivity are preserved.
	Varve chronology	X	=======+++					Requires either direct counting of varves back from present or construction of a chronology based on overlapping successions of continuous varved lake sediments. Subject to errors in matching separate sequences and to misidentification of annual layers.
Radiometric	Carbon-14	X to XXX	o•−+++====⇒?					Depends on availability of carbon. Based on decay of ^{14}C, produced by cosmic radiation, to ^{14}N. Subject to errors due to contamination, particularly in older deposits and in carbonate material (such as mollusk shells, marl, soil carbonate).
	Uranium series	XX			••− − − −++++−			Used to date coral, mollusks, bone, cave carbonate, and carbonate coats on stones. Potentially useful in dating travertine and soil carbonate. A variety of isotopes of the U-decay series are used during $^{230}Th/^{234}U$ (most common and method described to left), $^{234}U/^{238}U$ (with a range back to 600,000 yr), $^{231}Pa/^{235}U$ (10,000 –12,000 yr), U-He (0–2 m.y.), and $^{226}Ra/^{230}Th$ (<10,000 yr). Errors due to the lack of a closed chemical system are a common problem, especially in mollusks and bone.
	Potassium-argon	X				•••− − +++++		Directly applicable only to igneous rocks and glauconite. Requires K-bearing phases such as feldspar, mica, and glass. Based on decay of ^{40}K to ^{40}Ar. Subject to errors due to excess argon, loss of argon, and contamination.

	Method	Applicability	Age Range and Optimum Resolution					Basis of Method and Remarks
			10^2	10^3	10^4	10^5	10^6	
Numerical Dates (cont.) / **Radiometric (cont.)**	Fission track	X			●●●–┼–+++┤			Directly applicable only to igneous rocks (including volcanic ash); requires uranium-bearing material (zircon, sphene, apatite, glass). Based on the continuous accumulation of tracks (strained zones) caused by recoiling U fission products. Subject to errors due to track misidentification and to track annealing.
	Uranium trend	XXXX		∞●●–––––●●				Based on open-system flux of uranium through sediment and soil; ^{238}U, ^{234}U, ^{230}Th, and ^{232}Th must be measured on about five different samples from a given-aged deposit and an isochron constructed to determine age.
	Thermoluminescence (TL) and electron-spin resonance (ESR)	XXXX		∞●●●●●●–––––––––●●				Based on displacement of electrons from parent atoms by alpha, beta, and gamma radiation. Applicable to feldspar and quartz in sediments and carbonate in soils. TL based on amount of light released as sample is heated compared with that released after known radiation dose. TL precision better than indicated for ceramics in 400–10,000-year range.
	Cosmogenic isotopes other than carbon–14	X	–?–?–?–?–?–?–?–?–?					Dating methods analogous to ^{14}C-dating are based on the cosmogenic isotopes (half-life in years in parentheses), ^{32}Si (300), ^{41}Ca (1.3×10^5), ^{36}Cl (3.08×10^5), ^{26}Al (7.3×10^5), ^{10}Be (1.5×10^6), ^{129}I (1.6×10^7), and ^{53}Mn (3.7×10^6).
Relative Dates	Amino acid racemization	XX	●●●●●●●●●●●●●●●–––––––●●●					Requires shell or skeletal material. Based on release of amino acids from protein and subsequent inversion of their stereoisomers. Shells tend to be more reliable than bone, wood, or organic-rich sediment. Is strongly dependent on other variables, especially temperature and leaching history. Commonly used as a relative dating or correlation technique, but yields numerical ages when calibrated by other techniques.
	Obsidian hydration	X	––––––––––––––––●●●					Based on thickness of the hydrated layer along obsidian crack or surface formed during given event. Age proportional to the thickness squared. Calibration depends on experimental determination of hydration rate or numerical dating. Subject to errors due to temperature history and variation in chemical composition.
	Tephra hydration	X	∞∞∞∞∞∞∞●●●●●●●●					Requires volcanic ash. Based on the progressive filling of bubble cavities in glass shards with water. Subject to the same limits as obsidian hydration, plus others, including the geometry of ash shards and bubble cavities.
	Lichenometry	X to XXX	––––●●●●					Requires exposed, stable rock substrates suitable for lichen growth. Most common in alpine and arctic regions, where lichen thallus diameter is proportional to age. Subject to error due to climatic differences, lichen kill, and misidentification. The limit of the useful range varies considerably with climate and rock type.

Method	Applicability	Age Range and Optimum Resolution 10^2	10^3	10^4	10^5	10^6	Basis of Method and Remarks
Relative Dates (cont.)							
Soil development	XXXX	∞∞∞∞●●●●●●- - - -┤- - ●●┤					Encompasses a number of soil properties that develop with time, all of which are dependent on other variables in addition to time (parent material, climate, vegetation, topography). Is most effective when these other variables are held constant or can be evaluated. Precision varies with the soil property measured; for example, accumulation of soil carbonate locally yields age estimates within ±20 percent.
Rock and mineral weathering	XX	∞∞∞∞●●●●●●- - - - - - - - -●●●					Includes a number of rock- and mineral-weathering features that develop with time, such as thickness of weathering rinds, solution of limestone, etching of pyroxene, grussification of granite, and buildup of desert varnish. Has the same basic limitations as soil development. Precision varies with the weathering feature measured.
Progressive landform modification	XXX	∞∞∞∞∞∞∞∞●●●●●●●●∞∞∞					In addition to time, depends on factors such as climate and lithology. Depends on reconstruction of original landform and understanding of process resulting in change of landform, including creep and erosion.
Rate of deposition	XX	- -?- -?- -?- -?-?- -					Requires relatively constant rate of sedimentation over time intervals considered. Numerical ages based on sediment thickness between horizons dated by other methods. Quite variable in alluvial deposition.
Geomorphic position and incision rate	XXX	∞∞∞∞∞∞∞●●●●●●●?- ?-					Geomorphic incision rates depend on stream size, sediment load, bedrock resistance to erosion, and uplift rates or other base-level changes. If one terrace level is dated, other terrace levels may be dated assuming constant rate of incision.
Rate of deformation	XXX	●●?●?●?●?●●?●?●●?●?●●?●●?					Dating assumes deformation rate constant over interval of concern and requires numerical dating for calibration. At spreading centers and plate boundaries, nearly constant rates may be valid for intervals of millions of years.
Correlation							
Stratigraphy	XXXX						Based on physical properties and sequence of units, which includes superposition and inset relations. Depends on the establishment of time equivalence of units; deposition of Quaternary units normally occurs in response to cyclic climatic changes.
Tephrochronology	X	RESOLUTION DEPENDS ON RECOGNITION OF FEATURE AND ACCURACY OF DATING THAT FEATURE					Requires volcanic ash (tephra) and unique chemical or petrographic identification and (or) dating of the ash. Very useful in correlation because an ash eruption represents a virtually instantaneous geologic event.
Paleomagnetism	XX						Depends on correlation of remnant magnetic vector, which includes polarity, or a sequence of vectors with a known chronology of magnetic variation. Subject to errors due to chemical magnetic overprinting and physical disturbance.

	Method	Applicability	Age Range and Optimum Resolution					Basis of Method and Remarks
			10^2	10^3	10^4	10^5	10^6	
Correlation (cont.)	Fossils and artifacts	XX						Depends on the availability of fossils, including pollen, and artifacts. Resolution depends on the rate of evolution or change of organisms or cultures and on calibration by other techniques. Subject to errors due to misidentification and interpretation.
	Stable isotopes	X						Depends on correlation of the sequence of isotopic changes with an age-controlled master chronology. Oxygen isotopic changes with an age-controlled master chronology. Oxygen isotopic record is useful in deep-sea and ice-cap cores and perhaps in cave deposits.

RESOLUTION DEPENDS ON RECOGNITION OF FEATURE AND ACCURACY OF DATING THAT FEATURE

APPLICABILITY

XXXX, nearly always applicable   XX, often applicable
XXX, very often applicable   X, seldom applicable

OPTIMUM RESOLUTION

======, <2 percent   ••••••, 25–75 percent
++++++, 2–8 percent   ○○○○○○, 75–200 percent
------, 8–25 percent

(Modified slightly from Pierce, K.L., 1986. Dating methods. In *Active Tectonics*)

Glossary

accelograph: instrument that measures ground acceleration during seismic shaking.

accordant summits: a landscape in which all the peaks have approximately the same elevations.

active fault: a fault that has moved within a given period of time, typically the last 10,000 years.

aftershock: an earthquake that follows and is less powerful than the main shock.

aggradation: in a *fluvial system,* the accumulation of sediments in response to a rise in *base level* or other causes.

alignment (leveling) array: line of *control points* used to measure displacement across a fault.

alluvial fan: a conical, depositional landform found along many mountain fronts of arid and semiarid regions.

alluvium: loose sedimentary material deposited by rivers or streams.

antecedent stream: a stream that existed before geologic processes altered the landscape; such streams commonly cut through geologic structures or across high-standing topography of the modern landscape.

antithetic shear: a secondary fault with the opposite sense of displacement as the main fault.

Archimedes' principle: principle that the weight of a floating solid equals the weight of fluid that it displaces.

aseismic: describes an event or process that occurs without accompanying earthquake activity.

asthenosphere: plastic layer of the upper mantle that lies beneath the *lithosphere.*

asymmetry factor (AF): a geomorphic index used to detect active tilting.

balanced cross section: a geologic cross section in which strata are parallel, and individual layers maintain uniform or uniformly-varying thickness.

base level: the lowest elevation that a specific fluvial system drains to; the concept includes both *local base level* and *ultimate base level* (usually sea level).

beheaded stream: a stream that ends upstream at a fault and has been faulted away from its headwaters.

body waves: seismic waves that travel through the interior of the Earth.

brittle behavior: when a material responds to an applied stress by fracturing.

buried reverse fault: a compressional fault that does not or did not break the ground surface when it was last active.

capable fault: according to the U.S. Nuclear Regulatory Commission definition, a fault that has moved at least once in the past 50 k.y. or more than once in the past 500 k.y.

catastrophe: a disaster from which recovery is a long and involved process.

characteristic earthquake: an earthquake that strikes a given fault zone with approximately the same magnitude and with similar characteristics at approximately equal intervals.

coastal terrace: (sometimes called an *uplifted marine terrace*) a set of coastal landforms, typically either a *wave-cut platform* or a coral reef complex, that has been uplifted above the modern shoreline.

colluvial wedge: a deposit of colluvium at the base of a slope, thickest near the slope and progressively thinner farther away.

colluvium: loose sedimentary material deposited by gravity-driven processes, usually at the base of a slope.

complex response: a model that states that features within a system (such as individual landforms in a landscape system) may be caused by changes in *intrinsic variables*, and not the direct result of external stimuli.

conditional probability: probability of seismic risk based on available knowledge.

continental drift: movement of continents in response to sea-floor spreading. The most recent episode of continental drift began about 200 million years ago with the breakup of the supercontinent Pangaea.

control points: surveyed points in a geodetic net or array.

convergent plate boundary: boundary between two lithospheric plates in which one plate descends below the other (subduction).

coseismic: describes an event or process that coincides with an earthquake.

creep: see *tectonic creep*.

critical facilities: facilities that, if damaged or destroyed, might cause catastrophic loss of life, property damage, or disruption of society.

Cycle of Erosion: model of landscape development developed by W.M. Davis. In this model, landscapes go through three characteristic stages: youth, maturity, and old age.

décollement: a low-angle structure (typically a *thrust fault*) separating more deformed rocks above from less deformed rocks below.

deflected drainage: a stream that follows a strike-slip fault along some or all of its length.

degradation: in a *fluvial system*, the removal of sediments or erosion of the channel in response to a fall in *base level* or other causes.

dendritic drainage pattern: "finger-like" pattern of streams associated with homogeneous bedrock and gentle slopes.

dendrochronology: the study of tree rings.

denudation: regional erosion of the surface.

depth of compensation: in *isostasy*, it is the depth in a fluid at which pressure is everywhere equal.

detachment fault: a low-angle normal fault across which there is significant displacement.

diffusion equation: a mathematical expression that is used to quantitatively model *fault-scarp degradation.*

dilatancy: the development of cracks and pores in a material that is subjected to stress.

dilatancy-diffusion model: the theory that dilatancy in rocks near a fault zone leads to an influx of water that triggers an earthquake.

dip: the maximum slope angle on a sloping surface.

divergent plate boundary: boundary between lithospheric plates characterized by production of new lithosphere; found along oceanic ridges.

drainage basin: the area in which all rain that falls exits through the same stream.

drainage density: the number and length of channels per unit area.

drape fold: a fold that forms over, and as a result of, a buried normal fault.

dry-tilt net: three *control points* used to measure tilt of the ground surface.

dynamic equilibrium: a condition of stability that is created by self-adjustment of all processes operating within the system (see *intrinsic variables* and *threshold*).

earthquake: a sudden motion or trembling in the Earth caused by the abrupt release of strain on a *fault.*

Earthquake Cycle: model in which earthquakes are the result of the accumulation of elastic strain.

earthquake precursors: events or phenomena that precede an earthquake.

elastic behavior: deformation that is recovered fully and instantaneously when the driving force is removed.

elastic thickness: in *isostasy*, it is the depth to which the crust acts as a rigid or brittle solid.

emergence: motion of the land up relative to sea level, such that the coastline advances oceanward through time.

ephemeris: a mathematical model of the precise orbital path of a satellite.

epicenter: the point on the surface of the Earth directly above the *focus* of an earthquake.

erosion: general term describing the processes of *weathering* and transport of sediment.

exhumation: the unburial of rocks by erosion.

extrinsic variables: processes that originate or operate outside of a system.

fault: a fracture or fracture system that has experienced movement along opposite sides of the fracture.

fault-bend fold: a fold formed by a change in dip on an underlying fault.

fault gouge: a clay zone formed by pulverized rock during an earthquake, which may create a groundwater barrier.

fault-propagation fold: a fold that forms around the tip of a fault that does not rupture the ground surface (see *buried reverse fault*).

fault scarp: a steep slope formed by a fault rupturing the surface.

fault scarp degradation: erosion of a fault scarp. See *morphological dating*.

fault-valve mechanism: hypothesis that earthquakes may be closely linked with fluid pressures in the crust, and that discharges of crustal fluids may accompany earthquakes.

fault zone: a related group of faults in a subparallel belt or zone.

fill terrace: a type of *stream terrace* that consists of a thick accumulation of *alluvium* (in contrast to a *strath terrace*).

flexural-slip fault: a fault on a stratigraphic bedding plane or other plane of weakness caused solely by flexure associated with active folding.

flexural support: a process by which the weight of mountains or other loads atop the Earth's crust are carried by the strength of the crust.

flexure: bending of a material or a surface (see *warping*).

floodplain: flat land or valley floor that borders a stream or river, formed by migration of meanders and/or periodic flooding.

fluvial geomorphology: the study of river processes and landforms caused by river processes.

fluvial system: a river or stream.

focus: the location within the Earth at which an earthquake originates.

fold-and-thrust belt: a zone characterized by subparallel faults and folds that reflect active compression.

footwall: the side of a fault that lies beneath the inclined fault plane.

foreshock: an earthquake that precedes and is less powerful than the main shock.

geodesy: study and measurement of the shape of the Earth's surface.

geoid: a surface of equal-gravity potential around the Earth. In most applications, mean sea level is used.

geomorphic record: the sum of landforms and *Quaternary* deposits at a site or in an area.

geomorphology: the study of landforms and surface processes.

Global Positioning System (GPS): geodetic positioning technique that utilizes the array of satellites maintained by the U.S. Department of Defense.

graben: a structural block that is down-dropped by faults on both sides of it.

graded river: a river in which driving forces (e.g., gravity, slope, discharge) and resisting forces (e.g., volume of sediment transported, channel roughness, sinuosity) are equal along its entire length.

ground acceleration: a quantitative measurement of the intensity of seismic shaking (usually given as a percent of the acceleration of gravity).

ground-penetrating radar: a method of sending radar waves into the subsurface and measuring their reflections with the goal of determining geometry beneath the surface.

half graben: a structural block that is down-dropped by a fault on only one side of it.

Subsidence of a half graben must be accompanied by *tilting* or *warping*.

hanging wall: the side of a fault that lies above the inclined fault plane.

Holocene Epoch: the last 10,000 years.

horst: a structural block that is lifted up relative to the blocks on either side of it.

hypocenter: the point in the earth where an earthquake originates; also known as the *focus*.

hypsometric curve: a graphical representation of the elevation distribution of a given landscape.

hypsometric integral: area under the *hypsometric curve*.

hypsometry: measurement and analysis of the distribution of land area at different elevations.

incision: local or regional erosion by streams, typically causing an increase in *relief*.

induced seismicity: earthquakes caused by human activity such as building dams, injecting fluids into the subsurface, or underground testing of nuclear weapons.

intensity (of an earthquake): a relative measurement of the strength of shaking at any given location. Intensity generally decreases with increasing distance from the epicenter.

interseismic: describes an event or process that occurs between major earthquakes.

intraplate earthquakes: earthquakes that occur in the interior of a lithospheric plate, away from any plate boundary.

intrinsic variables: processes that originate or operate within a system.

isostasy: the principle by which thicker, more buoyant crust stands topographically higher than thinner, denser crust.

landform: a discrete element of the landscape, such as a hill, a terrace, or an alluvial fan.

landslide: any downslope motion under the force of gravity—sometimes a secondary effect of earthquake shaking.

liquefaction: transformation of water-saturated sediments from a solid to a liquid state in response to shaking.

listric fault: a *normal fault* that is curved.

lithosphere: the upper portion of the Earth, consisting of the crust and upper portion of the mantle, that is characterized by brittle behavior.

locked fault: a fault which does not exhibit *tectonic creep*; a fault on which stress accumulates.

longshore transport: movement of sediment parallel to the shoreline as a result of waves that strike the shore at an oblique angle.

Love wave: a type of surface wave that causes a sideways shaking (in contrast to *Rayleigh wave*).

magnitude (of an earthquake): an absolute measurement of the energy of a given earthquake.

marine terrace: (see *coastal terrace*)

material amplification: a local increase in the intensity of seismic shaking caused by near-surface material (usually loose sediments or artificial fill).

maximum credible earthquake: the largest earthquake likely to be generated by faults in a given area.

meander: one of a series of curves in a sinuous stream or river.

measurement interval bias: observation that the rates of many processes seem to be slower in the distant geologic past than they are in the present or in the recent past.

mid-ocean ridge: divergent spreading center at the center of an ocean, called a ridge because the newly formed ocean crust is relatively buoyant and causes the topography to stand high.

Modified Mercalli Scale: a system for estimating earthquake *intensity*.

moment magnitude: a system for measuring earthquake *magnitude* based on the total energy released by the earthquake (also see *seismic moment*).

monsoon: pattern of seasonal storms that strike India, Southeast Asia, western Africa, and northern Australia.

morphological dating: estimating the age of a landform based on its shape, usually estimating the amount of erosion that has occurred since the landform was formed.

morphometry: quantitative measurement and analysis of topography.

mountain front: steep escarpment that marks the boundary between mountainous topography and relatively flat topography.

mountain-front sinuosity (S_{mf}): a geomorphic index used to detect tectonic activity along mountain fronts.

normal fault: a fault across which there is extension.

numerical age control: estimates of the age of a material or feature in an absolute number of years (as opposed to *relative age control*).

offset stream: a stream the channel of which is displaced across a strike-slip fault.

orogeny: regional increase in topography caused by tectonic processes (mountain-building).

orthometric height: elevation above or below the geoid.

P-wave: compressional ("push-pull") *seismic waves*.

paleoseismology: the study of earthquakes that occurred in the geologic past.

peneplain: a low-relief plain that is the theoretical end-product of erosion without tectonic activity (see *Cycle of Erosion*).

piercing points: two points on opposite sides of a fault that were originally connected, but were offset by one or more ruptures along the fault.

plastic behavior: a permanent change in the shape of a material after a force is applied to it.

plate tectonics: a model of global tectonics that suggests that the outer layer of the earth known as the lithosphere is composed of several large plates that move relative to one another; continents and ocean basins are passive riders on these plates.

Pleistocene Epoch: the period of geologic time from about 2 million to 10,000 years ago. Much of the Pleistocene was characterized by the growth and decline of glaciers in many areas.

post-glacial rebound: uplift caused by *isostasy* that follows the melting of a large continental ice sheet.

postseismic: describes an event or process that occurs shortly after an earthquake.

preseismic: describes an event or process that occurs shortly before an earthquake.

pressure ridge: a hill along a strike-slip fault zone formed by upwarping at *restraining bends* or between two different strands of the fault.

process-response model: a model of the development of landforms and the landscape based on an understanding of the processes at work and how they shape the surface.

pull of the present: see *measurement interval bias*.

Quaternary: the latest period of geologic time up to and including the present. The Quaternary includes the *Pleistocene* and the *Holocene*, and ranges from approximately 1.65 million years ago to the present.

radiocarbon dating: a method that estimates the absolute age of a sample based on the ratio of radiogenic carbon (^{14}C) to stable carbon (^{12}C and ^{13}C).

Rayleigh wave: a type of surface wave that causes an elliptical motion like the rolling of an ocean wave (in contrast to *Love wave*).

recurrence interval: the average period of time between major earthquakes on a given fault (see *characteristic earthquake*).

rejuvenation: renewed uplift and erosion of a mature landscape (see *Cycle of Erosion*).

relative age control: estimates of the age of a material or feature compared with other features (as opposed to *numerical age control*).

relative spacing: a method for determining the ages of a sequence of *coastal terraces* based on their present-day elevations and knowledge of Quaternary sea-level history.

relative tectonic activity class designation: classification of a given area into one of several categories of ongoing tectonic activity.

releasing bend: a bend in a strike-slip fault that causes extension across the area of the bend.

relief: generally, the "ruggedness" of the topography; specifically, the highest elevation minus the lowest elevation in a given area.

response spectra: in earthquake engineering, it is the relationship between seismic-wave period and ground shaking.

restraining bend: a bend in a strike-slip fault that causes compression across the area of the bend.

retrodeformation: interpretation of a geologic cross section with the goal of understanding its geometry before deformation occurred.

reverse fault: a fault across which there has been convergence.

Richter magnitude: (also called *local magnitude*) a system for measuring earthquake *magnitude* based on the maximum amplitude and period of seismic waves recorded by a seismograph at a set distance from the earthquake *epicenter.*

rupture: breakage of a material under stress.

S-wave: *seismic waves* in which particle displacement is perpendicular to the direction of propagation of the wave.

sag pond: a pond along a strike-slip fault zone formed by downwarping between two different strands of the fault.

sand boil: sand extruded from the surface during seismic shaking, caused by high fluid pressure and *liquefaction.* Also called a "sand blow" or "sand volcano."

Satellite Laser Ranging (SLR): geodetic positioning technique based on the travel time of a laser pulse from a measuring station to a ranging satellite back to the measuring station.

satellite radar interferometry: geodetic positioning technique that uses pairs of radar images with the goal of measuring small changes in position over broad areas.

scarp: (short for escarpment) a slope steeper than the surrounding topography; related to a change in material, process, or geomorphic history.

seacliff: on a *coastal terrace* or a modern coastline, it is the steep slope cut by wave action at its base.

segmentation: subdivision of a fault zone into smaller units with discrete rupture histories.

seismic: refers to vibrations in the earth produced by earthquakes.

Seismic Deformation Cycle: a repeating pattern of *preseismic, coseismic, postseismic,* and *interseismic* deformation at or near a fault.

seismic gap: a portion of a fault zone, between two areas that have ruptured in historical or recent time, that has *not* ruptured.

seismic moment: a measurement of the total amount of energy released during an earthquake.

seismic reflection profiling: a method of inducing vibrations and measuring their reflections with the goal of determining geometry beneath the surface.

seismic risk: an estimate of the likelihood and the potential damage of an earthquake in a given area.

seismic waves: energy released from a fault rupture, subdivided into *body waves* (*P-waves* and *S-waves*) and *surface waves* (*Rayleigh waves* and *Love waves*).

seismic zoning: legal definition of land as appropriate or inappropriate for different uses based on proximity to active faults, presence of material that may amplify shaking, etc.

seismograph: instrument for measuring *seismic waves;* the record of seismic waves itself is called a *seismogram.*

seismometer: instrument used to measure seismic waves.

shoreline angle: on an erosional coastline, the line at which the wave-cut platform meets the seacliff.

shutter ridge: a ridge offset by a strike-slip fault such that the ridge below the fault is juxtaposed against the gully above the fault.

slip rate: long-term rate of motion on a fault or fault zone.

soil chronosequence: a series of soil profiles systematically arranged from youngest to oldest.

strain: deformation resulting from *stress*.

strain partitioning: observation that in areas with oblique strain, horizontal and vertical deformation often occur on distinct and separate structures.

strath terrace: a type of *stream terrace* that consists of a cut bedrock surface with little or no *alluvium* overlying it (in contrast to a *fill terrace*).

stream length-gradient index (SL): a geomorphic index used to identify possible areas of tectonic activity.

stream power: the rate at which a stream can do work.

stress: force per unit area; may be compressive, tensile, or shear.

strike: the orientation of a horizontal line on a sloping surface.

strike-slip fault: a fault across which displacement is predominantly horizontal.

subduction: process in which one lithospheric plate descends beneath another.

subduction zone: a convergent plate boundary at which dense crust of one of the plates sinks down into the mantle.

subenvelope map: a topographic map based on stream elevations, with the higher elevations between streams stripped away.

submergence: motion of the land down relative to sea level, such that the coastline advances landward through time.

subsidence: downwarping of an area of the Earth's surface.

surface uplift: a specific type of *uplift* in which the elevation of the surface increases through time.

surface waves: seismic waves that travel along the surface of the Earth or discrete boundaries within the Earth.

synthetic shear: a secondary fault with the same sense of displacement as the main fault.

tectonic creep: movement along a fault zone that does *not* occur at the same time as an earthquake.

tectonic cycle: the global process of the creation, evolution, and destruction of the crust of the Earth.

tectonic geomorphology: (1) the study of landforms shaped by tectonic process; (2) application of geomorphic principles to reveal the presence, pattern, or rates of tectonic processes.

tectonics: processes of deformation (motion) of the Earth's crust, and the structures and landforms that result from those processes.

terrace: an inactive bench, typically near an active stream or coastline. The term is applied to both the flat surface of the terrace (the *tread*) and the slope below (the *riser*).

threshold: a critical transition point, such as the maximum amount of change that a system can absorb before its *dynamic equilibrium* becomes unbalanced.

thrust fault: a type of *reverse fault* which is less steep than 45°.

tide gauge: an instrument used to measure sea level over time.

tilting: process by which a horizontal surface acquires a slope (usually without *warping*).

time history: properties of ground shaking over the duration of an earthquake at a particular site.

transform fault: type of fault associated with oceanic ridges; may form a plate boundary, such as the San Andreas Fault in California.

transport-limited slope: a hillslope on which erosion is limited only by the rate of sediment transport (also see *weathering-limited slope*).

transverse topographic symmetry factor (T): a geomorphic index used to detect active tilting.

trilateration net: network of *control points* used to measure vertical or horizontal deformation.

triple junction: the point at which three plate boundaries meet.

tsunami: a potentially damaging ocean wave triggered by a submarine earthquake or other large-scale shift on the ocean floor.

uniform-slip model: a model in which a fault or faults are characterized by constant displacement per rupture event, a constant long-term slip rate, and a characteristic earthquake magnitude.

unpaired stream terrace: a terrace segment on only one side of the modern stream; less likely to reflect a regional tectonic or climatic event than *paired terraces.*

uplift of rocks: any upward vertical motion, whether or not the elevation of the surface increases as a result.

uplift path: on a graph of sea-level history (elevation versus age), it is the line that traces how an individual *coastal terrace* was formed and uplifted to its present elevation.

variable-slip model: a model in which a fault or faults are characterized by irregular amounts of slip per rupture event, long-term slip rates, and/or earthquake magnitudes.

vertical extent of mortality (VEM): vertical width of a zone in which non-mobile marine organisms are killed by coseismic uplift.

Very Long Baseline Interferometry (VLBI): geodetic positioning technique that uses radio telescopes to measure signals from quasars.

volcanic tumescence: uplift caused by rising magma beneath the surface.

warping: process by which a planar surface becomes folded.

wave-cut platform: on a *coastal terrace* or a modern coastline, the subhorizontal surface cut by waves as well as by secondary biological and chemical processes.

wavelength: in *isostasy* and *flexural support,* the lateral dimensions of a load on the Earth's crust.

weathering: the sum of all physical and chemical processes that break down rock at and near the Earth's surface.

weathering-limited slope: a hillslope on which erosion is limited only by the rate of bedrock (or substrate) weathering (also see *transport-limited slope*).

winter berm: (also called the *storm berm*) on a sandy shoreline, it is the high ridge of sand that marks the highest limit of wave action during winter storms (also see *beach ridges*).

Index